T0301774

Zinc Oxide and Related
Materials — 2009

MATERIALS RESEARCH SOCIETY
SYMPOSIUM PROCEEDINGS VOLUME 1201

Zinc Oxide and Related Materials — 2009

Symposium held November 30 – December 3, 2009, Boston, Massachusetts, U.S.A.

EDITORS:

Steven M. Durbin
University of Canterbury
Christchurch, New Zealand

Holger von Wenckstern
Universität Leipzig
Leipzig, Germany

Martin W. Allen
University of Canterbury
Christchurch, New Zealand

Materials Research Society
Warrendale, Pennsylvania

CAMBRIDGE
UNIVERSITY PRESS

Shaftesbury Road, Cambridge CB2 8EA, United Kingdom

One Liberty Plaza, 20th Floor, New York, NY 10006, USA

477 Williamstown Road, Port Melbourne, VIC 3207, Australia

314–321, 3rd Floor, Plot 3, Splendor Forum, Jasola District Centre, New Delhi – 110025, India

103 Penang Road, #05–06/07, Visioncrest Commercial, Singapore 238467

Cambridge University Press is part of Cambridge University Press & Assessment, a department of the University of Cambridge.

We share the University's mission to contribute to society through the pursuit of education, learning and research at the highest international levels of excellence.

www.cambridge.org
Information on this title: www.cambridge.org/9781605111742

Materials Research Society
506 Keystone Drive, Warrendale, PA 15086
http://www.mrs.org

© Materials Research Society 2010

First published 2010
First paperback edition 2012

Single article reprints from this publication are available through University Microfilms Inc., 300 North Zeeb Road, Ann Arbor, MI 48106

CODEN: MRSPDH

A catalogue record for this publication is available from the British Library

ISBN   978-1-605-11174-2   Hardback
ISBN   978-1-107-40813-5   Paperback

CONTENTS

DEVICES

DOPING

*Invited Paper

GROWTH

*Invited Paper

POSTER SESSION II

PREFACE

Symposium H, "Zinc Oxide and Related Materials," held November 30–December 3 at the 2009 MRS Fall Meeting in Boston, Massachusetts, was the third such gathering of researchers from universities, research institutions, Government laboratories and industry to review progress in this area since 2006 — a reflection of the continuing worldwide interest in this emerging semiconductor system. The symposium consisted of 12 invited papers, 72 contributed oral papers, and 97 poster presentations covering sessions on devices, doping, electronic and optical properties, electric and magnetic properties, growth, nano-structures, optical properties and opto-electronic devices, transparent conducting oxides and thin-film transistors. The lively and wide-ranging debate during and after each session was stimulated by the high quality of research presented at the symposium, and by the challenges still remaining in gaining a more complete understanding of this highly functional material. This proceedings volume features selected papers presented during these sessions and gives an overview of the research presented during the course of the symposium.

We particularly wish to thank the symposium organizers Jürgen Christen, Leonard J. Brillson, Hiroshi Fujioka, and H. Hoe Tan for making the symposium such a great success. We are grateful for the generous financial support of the Air Force Office of Scientific Research. Finally, we thank all the session chairs, invited speakers, and contributed oral and poster presenters for continuing to make the MRS Fall Meeting a premier forum for research into Zinc Oxide and Related Materials.

Steven M. Durbin
Holger von Wenckstern
Martin W. Allen

February 2010

MATERIALS RESEARCH SOCIETY SYMPOSIUM PROCEEDINGS

MATERIALS RESEARCH SOCIETY SYMPOSIUM PROCEEDINGS

Prior Materials Research Society Symposium Proceedings available by contacting Materials Research Society

Devices

Mater. Res. Soc. Symp. Proc. Vol. 1201 © 2010 Materials Research Society   1201-H01-01

ZnO-Based MESFET Devices

M. Grundmann, H. Frenzel, A. Lajn, H. von Wenckstern, F. Schein, and M. Lorenz
Universität Leipzig, Institut für Experimentelle Physik II, Linnéstr. 5, 04103 Leipzig, Germany

ABSTRACT

We present transistors and inverters based on the MESFET principle. The channel consists of thin ZnO:Mg thin films on sapphire, deposited with pulsed laser deposition. The ohmic source and drain contacts are formed with sputtered gold. The Schottky gate electrode is formed by metal oxides providing high barrier height and a reliable contact. The voltage swing of our inverters is superior to any other reported oxide devices. Annealing studies show that our devices withstand temperatures up to 150°C, partly improving during annealing.

INTRODUCTION

Thin film field-effect transistors (TFT) and circuits based on them have many applications, the most prominent probably in flat panel display. Amorphous and poly- or microcrystalline silicon enables a viable technology. The typical performance values are a channel mobility below 1 cm^2/Vs and an on/off current ratio above 10^6 [1, 2]. Devices based on oxides provide a viable alternative, offering higher performance, lower fabrication temperatures, transparency and flexible substrates. Most oxide-based TFTs are based on the MISFET principle, i.e. a gate electrode with an insulating dielectric layer, either as bottom or top gate (Fig. 1a, b). While this strategy has been very successful in silicon technology due to the low density of defects at the Si/SiO_2 interface and the low leakage current through SiO_2, typical thickness of the dielectric in oxide MISFETs is 50-200 nm. Thus a fairly larger portion of the gate voltage drops over the insulator. Here, we show MESFETs, i.e. TFTs with a Schottky diode as gate electrode [3] (Fig. 1c). The high Schottky barrier height of typically 0.95 eV for Ag and 0.90 eV for Pt [4] provides rather low reverse currents of 5 nA/cm^2 (at -1V) and low voltage operation of the devices. Based on the MESFETs, we report inverters with the highest reported gain in oxide electronics.

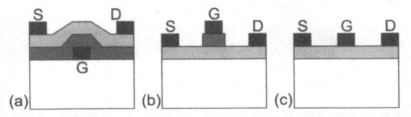

Fig. 1: Schemes of various thin film field-effect transistor geometries: (a) bottom gate MISFET, (b) top gate MISFET, and (c) MESFET. S: source, D: drain, G: gate, dark grey: dielectric, light grey: channel, white: substrate.

EXPERIMENTAL

The channel of the TFTs is formed by ZnO or $Mg_{0.003}Zn_{0.997}O$. The latter offers improved reproducibility. For the devices reported here, the material has been deposited by pulsed laser deposition (PLD) on sapphire at T=650 °C and oxygen pressure of p=0.02 mbar. The typical surface morphology of our ZnO channel layers on sapphire is shown in Fig. 2. We note that our process also yields working transistors on glass substrate [5]. The ohmic contacts to the channel are formed by sputtered gold. The Schottky gate is fabricated by reactive sputtering in an Ar/O_2 plasma. Both sputter processes are in principle suitable for room temperature, large area and mass production. As metals in the reactive sputtering process for the Schottky contact Ag, Pt and other metals have been investigated [6].

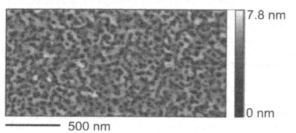

Fig. 2: Scanning force microscopy image of typical ZnO channel layer deposited using pulsed laser deposition on a-plane sapphire substrate. The layer thickness is 30 nm.

RESULTS

Fig. 3a shows transfer characteristics of various ZnO MESFETs with different W/L-ratios between 10.75 (denoted as standard geometry) and 700.

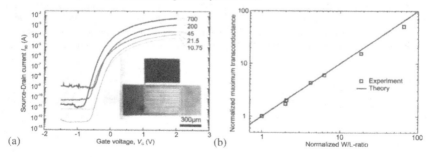

Fig. 3: (a) Transfer characteristics of our ZnO-MESFETs with different W/L-ratios from 700 to 10.75. Inset: Optical image of the MESFET with W/L=700 with interdigitated contact geometry. (b) W/L-ratio scaling behavior of the MESFETs. The transconductance and W/L is normalized with respect to our standard geometry with W/L=10.75.

The MESFETs with higher W/L were designed with interdigitated source-drain and meander-shaped gate contact. The forward current (and with that the transconductance, Fig. 3b) scales with the W/L-ratio over two orders of magnitude. Our MESFET with W/L=700 reaches an on-current of 15 mA and an on/off-ratio of $5-10^8$ making it suitable as driving transistor in e.g. OLED displays.

The performance of our MESFETs [3] is in the regime of the best MISFETs with regard to channel mobility and on/off ratio (Fig. 4a) [7].

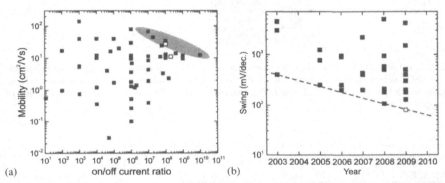

(a)   on/off current ratio   (b)   Year

Fig. 4: (a) On/off current ratio and channel (field effect) mobility for MISFETs (values from literature, filled squares) and our MESFETs (open squares). (b) Historic development of voltage swing (for MISFETs sub-threshold voltage swing) for MISFETs (from literature, filled squares) and our MESFET (open squares).

The channel mobility of a MESFET does not suffer from possible carrier scattering at the interface of the channel and the dielectric. This is an inherent advantage. However, the channel mobility is primarily determined by the material quality of the channel, its doping and the size and orientation of grains. The high on/off current ratio of over 10^8 is a consequence of the good rectification of our gate diodes.

In Fig. 4b the voltage swing (voltage difference at gate for increase of source-drain current by one order of magnitude at the steepest part of I_{SD} vs. V_G) of our MESFETs are compared to the sub-threshold voltage swing of MISFETs reported in the literature. Due to the low operation voltages of our MESFETs, they exhibit the smallest (i.e. best) value of about 80 mV/decade of all reported devices. This is close to the theoretical diffusion-determined minimum of $S=\ln(10)(kT/e)\approx60$mV/decade for FETs operated at room temperature [8].

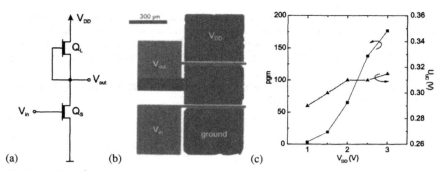

Fig. 5: (a) Schematic circuit diagram of an inverter. (b) Optical image ($Mg_{0.003}Zn_{0.997}O$ channels on Al_2O_3 substrate, Ag_xO gates, Au ohmic contacts and connections) with voltages overlaid. Gate width W=20 μm, W/L=21.5. (c) Peak maximum gain and uncertainty range as a function of supply voltage V_{DD}.

TFT type	pgm	U_{MG} (V)	U_{UC} (V)	U_{DD} (V)	
MESFET	175	0.3	~0.3	3	this work typ.
MESFET	>200	0.3	~0.3	3	this work best
MISFET	1.5	8	12	30	[9]
MISFET	1.7	5	8	18	[10]
MISFET	4.5	3	~2	10	[11]
MISFET	~11	3.3	1.9	10	[12]
MISFET	~56	2.8	~0.9	7	[13]

Tab. 1: Comparison of inverter device parameters for various oxide-channel integrated circuits. Peak maximum gain pgm, voltage at maximum gain U_{MG}, uncertainty level U_{UC}, at given supply voltage V_{DD}.

In Fig. 5b we depict an inverter based on our MESFET concept. The figures of merit for the input/output characteristic are the peak gain maximum pgm (max. of $|dV_{out}/dV_{in}|$) and the uncertainty voltage range U_{UC} between the points for which $|dV_{out}/dV_{in}|=1$. The first quantity should be as large as possible, the latter as small as possible; for the ideal inverter they are infinity and zero, respectively. In Fig. 5c pgm and U_{UC} are shown for a typical inverter as a function of the supply voltage V_{DD}. At 3V, pgm is about 175; our best devices exhibit values above 200. The uncertainty range remains fairly constant, about 0.3V. Comparing these results to inverters based on oxide MISFETs (Tab. 1), our values represent the best devices reported so far.

6

CONCLUSIONS

Oxide MESFETs have been demonstrated to exhibit very good performance and superior integrated circuits compared to oxide MISFET technology reported in the literature. Thus we consider our approach a viable alternative with great promise for high performance analogue and logic integrated circuits.

ACKNOWLEDGMENTS

We gratefully thank G. Biehne, H. Hochmuth and M. Lorenz for their technical support. This work was supported by Deutsche Forschungsgemeinschaft in the framework of Sonderforschungsbereich SFB 762 "Functionality of Oxidic Interfaces" and the Leipzig Graduate School of Natural Sciences "Building with Molecules and Nano-objects (BuildMoNa)". AL and HvW are grateful for support by the European Social Fund. AL is also supported by the Studienstiftung des deutschen Volkes.

REFERENCES

[1]   J.-J. Huang, C.-J. Liu, H.-C. Lin, C.-J. Tsai, Y.-P. Chen, G.-R. Hu, and C.-C. Lee, Journal of Physics D: Applied Physics 41, 245502 (2008).

[2]   H. Lee, G. Yoo, J.-S. Yoo and J. Kanicki, Journal of Applied Physics 105, 124522 (2009)

[3]   H. Frenzel, A. Lajn, M. Brandt, H. von Wenckstern, G. Biehne, H. Hochmuth, M. Lorenz, and M. Grundmann, Appl. Phys. Lett. 92, 192108 (2008).

[4]   H. Frenzel, A. Lajn, H. von Wenckstern, G. Biehne, H. Hochmuth and M. Grundmann, Thin Solid Films 518, 1119 (2009).

[5]   H. Frenzel, M. Lorenz, A. Lajn, H. von Wenckstern, G. Biehne, H. Hochmuth, M. Grundmann, Appl. Phys. Lett. 95, 153503 (2009).

[6]   A. Lajn, H. von Wenckstern, Z. Zhang, C. Czekalla, G. Biehne, J. Lenzner, H. Hochmuth, M. Lorenz, M. Grundmann, S. Wickert, C. Vogt, and R. Denecke, J. Vac. Sci. Technol. B 27, 1769 (2009).

[7]   M. Grundmann, H. Frenzel, A. Lajn, M. Lorenz, F. Schein, H. von Wenckstern, phys. stat. sol. (b) (2010), in press

[8]   S. Jit, P. K. Pandey and P. K. Tiwari, Solid-State Electronics 53, 57 (2009).

[9]   R.E. Presley, D. Hong, H.Q. Chiang, C.M. Hung, R.L. Hoffman, and J.F. Wager, Sol. Stat. Electron. 50, 500 (2006).

[10]   M. Ofuji, K. Abe, H. Shimizu, N. Kaji, R. Hayashi, M. Sano, H. Kumomi, K. Nomura, T. Kamiya, and H. Hosono, IEEE Electr. Dev. Lett. 28, 273 (2007).

[11]   J. Sun, D.A. Mourey, D. Zhao, and T.N. Jackson, J. Electr. Mat. 37, 755 (2008).

[12]   D.P. Heineck, B.R. McFarlane, and J.F. Wager, IEEE Electr. Dev. Lett. 30, 514 (2009).

[13]   J.H. Na, M. Kitamura, and Y. Arakawa, Appl. Phys. Lett. 93, 213505 (2008).

Mater. Res. Soc. Symp. Proc. Vol. 1201 © 2010 Materials Research Society   1201-H01-08

GaN/ZnO and AlGaN/ZnO Heterostructure LEDs: Growth, Fabrication, Optical and Electrical Characterization

J. Benz[1], S. Eisermann[1], P. J. Klar[1], B. K. Meyer[1], T. Detchprohm[2] and C. Wetzel[2]
[1]I. Physikalisches Instiut, Justus-Liebig Universität, Heinrich-Buff-Ring 16, 35392 Giessen, Germany
[2]Department of Physics, Applied Physics, and Astronomy, Rensselaer Polytechnic Institute, 110 Eighth Street, Troy, NY 12180-3590, U.S.A.

ABSTRACT

The wide bandgap polar semiconductors GaN and ZnO and their related alloys exhibit fascinating properties in terms of bandgap engineering, carrier confinement, internal polarisation fields, and surface terminations. With a small lattice mismatch of ~1.8 % between GaN and ZnO and the possibility to grow MgZnO lattice-matched to GaN, the system AlGaN/MgZnO offers the opportunity to design novel optoelectronic devices circumventing the problem of p-type doping of ZnO. In such AlGaN/MgZnO heterostructures with either hetero- or isovalent interfaces, tuning of band offsets is possible in various ways by polarisation fields, surface termination, strain, and composition. These aspects need to be fully understood to be able to make full use of this class of heterostructures. We report on the growth of ZnO films by chemical vapor deposition on p-type GaN and AlGaN grown by metal-organic vapor deposition on sapphire templates and on the fabrication of corresponding light-emitting diode (LED) structures. Electrical and optical properties of the n-ZnO/p-GaN and n-ZnO/p-AlGaN LEDs will be compared and the observed differences will be discussed in terms of the band alignment at the heterointerface.

INTRODUCTION

The wide bandgap semiconductor zinc oxide is a promising material for the production of blue and ultraviolet optoelectronic devices such as light emitting diodes (LEDs), laser diodes and photo diodes. Compared to GaN, the semiconductor mainly used in the optoelectronic industry for the production of short wavelength devices; ZnO offers several advantages, e.g. larger exciton binding energy (60 meV for ZnO versus 26 meV for GaN), which may lead to UV sources with higher brightness and lower power thresholds at room temperature. ZnO possesses higher radiation hardness than Si, GaAs, CdS and GaN, therefore it should be suitable for space applications. Last but not least, large area substrates of ZnO are available at relatively low material costs and ZnO offers a simplified processing, as it can be microstructured by conventional wet-chemical etching [1-4]. Despite all these advantages, there remains one obstacle to be overcome before reliable, entirely ZnO-based optoelectronic devices become reality: the problem of p-type doping of ZnO. So far, there is no way to reliably produce stable and high quality p-type ZnO. Considering the similarity of the physical properties of ZnO and GaN, which both crystallize in wurtzite structure with a lattice mismatch of ~1.8%, and the availability of high quality p-GaN, a natural way of circumventing the doping issue is to grow n-MgZnO/p-AlGaN based devices. Several groups have started to work in this field growing n-ZnO/p-GaN heterostructures. Such heterostructure devices are supposed to exhibit improved current confinement compared to homojunctions, which could lead to higher recombination rates at the interface and thus higher device efficiency [1-4]. These heterostructures must be considered a first step towards light-emitting devices with more complicated active regions consisting of MgZnO/ZnO quantum wells.

EXPERIMENT

Schematic diagrams of the two heterostructures investigated are shown in figure 1. The growth of the GaN templates was done at Rensselaer Polytechnic Institute. The templates were grown by metal-organic vapor-phase epitaxy (MOCVD). A 3.7 µm thick GaN buffer layer was deposited on a sapphire substrate, followed by a layer of weakly Mg doped GaN of 325 nm thickness and a highly doped GaN:Mg, AlGaN:Mg layer of 25 nm thickness, respectively. The unintentionally doped n-type ZnO layer was grown on to these templates by chemical vapor deposition (CVD) in Giessen. The hole density of the GaN templates was determined to be $p \sim 10^{18}$ cm^{-3} at room temperature and for the electron density of the ZnO layer a rough approximation is $n \sim 10^{17}$ cm^{-3}. Photolithography followed by a wet-chemical etching step in a solution of water (H_2O), phosphoric acid (H_3PO_4) and acetic acid ($C_2H_4O_2$) in parts of 30:1:1 by volume was used for the fabrication of mesa structures. The square shaped ZnO mesas have edge lengths of about 1200 µm. The contact areas on top of the mesa and surrounding the mesa were fabricated by evaporating Au on thin layers of Cr and Ni to improve adhesion, respectively. The inset in figure 2 shows a contacted sample.

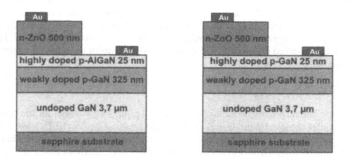

Figure 1. Schematic diagram of the ZnO/GaN and ZnO/AlGaN heterostructure.

DISCUSSION

The measured I-V characteristics at room temperature presented in figure 2 show a clearly rectifying, diode-like behavior with threshold voltages of ~3.4 V for ZnO/GaN and ~4.8 V for ZnO/AlGaN, respectively. The reverse breakdown voltages are estimated to be ~8.6 V for the ZnO/GaN structure and ~12.1 V for the ZnO/AlGaN structure, respectively. The ideality factor n and the saturation current I_s are calculated from a modified diode equation (see equation 1), which takes the series resistance R_s of the layers and contacts into account.

$$U=(nkT/e)\ln(I/I_s)+R_sI \tag{1}$$

The ideality factor is determined to be $n = 26$, the saturation current to $I_s = 4.7 \times 10^{-7}$ A and the series resistance to $R_s = 11$ kΩ for ZnO/GaN and to $n = 36$, $I_s = 2.6 \times 10^{-7}$ A and $R_s = 7$ kΩ for ZnO/AlGaN. We assume that these high values for n arise from the interplay of multiple transport mechanisms at the heterointerface. The comparatively high values for the series resistances and for the threshold voltages compared to the literature [2,4] indicate that the contacts need to be improved.

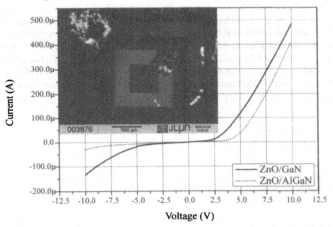

Figure 2. *I-V* characteristics of the ZnO/GaN and ZnO/AlGaN heterostructure. The inset shows a contacted sample.

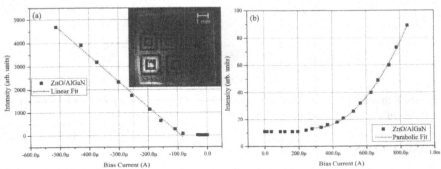

Figure 3. Integral EL of a ZnO/AlGaN sample as a function of (a) reverse and (b) forward bias. The inset shows the bright green EL obtained under reverse bias.

11

The integral electroluminescence (EL) was measured by placing the sample in front of a photomultiplier. Figure 3 shows the intensity as a function of the applied current. We obtained a linear relationship under reverse bias, corresponding to recombination of free carriers at localized defects and impurities. Under forward bias the relationship is parabolic, which we attribute to radiative band to band emission in the interface region. The intensity under reverse bias is one order of magnitude higher than that under forward bias, thus we were not able to measure the spectral distribution of the intensity under forward bias, while the intensity under reverse bias was bright enough to be seen with the naked eye in a dark room and therefore could be further analyzed spectrally. Figure 4 shows the spectral EL of a ZnO/AlGaN sample under a reverse bias of 15 V and obtained using a UV spectrometer and lock-in technique. The spectrum consists of a broad peak centered at 550 nm. The band may originate from defect levels, probably Cu, in ZnO which act as efficient radiative traps for the minority holes. Alternatively it may also be due to deep defect levels in GaN [5]. The inset shows the *I-V* characteristics before and after applying a reverse bias of 15 V for several hours to measure the EL, the value for the reverse breakdown voltage decreases significantly indicating that the device structure has changed under high DC bias.

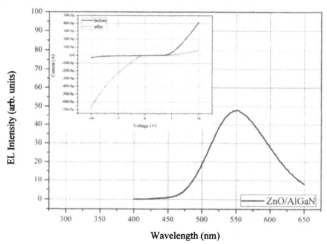

Figure 4. Spectral EL of a ZnO/AlGaN sample. The inset shows the I-V characteristics before and after applying a reverse bias of 15 V for several hours to measure the EL.

CONCLUSIONS

Here, we reported on the fabrication of n-ZnO/p-GaN and n-ZnO/p-AlGaN heterostructure diodes. We investigated the current-voltage characteristics and achieved a clear rectifying behavior. We found a weak electroluminescence (EL) under forward bias as well as reverse bias for both heterosystems. The EL under reverse bias very likely originates from defect-related minority carrier recombination in ZnO or GaN, whereas the weaker EL under forward bias is partly due to band edge emission. Further improvement is anticipated by including MgZnO/ZnO quantum wells in the active region.

REFERENCES

1. D. C. Look, Mater. Sci. Eng., B **80**, 383 (2001)
2. Y. I. Alivov, J. E. Van Nostrand, D. C. Look, M. V. Chukichev, B. M. Ataev, Appl. Phys. Lett., **83**, 2943 (2003)
3. D. J. Rogers, F. Hosseini Teherani, A. Yasan, K. Minder, P. Kung, M Razeghi, Appl. Phys. Lett., **88**, 141918 (2006)
4. Y. I. Alivov, E. V. Kalinina , A. E. Cherenkov, D. C. Look, B. M. Ataev, A. K. Omaev, M. V. Chukichev, D. M. Bagnall, Appl. Phys. Lett., **83**, 4719 (2003)
5. I. E. Titkov, A. S. Zubrilov, L. A. Delimova, D. V. Mashovets, I. A. Liniichuk, I. V. Grekhov, Semiconductors, **41**, 564 (2007)

Mater. Res. Soc. Symp. Proc. Vol. 1201 © 2010 Materials Research Society   1201-H01-09

Cyan electroluminescence from n-ZnO/i-CdZnO/p-Si heterojunction diodes

Lin Li, Zheng Yang and Jianlin Liu
Quantum Structures Laboratory, Department of Electrical Engineering, University of California,
Riverside, California 92521, USA

ABSTRACT

n-ZnO/i-CdZnO thin film was grown on p-type Si substrate by plasma-assisted molecular-beam epitaxy (MBE). Rectifying I-V curves show typical diode characteristics. Cyan electroluminescence emissions at around 473 nm were observed when the diodes were forward-biased at room temperature. The emission intensity increases with the increase of the injection current. Room temperature photoluminescence verifies the electroluminescence emissions come from CdZnO layer.

INTRODUCTION

ZnO is a promising optoelectronic material for light emitting devices in ultraviolet and visible wavelength [1] because of its large wide direct band gap of 3.3 eV at room temperature and large exciton binding energy of 60meV. The bandgap of ZnO can be extended wider by alloying with MgO [2], while it can also be adjusted narrower to 1.8 eV by alloying with CdO [3-6]. This property makes ZnO based light sources emitting light from ultraviolet region to green band or even further. However, reliable p-type ZnO with high carrier concentration has always been an issue for ZnO light emitting devices application. Here, we demonstrate ZnO based double heterojunction LED devices using p-type Si as p-layer, with inserted CdZnO as active layer. Visible cyan emissions were observed and studied.

EXPERIMENT

ZnO based heterojunction diodes were grown by plasma-assisted MBE on p-type (1–10Ωcm) Si (100) substrates. Elemental Zn (6N), Cd (6N), Ga (6N) were heated by effusion cells and used as metal sources. Oxygen (5N) plasma generated by a radiofrequency plasma generator was used as the oxygen source. Fig. 1 shows the structure of the sample. A 100 nm CdZnO layer was first deposited at 150 °C on p-type Si(100) substrate, followed by 800 °C in situ annealing for 5 min under vacuum, then a 350 nm Ga-doped ZnO layer was deposited at 500 °C as n-type layer. Room temperature (RT) PL measurements were carried out using a home-built PL system, with a 325-nm He-Cd laser as excitation source and a photomultiplier tube behind the monochrometer as detector. Samples were etched by diluted hydrochloride acid to different depth to investigate the PL emission from different layers. Heterojunction diodes were fabricated by standard photolithography techniques. Etching was done using diluted hydrochloride acid to reach down to the substrate. Mesas with size of 800μm×800μm were formed on the samples. Metal contacts were deposited by E-beam evaporator. Au/Ti contacts with thicknesses of 200/10nm were used for contacts of both Ga-doped ZnO and p-type Si. The contacts were subjected to rapid thermal annealing (RTA) under nitrogen ambient at 600 °C for 60 seconds to form ohmic contacts. Current-voltage (I-V) characteristics were measured using Agilent 4155C semiconductor parameter analyzer. EL measurements were carried out in the same system as PL measurement.

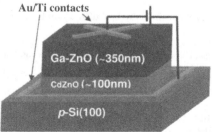

FIG. 1. (color online) Device structure of the sample: Ga-doped ZnO/CdZnO/p-Si.

DISCUSSION

To study the optical properties of Ga-doped ZnO layer and CdZnO layer, Room temperature (RT) PL measurements of each layer were studied. PL measurement of Ga-ZnO layer was carried out from the surface of the sample, while another piece of sample was etched to about 100 nm from substrate to investigate the PL emission of CdZnO layer. As shown in Fig. 2, the PL emission of Ga-doped ZnO layer shows strong and sharp ZnO near band edge (NBE) emission at around 380 nm, although there is deep level emission centered around 522 nm, which comes from the defects. On the other hand, the RT PL emission of CdZnO layer is dominated by the NBE emission of CdZnO at around 451 nm (2.75 eV). There is also a very weak emission peak at 382 nm (3.25 eV), which comes from residual GaZnO in the etched sample.

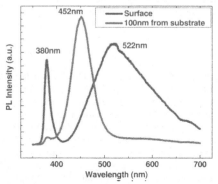

FIG. 2. (color online) Room temperature PL of the sample from surface (blue line) and 100nm from substrate (i.e. CdZnO layer) (red line) respectively.

Fig. 3 shows the *I-V* curve of the fabricated LED device with the voltage configuration as shown in Fig. 1. It shows typical diode characteristic under forward bias. The inset shows the

16

Ohmic behavior of n-n contacts and p-p contacts. The energy band diagram of the sample is presented in Fig. 4. The room temperature PL maximum position is approximately used as the band gap of CdZnO layer, which is about 2.74 eV. The Cd concentration is calculated using the relation $E_g(x) = 3.37 - 2.82x + 0.95x^2$ from Ref. 7. And x is calculated to be 0.24. The electron affinity of CdZnO layer is assumed to be linear distributed between 4.3 eV (ZnO) and 4.5 eV (CdO) [8, 9]. Thus it results in 4.35 eV. So there is a 0.05 eV difference in electron affinity between CdZnO layer and Ga-doped ZnO layer on the top. The electron affinity between ZnO and Si substrate is 0.3 eV [10], so a 0.35 eV electron affinity is observed at the CdZnO/Si interface. Therefore, a double heterojunction is formed clearly as shown in the band alignment. Room temperature EL spectra of the sample at different injection current are shown in Fig. 5. The room temperature EL emissions are dominated by the NBE emission from CdZnO active layer at around 473 nm, which corresponds to the room temperature PL emission from CdZnO layer. The difference between PL and EL peak positions may come from the non-uniformity of the CdZnO layer and heat induced bandgap shrinkage under the high injection currents. There is also a weaker shoulder at around 405nm in each injection current; it may come from the Cd weak phase near the CdZnO/Ga-doped ZnO interface.

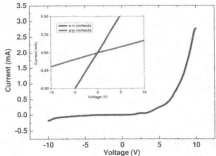

FIG. 3. (color online) *I-V* characteristics of the sample, with voltage configurations as shown in Fig 1, which shows typical rectifying characteristics.

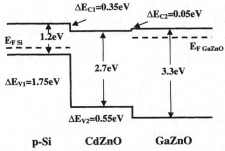

FIG. 4. Band alignment structure of the sample.

17

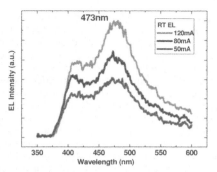

FIG. 5. (color online) Room temperature EL characteristics of the sample under different injection currents.

CONCLUSIONS

ZnO based heterojunction using Ga-doped ZnO as n-layer, CdZnO as an active layer, and p-type Si substrate as p-layer was grown by MBE. Dominant cyan EL emissions were observed at RT. RT PL measurements at different thicknesses and temperature dependent EL confirmed that these emissions come from the radiative recombinations in CdZnO active layers in the film. ZnO based cyan emitting LED devices are demonstrated.

ACKNOWLEDGMENTS

This work was supported by DOE (DE-FG02-08ER46520) and NSF (ECCS-0900978).

REFERENCES

1. 1. D. C. Look, Mater. Sci. Eng., B **80**, 383 (2001).
2. A.Ohtomo, M. Kawasaki, T. Koida, K. Masubuchi, H. Koinuma, Y. Sakurai and Y. Yoshida, T. Yasuda and Y. Segawa, Appl. Phys. Lett. **72**, 2466 (1998).
3. T. Makino, Y. Segawa, M. Kawasaki, A. Ohtomo, R. Shiroki, K. Tamura, T. Yasuda, and H. Koinuma, Appl. Phys. Lett. **78**, 1237 (2001).
4. S. Shigemori, A. Nakamura, J. Ishihara, T. Aoki, and J. Temmyo, Jpn. J. Appl. Phys. **43**, L1088 (2004).
5. S. Sadofev, S. Blumstengel, J. Cui, J. Puls, S. Rogaschewski, P. Schafer, and F. Henneberger, Appl. Phys. Lett. **89**, 201907 (2006).
6. Z. Yang, L. Li, Z. Zuo, and J. L. Liu, J. Cryst. Growth, (in press).
7. X. J. Wang, I. A. Buyanova, W. M. Chen, M. Izadifard, S. Rawal, D. P. Norton, S. J. Pearton, A. Osinsky, J. W. Dong, and A. Dabiran, Appl. Phys. Lett. **89**, 151909 (2006).
8. A. Nakamura, T. Ohashi, K. Yamamoto, J. Ishihara, T. Aoki, J. Temmyo and H. Gotoh, Appl. Phys. Lett. **90**, 093512 (2007).
9. R. Ferro and J. A. Rodríguez, Sol. Energy Mater. Sol. Cells **64**, 363 (2000)

10. L. J. Mandalapu, Z. Yang, F. X. Xiu, D. T. Zhao, and J. L. Liu, Appl. Phys. Lett. 88, 112108 (2006).

Doping

Mater. Res. Soc. Symp. Proc. Vol. 1201 © 2010 Materials Research Society   1201-H02-03

Divacancy-Hydrogen Complexes in Zinc Oxide

J. Kuriplach [1], G. Brauer [2], O. Melikhova [1], J. Cizek [1], I. Prochazka [1] and W. Anwand [2]

[1] Charles University, Faculty of Mathematics and Physics, Department of Low Temperature Physics, CZ-18000 V Holešovičkách 2, Prague, Czech Republic

[2] Institut für Strahlenphysik, Forschungszentrum Dresden-Rossendorf, Postfach 510119, D-01314 Dresden, Germany

ABSTRACT

In the present work we study Zn+O divacancies filled up with varying amount of hydrogen atoms. Besides the structure and energy-related properties of such defects, we also investigate their capability to trap positrons taking into account positron induced forces. We show that the Zn+O divacancy may trap positrons when up to two hydrogen atoms are located inside the divacancy. The calculated properties are discussed in the context of other computational and experimental studies of ZnO.

INTRODUCTION

The nature of point defects in zinc oxide is not yet fully understood, but this subject attracts wide attention as it is of vital importance for various technological applications mainly in electronics. An important class of point defects in ZnO is represented by hydrogen-related defects that are also often considered in relation to the n-type conductivity of ZnO materials [1,2]. Recently, it has been found that nominally undoped commercially available ZnO single crystals contain an appreciable amount of hydrogen at a level of at least 0.3 at.% [3]. The forms in which such hydrogen atoms are incorporated into the ZnO lattice are not precisely known at present, and there are several possibilities, in principle. In particular, hydrogen may occupy interstitial positions and/or may be present in the form of H_2 molecules in 'channels' along the c-axis of the hexagonal ZnO lattice [1]. Hydrogen may also form complexes with other impurities though their concentrations are usually much smaller than that of H, as found in [3]. Furthermore, hydrogen atoms can enter open volume defects, like oxygen and zinc vacancies (see e.g. [4]) and related defects. Positron annihilation techniques are sensitive to such kind of defects and, as shown in [3], positron trapping in hydrothermally-grown ZnO materials occurs in Zn-vacancy-hydrogen related complexes. Namely, a suggestion has been given that positrons may get trapped in a Zn-vacancy occupied by one hydrogen atom. In the present work, we extend our study to Zn+O divacancies containing hydrogen atoms ($V_{Zn+O}+n$H complexes). We investigate their structure, energy-related properties, and also positron characteristics taking into account positron induced forces acting on the ions surrounding studied defects. Our previous positron studies as well as further motivations, explanations and references to other related literature can be found in refs. [3,5].

COMPUTATIONAL METHODS

Realistic configurations of studied defects were obtained by means of relaxing the total energy of corresponding supercells with respect to atomic positions. In particular, the Vienna ab initio simulation package (VASP) [6] was employed for this purpose. In the course of calculations, projector augmented-wave potentials within the local-density approximation (LDA) were used [7]. 96 atom-based supercells containing defects were utilized.

ZnO exhibits the wurtzite crystal structure and it can be considered as a network of ZnO_4 tetrahedra interconnected at their corners (see e.g. figure 1 in ref. [3]). A $V_{Zn+O}+n$H complex can be considered as a part of such a tetrahedron with H atoms inside, missing Zn central atom and one O corner atom and, thereby, $V_{Zn}(V_OO_3)+n$H 'formula' describes the situation, which will be useful for defect structure explanations (see figure 1). We studied defects based on one of possible two configurations of the V_{Zn+O} divacancy in which the removed/missing nearest neighbor Zn+O pair has a non-zero projection to the basal plane of the ZnO hexagonal wurtzite lattice.[1] For this configuration we placed $n = 0$, 1, and 2 hydrogen atoms into various possible positions in V_{Zn}, as described in [3], and also in the center of V_O [4]. In particular, for $n = 1$ we have three non-equivalent configurations denoted further by $V_{Zn+O}+1$H OV, $V_{Zn+O}+1$H Oab, and $V_{Zn+O}+1$H Oc. Here the suffix OV marks the configuration with a H atom in the center of V_O, and suffixes Oab and Oc mark, respectively, configurations with a H atom close to an O atom around V_{Zn} with the O-H pair having a non zero projection to the hexagonal basal plane and lying along the c-axis direction (see figure 1a). In the case of $n = 2$, there are totally four configurations $V_{Zn+O}+2$H Oab$V$ (figure 1b), $V_{Zn+O}+2$H OcV, $V_{Zn+O}+2$H Oab, and $V_{Zn+O}+2$H Oabc. The meaning of these abbreviations can be derived from that given above for $V_{Zn+O}+1$H complexes, considering that there are always two H atoms that cannot occupy the same position inside V_{Zn+O}.

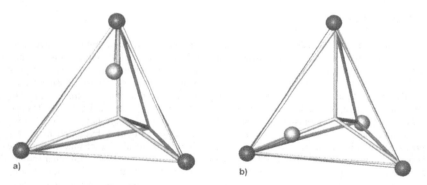

a)   b)

Figure 1: Relaxed configurations of a) $V_{Zn+O}+1$H Oc and b) $V_{Zn+O}+2$H Oab$V$ complexes. Dark (light) spheres represent O (H) atoms. The original (nonrelaxed) ZnO_4 tetrahedra are plotted by thick solid lines. Thinner (darker) lines join relaxed positions.

---

[1] The second configuration with the Zn-O pair oriented along the c-axis direction will be the subject of another study.

Formation energies ($E_f(D; q, E_F)$) for various defects (D) with and without hydrogen in dependence on the Fermi level position (E_F) and defect charge state (q) have been estimated. For this purpose, the formula from ref. [8], where we added also terms related to hydrogen (following ref. [1]), was employed

$$E_f(D; q, E_F) = E(D; q, n_{Zn}, n_O, n_H) - (n_{Zn}-n_O)\mu_{Zn} + n_O E_{ZnO} - n_H \mu_H + q E_F . \qquad (1)$$

Here $E(D; q, n_{Zn}, n_O, n_H)$ and E_{ZnO} are, respectively, the energy of the corresponding supercell containing the defect D considered (in charge state q) and the energy per one ZnO pair in bulk ZnO material. n_{Zn}, n_O, and n_H specify the number of Zn, O, and H atoms in the supercell, respectively. μ_{Zn} and μ_H are the chemical potentials of Zn and H. We will not repeat the whole discussion [8] here and mention only that in binary systems/compounds chemical potentials of constituent atoms (i.e. Zn and O in our case) are not fixed, but depend on the conditions to which the material is exposed. For this reason one can specify a lower and an upper limit for both chemical potentials (μ_{Zn} and μ_O) and one can thus principally distinguish between Zn-rich (O-poor) and Zn-poor (O-rich) conditions. Hydrogen is treated as an impurity [1] and its chemical potential is taken to be one half of the free H_2 molecule energy. Formation energies were evaluated on the basis of energies obtained within the LDA framework so, in principle, they suffer from the well know LDA band gap underestimation issue (cf. ref. [9]), which we intend to address in a future work.

Positron calculations have been performed using a real space method as implemented in the atomic superposition (ATSUP) method [10]. Positron-induced forces acting on atoms around the defect were implemented according to the scheme developed by Makkonen et al. [11]. In such calculations, the so-called conventional scheme [11,12] for positron calculations is considered, which means that the electronic structure is not primarily influenced by the presence of the positron trapped in a defect. On the other hand, the positions of ions (Zn, O, and H) are influenced by the presence of the positron through the positron-induced forces (PFs). In the following, we shall distinguish between two schemes for positron calculations. In both of them, we use self-consistent (SC) and relaxed (RE) electron charge densities and Coulomb potentials. The first scheme neglects PFs (hence SC RE), whereas the second one takes them into account (SC RE PF). In positron calculations, the approach of Boroński and Nieminen [13] to the electron-positron correlations and enhancement factor – with the correction [14] for incomplete positron screening with the high frequency dielectric constant $\varepsilon_\infty = 4$ – has been used. We further refer readers to refs. [11,12] concerning theoretical background of positron calculations and to refs. [3,5] regarding our previous calculations for ZnO.

RESULTS AND DISCUSSION

We shall now present and discuss the results for the studied $V_{Zn+O}+n$H defects from the viewpoint of structural, energy-related, and positron properties.

Structure of defects

Examples of relaxed defect configurations are shown in figure 1. Hydrogen atoms that were initially placed in the center of the V_O-part of the divacancy remain roughly at their positions after relaxation, i.e. they are surrounded by three Zn atoms like in the oxygen vacancy – just one Zn atom is missing due to V_{Zn} (figure 1b). Also O-H pairs preserve their character in relaxed configurations keeping the O-H distance at approximately 1 Å (figure 1a).

Table I gives energies of charge neutral configurations of $V_{Zn+O}+1H$ and $V_{Zn+O}+2H$ complexes. These energies are calculated with respect to the lowest energy in either group of complexes. One can see that complexes with H atoms located in the V_O-part of the V_{Zn+O} divacancy have apparently larger energies than those containing O-H pairs only. This stems from the fact that V_O is positively charged relative to the lattice (even though the whole supercell is neutral), as is also the H atom because it is losing a 'part' of its electron. Thus, V_O and H atom are mutually repulsive. This indicates that lower energy configurations (i.e. without any O...V part) in either group of studied complexes should preferably be found in reality.

Table I. Calculated energy-related and positron characteristics for studied $V_{Zn+O}+nH$ complexes. E_r is the energy of a given configuration with respect to the lowest energy configuration (see the text). τ and E_b denote, respectively, the positron lifetime and positron binding energy to defects.

Defect	E_r (eV)	τ (ps)	E_b (eV)	τ (ps)	E_b (eV)
V_{Zn+O}	—	265	1.1	224	1.0
$V_{Zn+O}+1H$ OV	1.78	246	1.0	218	0.9
$V_{Zn+O}+1H$ Oab	0.00	223	0.4	208	0.3
$V_{Zn+O}+1H$ Oc	0.08	217	0.4	207	0.3
$V_{Zn+O}+2H$ Oab$V$	0.20	199	0.3	185	0.2
$V_{Zn+O}+2H$ OcV	0.24	199	0.3	183	0.2
$V_{Zn+O}+2H$ Oab	0.00	154	~0	154	~0
$V_{Zn+O}+2H$ Oabc	0.05	154	~0	154	~0
		SC RE		SC RE PF	

Formation energies

Figure 2 shows resulting formation energies for studied complexes and other selected defects calculated using equation 1 for Zn-rich, 'intermediate', and Zn-poor conditions. By intermediate conditions we understand the case when the chemical potential μ_{Zn} is just in the middle between its upper and lower limit. When H-related defects are not considered for the moment, formation energy diagrams follow those shown in refs. [8,9].

As for $V_{Zn}+nH$ defects, the corresponding energies to calculate formation energies were taken from [3]. One can see that these defects have lowest formation energies from all the defects considered for n-type materials under O-rich (Zn-poor) and intermediate conditions. This suggests that $V_{Zn}+nH$ complexes exist in ZnO under broad growth conditions.

Regarding $V_{Zn+O}+nH$ complexes, we considered only neutral defects at this stage of research. Corresponding formation energies are positioned quite high (figure 2) and, therefore, should not be populated to a large extent when compared to other defects considered, depending also on the Fermi level position and conditions (Zn-rich, Zn-poor). However, charge states need to be investigated too in order to obtain the full picture. Though the creation of $V_{Zn+O}+nH$ complexes could be hindered due to energy reasons, we note that V_{Zn+O} divacancies can be introduced into ZnO materials e.g. by irradiation.

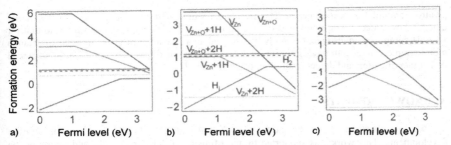

Figure 2: Formation energies of various defects in ZnO for a) Zn-rich, b) intermediate, and c) O-rich conditions in dependence on the Fermi level position. The zero Fermi level corresponds to the valence-band maximum and the maximum Fermi level (3.4 eV) corresponds to the conduction band maximum. The slope of some segments indicates a charge state (see refs. [1,8] for details).

<u>**Positron characteristics**</u>

Positron lifetimes and binding energies to defects for studied complexes are listed in table I, both for SC RE and SC RE PF computational schemes. The SC RE PF scheme yields systematically lower lifetimes and binding energies due to the attractive interaction between the positron and surrounding oxygen ions, which results in lowering the free volume available to the positron compared to the case when positron induced forces are not considered. This indicates that positron induced forces are inevitable in order to obtain realistic positron characteristics.

The calculated bulk positron lifetime is 154 ps [3] and agrees well with the experimental value (151 ± 2) ps given in ref. [5]. We refer to [3] for a detailed summary of the ZnO bulk lifetime measurements and calculations. The positron lifetime and binding energy for the V_{Zn+O} divacancy show that this defect is a deep positron trap as is also the $V_{Zn+O}+1H$ OV complex. But other $V_{Zn+O}+1H$ complexes are shallower traps due to the fact that the V_O part of the divacancy repulses positrons and the free volume of the V_{Zn}-part is reduced due to the presence of the H atom. Analogous considerations can be applied to $V_{Zn+O}+2H$ complexes for which the positron lifetime is further lowered (table I). $V_{Zn+O}+2H$ Oab and $V_{Zn+O}+2H$ Oabc complexes even do not trap positrons as the remaining free volume is not enough to trap positrons and their positron lifetimes approach that for delocalized positrons.

CONCLUSIONS

We have studied selected V_{Zn+O}-hydrogen complexes in ZnO and determined their structure, energy-related characteristics and positron properties. In such complexes, H atoms can occupy either the V_O-part of the divacancy, which is not energetically favorable, or more likely they are bound to an O atom in the vicinity of the V_{Zn}-part of the divacancy. $V_{Zn+O}+nH$ complexes seem to exhibit quite a high formation energy though charged states were not studied yet. In contrast, $V_{Zn}+nH$ complexes have lower formation energies and should exist in large concentrations under O-rich (Zn-poor) conditions for n-type ZnO materials, which seems to be supported by recent positron annihilation experiments. The positron lifetime and binding energy for $V_{Zn+O}+n$H defects show a decreasing trend with increasing number of H atoms, which is similar to that for $V_{Zn}+n$H complexes investigated recently. Further calculations are in progress in order to obtain a more complete and accurate description of $V_{Zn+O}+n$H complexes, which will assist in advanced defect identification and defect resolution in ZnO materials.

ACKNOWLEDGMENTS

We are grateful to M. J. Puska for his ATSUP code that served as a basis for further developments. This work was supported by the Ministry of Schools, Youths and Sports of the Czech Republic through the research plan MSM 0021620834 and the project COST OC 165. Financial support of the Academy of Science of the Czech Republic (project KJB101120906) is appreciated. The support provided by the computational facility LUNA (Academy of Sciences of the Czech Republic, VUMS Computers) is also acknowledged.

REFERENCES

1. C. G. Van de Walle, *Phys. Rev. Lett.* **85**, 1012 (2000).
2. A. Janotti and C. G. Van de Walle, *Rep. Prog. Phys.* **72**, 126501 (2009).
3. G. Brauer, W. Anwand, D. Grambole, J. Grenzer, W. Skorupa, J. Cizek, J. Kuriplach, I. Prochazka, C. C. Ling, C. K. So, D. Schulz, and D. Klimm, *Phys. Rev. B* **79**, 115212 (2009).
4. A. Janotti and C. G. Van de Walle, *Nature Mater.* **6**, 44 (2007).
5. G. Brauer, W. Anwand, W. Skorupa, J. Kuriplach, O. Melikhova, C. Moisson, H. von Wenckstern, H. Schmidt, M. Lorenz, and M. Grundmann, *Phys. Rev. B* **74**, 045208 (2006).
6. G. Kresse and J. Hafner, *Phys. Rev. B* **47**, 558 (1993), ibid. **49**, 14251 (1994); G. Kresse and J. Furthmüller, *Comput. Mater. Sci.* **6**, 15 (1996); *Phys. Rev. B* **54**, 11169 (1996).
7. G. Kresse and J. Hafner, *J. Phys.: Condens. Matter* **6**, 8245 (1994); G. Kresse and D. Joubert, *Phys. Rev. B* **59**, 1758 (1999).
8. A. F. Kohan, G. Ceder, D. Morgan, and C. G. Van de Walle, *Phys. Rev. B* **61**, 15019 (2000).
9. A. Janotti and C. G. Van de Walle, *Phys. Rev. B.* **76**, 165202 (2007).
10. M. J. Puska and R. M. Nieminen, *J. Phys. F: Met. Phys.* **13**, 333 (1983); A. P. Seitsonen, M. J. Puska, and R. M. Nieminen, *Phys. Rev. B* **51**, 14057 (1995).
11. I. Makkonen, M. Hakala, and M. J. Puska, *Phys. Rev. B* **73**, 035103 (2006).
12. M. J. Puska and R. M. Nieminen, *Rev. Mod. Phys.* **66**, 841 (1994).
13. E. Boroński and R. M. Nieminen, *Phys. Rev. B* **34**, 3820 (1986).
14. M. J. Puska, S. Mäkinen, M. Manninen, and R. M. Nieminen, *Phys. Rev. B* **39**, 7666 (1989).

Mater. Res. Soc. Symp. Proc. Vol. 1201 © 2010 Materials Research Society   1201-H02-05

Transportation of Na and Li in Hydrothermally Grown ZnO

Pekka T. Neuvonen[1], Lasse Vines[1], Klaus M. Johansen[1], Anders Hallén[2], Bengt G. Svensson[1] and Andrej Yu. Kuznetsov[1]

[1]Centre for Material Science and Nanotechnology, Department of Physics, University of Olso, P.O. Box 1048 Blindern, N-0316 Oslo, Norway

[2]Royal Institute of Technology, School of ICT, Dept. of Microelectronics and Applied Physics, P.O. Box Electrum 229, SE-164 40 Kista, Sweden

ABSTRACT

Secondary ion mass spectrometry has been applied to study the transportation of Na and Li in hydrothermally grown ZnO. A dose of 10^{15} cm^{-2} of Na$^+$ was implanted into ZnO to act as a diffusion source. A clear trap limited diffusion is observed at temperatures above $550\,°C$. From these profiles, an activation energy for the transport of Na of ~1.7 eV has been extracted. The prefactor for the diffusion constant and the solid solubility of Na cannot be deduced independently from the present data but their product estimated to be $\sim 3 \times 10^{16}$ cm^{-1}s^{-1}. A dissociation energy of ~2.4 eV is extracted for the trapped Na. The measured Na and Li profiles show that Li and Na compete for the same traps and interact in a way that Li is depleted from Na-rich regions. This is attributed to a lower formation energy of Na-on-zinc-site than that for Li-on-zinc-site defects and the zinc vacancy is considered as a major trap for migrating Na and Li atoms. Consequently, the diffusivity of Li is difficult to extract accurately from the present data, but in its interstitial configuration Li is indeed highly mobile having a diffusivity in excess of 10^{-11} cm^{2}s^{-1} at 500 $°C$.

INTRODUCTION

Zinc oxide (ZnO) is a wide (~3.4 eV) and direct energy-bandgap semiconductor with high exciton binding energy (~60 meV) [1], which makes it highly desirable material for optoelectronic applications [2, 3]. In addition, large scale manufacturing of ZnO wafers with good crystal quality has been accomplished.

ZnO has a tendency to be natively n-type. The origin of the native n-type conductivity still remains a topic of discussion. Previously, it has been suggested to be zinc interstitials (Zn$_i$) or oxygen vacancies (V$_o$) [4-6], but recent theoretical estimates [7] show, that the formation energies of these native defects are too high, or the energy level lies too deep in the bandgap, for these defects to be of main responsibility for the n-type conductivity. The role of hydrogen [8] and other impurities, like Al and Ga [9], has been widely studied, but unambiguous evidence for any of these individual impurities as being the only cause for the n-type conductivity, has not been presented. Partly due to this native behavior, the p-type conductivity has proven difficult to achieve. For instance, elements exhibiting shallow enough acceptor levels in the bandgap are scarce, or the solubility of the dopant is too low to overcome the native n-type conductivity. In addition, several acceptor-like impurities have a tendency to change configuration when the Fermi-level position (E$_F$) varies and become donor-like as E$_F$ approaches the valence band edge [10]; thus, E$_F$ is readily pinned to the middle of bandgap as the concentration of such impurities is sufficiently high.

In this work, group-Ia elements are studied. They are common impurities in hydrothermally grown (HT) ZnO, but diffusion properties and mutual interactions are still poorly understood. For instance, they have been regarded as possible p-type dopants, due to the expectation of reasonably shallow acceptor levels, but stable p-type conductivity has not been demonstrated yet.

EXPERIMENT

High resistivity (n-type, $\sim 1k\Omega cm$) hydrothermally grown ZnO wafers containing $\sim 4\times 10^{17}$ cm^{-3} Li, have been implanted at room temperature with Na$^+$ to a dose of 10^{15} cm^{-2}, with an energy of 150 keV, resulting in projected range of ~ 230 nm. A tilt angle of 7° was used to reduce channeling effects. After the implantation the samples were annealed in air at 550, 575 and 600 °C for 60, 50 (20+30) and 10 minutes, respectively. The annealings at 550 and 600 °C have been performed sequentially to the same wafer and the 575 °C was undertaken in two steps, for 20 and 30 minutes, respectively.

Na and Li concentrations versus depth profiles were measured by secondary ion mass spectrometry (SIMS) with a Cameca IMS7f microanalyzer. 10 keV O$_2^+$ ions were used as a primary beam and rastered over a surface area of 125 x 125 μm^2. The secondary ions were collected from the central part of the sputtered craters with a detection limit in the low 10^{14} cm^{-3} range for both Li and Na. Crater depths were measured with a Dektak 8 stylus profilometer, and the erosion rate was assumed to be constant when converting sputtering time to sample depth. Calibration of the Na and Li concentrations were performed using implanted reference samples.

RESULTS AND DISCUSSION

Na – Li interaction

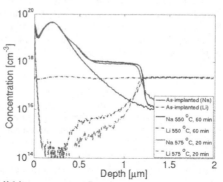

Figure 1. Sodium and lithium concentration versus depth profiles after annealings at 550 and 575 °C. The interaction between Na and Li, and the trap limited diffusion of Na are clearly seen in the picture.

The Na and Li concentration versus depth profiles after treatment at 550 and 575°C can be seen in the Figure 1. The Na profile shows clear trap limited diffusion characteristics and in accordance with recent results by Neuvonen et al. [11], Li is efficiently depleted from the Na-rich regions when annealing at temperatures ≥500 °C. Above 550 °C, Na starts to diffuse into the sample and Li is depleted from the corresponding regions. Apparently, Na and Li compete for the same trapping sites, and the results in Fig. 1 are attributed to lower formation energies of Na on Zn site (Na$_{Zn}$) and interstitial Li (Li$_i$) relative to that of Li on Zn site (Li$_{Zn}$) and interstitial Na (Na$_i$) in the studied samples [11]. Indeed, according to calculations performed by Wardle et al. [10], both Li and Na prefer to be on the substitutional zinc site under n-type conditions with the formation energy (E_{Form}) of Na_{Zn} being lower than that of Li_{Zn}. When the Fermi-level is moving towards mid-gap, both elements start, however, to prefer the interstitial site. The results in Fig. 1 suggest a scenario where Na_i is released from the implanted region and then replaces Li_{Zn} through the reaction $Na_i + Li_{Zn} \rightarrow Na_{Zn} + Li_i$ which is highly energetically favorable since $\{E_{From}(Na_{Zn}) + E_{Form}(Li_i)\} < \{E_{From}(Na_i) + E_{Form}(Li_{Zn})\}$ [10,11]. Gradually, all the traps are filled with Na and Li_i diffuses rapidly out of the Na-rich regions. In addition to Li_{Zn}, also V_{Zn} is considered as a major trap for the migrating Na_i and from positron annihilation spectroscopy the combined concentration of Li_{Zn} and V_{Zn} is estimated to be in the $10^{17} - 10^{18}$ cm^{-3} range for the present samples [12].

Transport of sodium and lithium

As stated previously, the Na concentration profiles in Figure 1 show characteristic trap limited diffusion and a numerical method to solve the Fick's law including trapping was applied (Eq. 1)

$$\frac{\partial[A]}{\partial t} = \frac{\partial}{\partial x}\left(D_A \frac{\partial[A]}{\partial t}\right) - \frac{\partial[AB]}{\partial t}$$

$$\frac{\partial[AB]}{\partial t} = K[A][B] - \nu[AB] \quad , \tag{1}$$

$$[B] = B_{tot} - [AB]$$

$$K = 4\pi R D_A$$

where, [A] and D_A are the concentration and diffusivity of the diffusing element, respectively, while B_{tot}, [B] and [AB] are the total concentration-, available concentration- and the concentration of the occupied traps, respectively. K and ν are the capturing- and dissociation rate of the trap, respectively. Assuming, that the trap and the diffusing element are charged, e.g. Na_i^+ and Li_{Zn}^-, the capture radius R of the trap can be calculated from Coulomb attraction and thermal energy (Eq. 2)

$$\frac{e^2}{4\pi R} = k_B T \quad , \tag{2}$$

where, e is the elementary charge, T is absolute temperature and k_B is the Boltzmann constant. The equation is in SI-units.

In addition, a solid solubility (SS) limitation (Eq. 3) has been added to restrict the concentration of sodium available (free) to migrate, as was done previously by Johansen et al. [13] for deuterium diffusion. The actual pre-factor for the diffusion (D_0) can not be extracted independently from the presented data, due to the lack of exact knowledge of SS of Na in ZnO. However, Janson et al. [14] have proven previously, that the product between the concentration of an element free to diffuse and the D_0 can be determined. In our model, [SS] is concentration of Na, which is free to diffuse and thus approximated by

$$[SS] = \frac{SS \cdot [A]}{SS + [A]} . \tag{3}$$

A major contribution to the trap concentration B_{tot} is considered to be V_{Zn} [12], as discussed in the previous section.

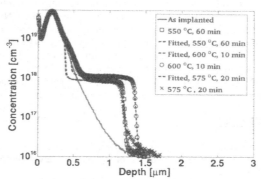

Figure 2. Measured (symbols) and simulated (dashed lines) profiles used in Na diffusion study. Solid line is an as-implanted sample.

Table I. Summary of the parameters used for the diffusion simulations and the extracted coefficients. The data has been grouped by the SS used in the simulation.

T [°C]	SS [cm^{-3}]	B_{tot} [cm^{-3}]	D_A [cm^{2}s^{-1}]	v [s^{-1}]
550	1x10^{16}	9.0x10^{17}	6.05x10^{-11}	1.55x10^{-2}
575	1x10^{16}	8.7x10^{17}	1.55x10^{-10}	7.50x10^{-2}
600	1x10^{16}	1.1x10^{18}	2.45x10^{-10}	5.50x10^{-2}
550	5x10^{16}	9.0x10^{17}	1.28x10^{-11}	1.48x10^{-2}
575	5x10^{16}	8.7x10^{17}	3.25x10^{-11}	7.38x10^{-2}
600	5x10^{16}	1.1x10^{18}	5.35x10^{-11}	5.20x10^{-2}
550	1x10^{17}	9.0x10^{17}	6.82x10^{-12}	1.35x10^{-2}
575	1x10^{17}	8.7x10^{17}	1.72x10^{-11}	6.80x10^{-2}
600	1x10^{17}	1.1x10^{18}	2.90x10^{-11}	5.10x10^{-2}

Figure 2 represents the measured and modelled profiles of Na for all the samples studied in this work (except 50 min at 575 °C). The numerical values deduced from the modelling are given in Table I. Since the value of SS is not known accurately, three simulations were carried out assuming SS equal to 1×10^{16}, 5×10^{16} and 1×10^{17} cm^{-3}, respectively. On the basis of the comparison with the experimental, it is not possible to distinguish between the validity of these three cases, however, increasing SS above 1×10^{17} cm^{-3} gives poor agreement in the beginning of the plateau region (~0.6-0.7 µm in depth) indicating an upper limit of the SS of $\leq1\times10^{17}$ cm^{-3}. Figure 3(a) shows an Arrhenius plot for extracted diffusion constant (D_A) of Na (SS = 10^{16} cm^{-3}) giving an activation energy of ~1.7 eV, with a rather large uncertainty of ~0.5 eV. One important contribution to the uncertainty is the relatively narrow temperature interval studied yielding a dynamic range for D_A of less than one order of magnitude, calling for future experiments [15] with more accurate determination of D_A in a wider temperature range. Furthermore, the extracted activation energy of ~1.7 eV may not necessarily reflect the diffusion of Na itself but rather the dissociation of Na from the traps in the implanted peak region, i.e., the observed transport of Na into the bulk of the samples may be controlled by the diffusion source rather than by the diffusion process itself.

Nevertheless, as revealed by Table I, for a given temperature the product between SS and D_A stays essentially constant, irrespective of the value assumed for SS, in accordance with that predicted by Janson et al. [14] for trap limited diffusion process. With an activation energy of 1.7 eV for D_A, this gives a value of 3×10^{16} cm^{-1}s^{-1} for the product between the pre-exponential factor for D_A and SS. If the pre-factor assumes a typical value of ~1 cm^{2}s^{-1} it implies SS $\approx 3 \times 10^{16}$ cm^{-3}, which appears to be realistic value consistent with the conclusions in the previous paragraph.

Assuming a characteristic vibrational frequency of 1×10^{13} s^{-1}, the dissociation of the trapped Na exhibits an activation energy of ~2.4 eV, Fig. 3(b). Since V_{Zn} is most likely the main trap for the migrating Na atoms, this value can be attributed to the energy barrier for the reaction $Na_{Zn} \rightarrow V_{Zn} + Na_i$, i.e., dissociation of Na_{Zn} into V_{Zn} and Na_i. If the activation energy of ~1.7 eV deduced for D_A is determined by diffusion of Na_i, these values imply a binding energy of ~0.7 eV of Na_{Zn} relative to Na_i in the present samples.

Even though no effect of hydrogen is seen in these samples, it may, in general, influence the transportation of Na and Li. Hydrogen is known to form complexes with Li [16], and presumably, similar complexes with Na, and thus hydrogen may stabilize the trapped Li(Na), or counteract the Fermi-level changes in the samples.

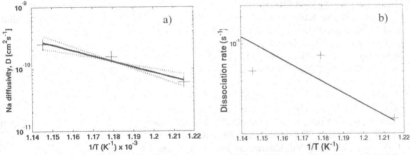

Figure 3. Arrhenius plot for Na diffusion constant (a) and dissociation of corresponding trap (b). The slope corresponds to activation energy for Na diffusion of ~1.7 ± 0.6 eV and a dissociation energy of trapped Na of ~2.4 eV.

Finally, as illustrated by the data in Fig. 1, the diffusion of Li in the Na-rich regions is very rapid. This diffusion is attributed to migration of Li_i and found to occur with a similar rate or higher than that of Na_i. From the present data, it is not possible to extract a more detailed estimate for the diffusion constant of Li_i than a lower limit of ~10^{-11} cm^{2}s^{-1} at 550 °C.

CONCLUSIONS

Na and Li are found to compete for the same trapping sites in hydrothermal ZnO. In samples implanted with Na at room temperature and then annealed at temperatures above 500 °C, Li is efficiently depleted from the Na-rich regions as Na is released from the implantation induced defects and gradually diffuses into the bulk of the samples. The exchange of Li with Na is attributed to that $\{E_{From}(Na_{Zn}) + E_{Form}(Li_i)\} < \{E_{From}(Na_i) + E_{Form}(Li_{Zn})\}$, in accordance with recent theoretical and experimental results published in the literature [10,11]. The transport of Na can be accurately described by a trap-limited diffusion process exhibiting an activation energy of ~1.7 eV for the migration and ~2.4 eV for the dissociation of the trapped Na atoms. The former value may not necessarily reflect the migration process itself of the mobile Na atoms (presume Na_i) but rather the release of these atoms from the implantation-induced defects. V_{Zn} is found to be a major trap in the studied samples and the latter value is ascribed to the energy barrier for dissociation of Na_{Zn} into V_{Zn} and Na_i. Further, the experimental data combined with modelling suggest a solid solubility limit of less than 10^{17} cm^{-3} for Na in the temperature range of 500 to 600 °C. Finally, Li displays a very fast diffusion, which is most likely due to Li_i, with a lower limit of 10^{-11} cm^2/s for the diffusion constant at 550 °C.

ACKNOWLEDGMENTS

The authors gratefully acknowledge support from the Norwegian Research Council through the NANOMAT- and FRINAT-programs for the fundings as well as NordForsk grants providing resources for the international collaboration.

REFERENCES

1. D. G. Thomas, J. Phys. Chem. Solids **15**, 86 (1960).
2. Y. R. Ryu, T. S. Lee, J. A. Lubguba, H. W. White, Y. S. Park and C. J. Youn, Appl. Phys. Lett. **87**, 153504 (2005).
3. A. Tsukazaki, M. Kubota, A. Ohtomo, T. Onuma, K. Ohtani, H. Ohno, S. F. Chichibu and M. Kawasaki, Jpn. J. Appl. Phys. **44**, L643 (2005).
4. S. E. Harrison, Phys. Rev. **93**, 52 (1954).
5. A. R. Hutson, Phys. Rev. **108**, 222 (1957).
6. A. F. Kohan, G. Ceder, D. Morgan and C. G. Van de Walle, Phys. Rev. B **61**, 15019 (2000).
7. A. Janotti and C. G. Van de Walle, Phys. Rev. B **76**, 165202 (2007).
8. S. F. J. Cox, E. A. Davis, S. P. Cottrell, P. J. C. King, J. S. Lord, J. M. Gil, H. V. Alberto, R. C. Vilão, J. Porto Duarte, N. Ayres de Campos, A. Weidinger, R. L. Lichti and S. J. C. Irvine, Phys. Rev. Lett. **86**, 2601 (2001).
9. E. V. Monakhov, A. Yu. Kuznetsov and B. G. Svensson, J. Phys. D **42**, 153001 (2009).
10. M. G. Wardle, J. P. Goss and P. R. Briddon, Phys. Rev. B **71**, 155205 (2005).
11. P. T. Neuvonen, L. Vines, A. Yu. Kuznetsov, B. G. Svensson, X. L. Du, F. Tuomisto and A. Hallén, Appl. Phys. Lett. **95**, 242111 (2009).
12. F. Tuomisto, A. Zubiaga, K. Kuitunen, P. T. Neuvonen, A. Yu. Kuznetsov, B. G. Svensson, MRS09 Fall Meeting, H2.2 (2009)
13. K. M. Johansen, J. S. Christensen, E. V. Monakhov, A. Yu. Kuznetsov and B. G. Svensson, Appl. Phys. Lett. **93**, 152109 (2008).
14. M. S. Janson, M. K. Linnarsson, A. Hallén and B. G. Svensson, Phys. Rev. B **64**, 195202 (2001).
15. Neuvonen et al., to be published
16. L. E. Halliburton, L. Wang, L. Bai, N. Y. Garces, N. C. Giles, M. J. Callahan and B. Wang, J. Appl. Phys. **96**, 7168 (2004).

Mater. Res. Soc. Symp. Proc. Vol. 1201 © 2010 Materials Research Society

David C. Look[1], Kartik Ghosh[2], and Kevin D. Leedy[3]

[1]Semiconductor Research Center, Wright State University, Dayton, OH 45435, U.S.A.

[2]Department of Physics, Astronomy, and Materials Science, Missouri State University, Springfield, MO 65897, U.S.A.

[3]Sensors Directorate, Air Force Research Laboratory, Wright-Patterson Air Force Base, OH 45433, U.S.A.

ABSTRACT

Temperature-dependent Hall-effect measurements have been performed on three Ga-doped ZnO thin films of various thicknesses (65, 177, and 283 nm), grown by pulsed laser deposition at 400 °C and annealed at 400 °C for 10 min in Ar, N_2, or forming-gas (5% H_2 in Ar). The donor N_D and acceptor N_A concentrations as a function of sample thickness and annealing conditions are determined by a new formalism that involves only ionized-impurity and boundary scattering. Before annealing, the samples are highly compensated, with $N_D = (2.8 \pm 0.3) \times 10^{20}$ cm^{-3} and $N_A = (2.6 \pm 0.2) \times 10^{20}$ cm^{-3}. After annealing in Ar the samples are less compensated, with $N_D = 3.7 \pm 0.1 \times 10^{20}$ cm^{-3} and $N_A = 2.0 \pm 0.1 \times 10^{20}$ cm^{-3}; furthermore, these quantities are nearly independent of thickness. However, after annealing in N_2 and forming-gas, N_D and N_A are thickness dependent, partly due to depth-dependent diffusion of N_2 and H, respectively.

INTRODUCTION

Highly conducting thin films of ZnO are being considered for many applications, including: (1) transparent electrodes for flat-panel displays and photovoltaic cells; (2) low-emissivity windows; (3) window defrosters; and (4) light-emitting and laser diodes [1-4]. At present, there is a surge of interest in developing ZnO as a replacement for indium tin oxide

(ITO), because In has become very expensive and is somewhat toxic. A key figure of merit for conductive films is the resistivity ρ; however, from a more basic point of view, ρ is determined by donor N_D and acceptor N_A concentrations [5], and these quantities must be determined if the material is to be developed to its fullest potential. In this work, we employ a simple, analytical method to determine N_D and N_A from mobility μ and carrier concentration n in degenerate semiconductor materials.

EXPERIMENTAL DETAILS

Three Ga-doped ZnO films, of thicknesses 65, 177, and 283 nm, were grown by pulsed laser deposition (PLD) using a 99.99%-pure ZnO target containing 3 wt% Ga_2O_3 [6]. The substrate was Si, coated with a 1-μm-thick layer of SiO_2, for electrical isolation, and the substrate temperature during growth was held at 400 °C. After growth, a piece of each wafer was subjected to a 400-°C, 10-min anneal in either forming-gas (5% H_2 in Ar), pure Ar, or pure N_2. The forming-gas (FG) and N_2 anneals were performed in a rapid thermal annealing system while the Ar anneal was carried out in a tube furnace. Temperature-dependent Hall-effect measurements were performed over the range 15 – 320 K with a LakeShore 7507 apparatus. Ohmic contacts were achieved by soldering small dots of indium onto the corners of 5-mm x 5-mm samples. The measured carrier concentration n was independent of temperature while the mobility was weakly temperature dependent, but only above about 150 K. Thus, for T < 150 K, we could analyze the data by applying degenerate Brooks-Herring ionized-impurity scattering theory along with a much weaker contribution from boundary scattering, as described in detail elsewhere [7]. Specifically, we analyzed mobility μ and carrier concentration n data at 20 K, although these quantities would have been nearly identical at 150 K and not much different even at 300 K.

THEORY

For ZnO, the degenerate Brooks-Herring formula can be written in the form [5,7]:

$$\mu_{ii}(K,n) = \mu_{max}(n)\frac{1-K}{1+K}$$

where $K = N_A/N_D$ and

$$\mu_{max}(n) = \frac{146.9}{\ln\left(1 + 6.46n_{20}^{1/3}\right) - \frac{6.46n_{20}^{1/3}}{1 + 6.46n_{20}^{1/3}}} \qquad (2)$$

Here n_{20} is a normalized value of n such that $n = n_{20} \times 10^{20}$ cm^{-3}. Note that the denominator of equation 2 is slowly varying with n and moreover is approximately unity for $n = 1 \times 10^{20}$ cm^{-3}. The boundary scattering mobility can be approximately written, *for our ZnO samples*, as [7]

$$\mu_{bdry}(d,n) = \frac{e}{m^*} \frac{d/4}{v_{Fermi}(n)} = \frac{e}{\hbar} \frac{d/4}{(3\pi^2 n)^{1/3}} = 2.645 \frac{d_{nm}}{n_{20}^{1/3}} \frac{cm^2}{Vs} \qquad (3)$$

where d_{nm} is the thickness in nm. Since the electrons are degenerate, the two scattering mechanisms can be combined by means of Matthiessen's Rule [5]:

$$\mu(d,n,K) = \left[\mu_{ii}(K,n)^{-1} + \mu_{bdry}(d,n)^{-1}\right]^{-1} \qquad (4)$$

For the convenient determination of N_D and N_A, we define a dimensionless quantity $Q = \mu_{max}(n_{expt})[\mu_{expt}^{-1} - n_{expt}^{1/3}/2.645d_{expt}]$, and then $K = (Q-1)/(Q+1)$, and finally $N_D = n_{expt}/(1-K)$ and $N_A = n_{expt}K/(1-K)$, where n_{expt} is in units of 10^{20} cm^{-3} and d_{expt} in nm, as before. Here, n_{expt}, μ_{expt}, and d_{expt} denote the measured values of n, μ, and d, respectively. Equations 1 – 4 provide a good description of uniform thin films, and we will analyze our three samples in that context. However, we will find that the two thicker samples are not uniform in depth after the FG and N_2 anneals.

DISCUSSION

Plots of N_D and N_A as functions of thickness and annealing ambient are presented in figure 1. (The unannealed 177-nm sample was not truly degenerate and thus its data are not included in figure 1.) For the unannealed samples, $N_D \sim (2.8 \pm 0.3) \times 10^{20}$ cm^{-3} and $N_A \sim (2.6 \pm 0.2) \times 10^{20}$ cm^{-3}, so $K > 0.9$. This high value of K is a manifestation of self compensation, in which the lattice lowers its energy by incorporating acceptors in response to the doping with Ga donors. Because the acceptor concentrations are very high, in the 10^{20}-cm^{-3} range, they must be mainly composed of either pure defect-type acceptors, such as Zn vacancies V_{Zn} or O interstitials O_I, or

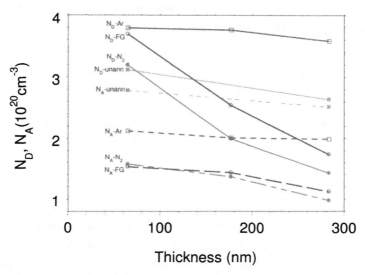

Figure 1. Donor N_D and acceptor N_A concentrations as functions of sample thickness for various annealing conditions.

they must involve Ga, since it also has a concentration $> 10^{20}$-cm^{-3}. Possible Ga-related acceptors are Ga_O, which is unlikely, and Ga_{Zn}-V_{Zn}. Since V_{Zn} has a low formation energy in n-type ZnO, and indeed has been shown to be the dominant acceptor in some cases [8], the most likely acceptors in our samples are V_{Zn} and Ga_{Zn}-V_{Zn}. If O_I is present, it is likely in the neutral species O_I-Ga_{Zn}.

    We first consider the 65-nm sample. Note that for the Ar anneal, N_D increases and N_A decreases, by about the same amount; i.e., $\Delta N_D \sim -\Delta N_A$. We conjecture that if Zn interstitials are available, say in the surface region or as components of extended defects, then they might be able to break away to the bulk and take part in the reaction $Zn_I + Ga_{Zn}\text{-}V_{Zn} \rightarrow Ga_{Zn} + Zn_{Zn}$, thus creating bulk donors and destroying bulk acceptors in equal numbers, as observed. Another potential reaction is $V_O + O_I\text{-}Ga_{Zn} \rightarrow Ga_{Zn} + O_O$, which has the net effect of changing a deep neutral donor V_O into an ionized shallow donor Ga_{Zn}. For the FG anneal, N_D increases about the same amount as that for the Ar anneal, but N_A *decreases* more. Here, we believe that the H in the FG is passivating acceptors through the reactions $H + Ga_{Zn}\text{-}V_{Zn} \rightarrow H\text{-}Ga_{Zn}\text{-}V_{Zn}$ and $2H + V_{Zn} \rightarrow 2H\text{-}V_{Zn}$. Note that the $2H\text{-}V_{Zn}$ complex has been identified in IR absorption spectra [9]. Finally, for the N_2 anneal, ΔN_D is small, while ΔN_A is about the same as that observed in the FG anneal. It is possible that N_2, which theoretically acts as a donor in ZnO [10], can also passivate acceptors such as $Ga_{Zn}\text{-}V_{Zn}$ and V_{Zn}.

For the thicker samples, caution must be exercised when analyzing certain types of anneals, because the electrical properties can be depth dependent, as discussed elsewhere [7]. That is, for the Ar anneals, N_D and N_A do not vary strongly with thickness, but for the FG and N_2 anneals, they do, as seen in figure 1. Consider the 283-nm FG-annealed film, which has $n_{expt} = 0.615 \times 10^{20}$ cm^{-3} and $\mu_{expt} = 29.8$ cm^2/V-s, at 20 K (and nearly the same values at room temperature). If we treat this layer as homogeneous, then equations $1 - 4$ give $N_D = 1.74 \times 10^{20}$ cm^{-3} and $N_A = 1.12 \times 10^{20}$ cm^{-3}, as plotted in figure 1. On the other hand, the 65-nm FG-annealed film is fitted with $N_D = 3.69 \times 10^{20}$ cm^{-3} and $N_A = 1.53 \times 10^{20}$ cm^{-3}. Why are the N_D's and N_A's so thickness-dependent for the FG anneal, but not for the Ar anneal? We hypothesize that the H in the FG does not penetrate the full thickness of the 283-nm layer and thus it creates a sample with nominally two layers of differing electrical properties. This situation can be analyzed with a 2-layer Hall-effect formulation [5,7], and we simplify the problem by assuming that $N_{D2} = N_{D1} = 3.69 \times 10^{20}$ cm^{-3}, based on the premise that H will not form complexes with donors since both species are positively charged. We then have two fitting parameters, N_{A2} and d_2 (or d_1, since $d_1 = 283 - d_2$), which can be varied in an attempt to reproduce the values measured for the total layer: $n_{expt} = 0.615 \times 10^{20}$ cm^{-3} and $\mu_{expt} = 29.8$ cm^2/V-s. An exact fit to both n_{expt} and μ_{expt} is achieved by choosing the parameters $N_{A2} = 3.38 \times 10^{20}$ cm^{-3} and $d_1 = 54.5$ nm. We can also apply this same analysis to the 177-nm film, and the results are $N_{A2} = 3.50 \times 10^{20}$ cm^{-3} and $d_1 = 59.0$ nm. The respective values of N_{A2} and d_1 for these two samples are consistent with each other and moreover imply an H penetration depth (d_1) of 57 ± 2 nm and a passivated acceptor concentration ($N_{A2} - N_{A1}$) in region 1 of about $1.9 \pm 0.1 \times 10^{20}$ cm^{-3}. For the N_2 anneal, we get

$N_{A2} = 2.98 \times 10^{20}$ cm^{-3} and $d_1 = 53.7$ nm for the 283-nm layer, and $N_{A2} = 2.86 \times 10^{20}$ cm^{-3} and d_1 = 42.4 nm for the 177-nm film. In this case, the penetration depth is 48 ± 6 nm, and the concentration of passivated acceptors in region 1 is $1.3 \pm 0.1 \times 10^{20}$ cm^{-3}. From a comparison of these results we conclude that the N_2 diffusion depth is about 10 nm smaller than the H diffusion depth, under our annealing conditions, and also that N_2 is about 30 % less effective than H in passivating acceptors.

The success of the 2-layer model in describing the depth-dependent electrical properties associated with the FG and N_2 anneals allows us to answer the question, posed earlier, regarding the sharp drop of N_D with thickness for these anneals. This drop is artificial and results from the inadequacy of the 1-layer model in describing FG and N_2 anneals in thick samples. On the other hand, the 1-layer model is entirely adequate for Ar anneals.

The formulation presented here for determining N_D and N_A is quite general and can be applied to any other degenerate semiconductor material by modifying equation 2 as follows: multiply the factor "146.9" by $(0.3/m^*)^2(\varepsilon_0/8.12)^2$, and the factor "6.46" by $(0.3/m^*)(\varepsilon_0/8.12)$, where m^* and ε_0 are the relative effective mass and static dielectric constant of the material, respectively. Also, the factor "2.645" in equation 3 may need to be modified, depending on the surface and interface properties. However, equation 3 as it stands can probably serve as a predictor of whether or not surface/interface scattering needs to be considered at all.

SUMMARY

We have developed a simple analytical model to determine donor N_D and acceptor N_A concentrations in degenerately-doped ZnO layers of different thicknesses, and have applied it to PLD-grown films subjected to annealing at 400 °C for 10 min in Ar, N_2, or forming-gas (5% H_2 in Ar). Anneals in Ar produce thickness-independent values of N_D and N_A, but anneals in FG or N_2 do not, evidently because the passivation of acceptors by H and N_2 occurs only in the top 50 – 60 nm of the film. When the FG and N_2 anneals are analyzed with a 1-layer (i.e., uniform-film) Hall-effect model, it appears that N_D drops sharply with thickness; however, this effect is artificial due to the inadequacy of the 1-layer model for samples with d >> 60 nm. However, analysis with a 2-layer model gives good results for our whole thickness range, 65 – 283 nm. The model presented here for determining N_D and N_A is quite general and can be applied to any other degenerate semiconductor material.

ACKNOWLEDGMENTS

We wish to thank T.A. Cooper for the Hall-effect measurements and B. Claflin for helpful discussions. Support is gratefully acknowledged from the following sources: AFOSR Grant FA9550-07-1-0013 (K. Reinhardt), NSF Grant DMR0513968 (L. Hess), and DOE Grant DE-FG02-07ER46389 (R. Kortan).

REFERENCES

1. T. Manami, Semicond. Sci. Technol. **20**, S35 (2005).

2. D.C. Look, Mater. Sci. and Eng. B **80**, 383 (2001).

3. S.J. Pearton, D.P. Norton, K. Ip, Y.W. Heo, and T. Steiner, Prog. In Mater. Sci. **50**, 293 (2005).

4. U. Ozgur, Y.I. Alivov, C. Liu, A. Teke, M.A. Reshchikov, S. Dogan, V. Avrutin, S.J. Cho, and H. Morkoc, J. Appl. Phys. **98**, 041301 (2005).

5. D.C. Look, *Electrical Characterization of GaAs Materials and Devices* (Wiley, New York, 1989).

6. B. Bayraktaroglu, K. Leedy, and R. Bedford, Appl. Phys. Lett. **93**, 022104 (2008).

7. D.C. Look, K.D. Leedy, D.H. Tomich, and B. Bayraktaroglu, Appl. Phys. Lett., accepted for publication.

8. F. Tuomisto, V. Ranki, K. Saarinen, and D.C. Look, Phys. Rev. Lett. **91**, 205502 (2003).

9. E.V. Lavrov, J. Weber, F. Börrnert, C.G. Van de Walle, and R. Helbig, Phys. Rev. B **66**, 165205 (2002).

10. S. Limpijumnong, X. Li, S-H. Wei, and S.B. Zhang, Physica B **376-377**, 686 (2006).

Mater. Res. Soc. Symp. Proc. Vol. 1201 © 2010 Materials Research Society   1201-H02-08

P-Type Nitrogen Doped ZnO Films Grown By Thermal Evaporation

Wei Mu, Lei Guo and Lei L. Kerr[1]

Dept. of Paper and Chemical Engineering, Miami University, Oxford, OH, 45056

David C. Look

Semiconductor Research Center, Wright State University, Dayton, OH, 45435

We report the formation of p-type nitrogen doped ZnO (ZnO:N) grown by thermal-evaporation deposition. The effects of nitrogen precursors on the electrical and optical properties of ZnO:N were investigated. This study shows that growth process plays a critical role in the electrical properties of ZnO:N. The chemical reaction mechanism was analyzed.

Introduction

For many applications, ZnO has certain advantages over GaN, such as a high exciton binding energy (60 meV), low cost, and the availability of large-area substrates [1-3]. However, the development of ZnO for optoelectronic devices is hindered due to the difficulty of synthesizing p-type ZnO to make a p-n homojunction. At this time, in spite of great efforts, there is no recipe for making good p-type ZnO in a reproducible manner [4-10]. Group-V elements are predicted to be the most promising dopant candidates, especially those involving nitrogen [11-18]. Many techniques have been explored to deposit ZnO:N including MOCVD [19-23], sputtering [24-27], pulsed laser deposition [28], molecular beam epitaxy [29], etc.

In this work, a low-cost thermal evaporation technique is developed to grow ZnO:N. The nitrogen doping chemistry in ZnO:N is unknown. The origin of the typically weak p-type electrical properties is far from being fully understood, partly because the results obtained from different groups are rather inconsistent [30-31]. It has been suggested that the NO or NO_2 molecule is a more efficient N source than the N_2 or N_2O molecule. This is because the diatom precursors contains pairs of N and can easily incorporate $(N-N)_O$ into the film [30], possibly contributing to the n-type behavior of ZnO:N. However, Xu et. al. observed that native donor-type defects (e.g. Zn_i or V_O) can be reduced by using an appropriate amount of N_2O and thus lead to an improvement of the p-type conductivity [31]. Therefore, in this study, two different N doping precursors (pure NO and 5 vol% NO and N_2 mixture) were used to investigate the effects of single-atom and diatom N on film conductivity.

Experiments

ZnO is a naturally n-type material, mainly due to donor-type impurities such as Al, but also partially due to native defects, in particular, the zinc interstitial [32-33].

[1] Corresponding author: Dr. Lei L. Kerr, Dept. of Paper and Chemical Engineering, Miami University, Oxford, OH 45056. (Tel) 513-529-0768. (Fax) 513-529-0761. kerrll@muohio.edu.

Thus, theoretically, if such impurities and defects could be eliminated, and if acceptors such as N could be incorporated, then a high hole concentration would be possible. In this study, two separate processes of thermal evaporation deposition were designed to determine the effects of oxygen nonstoichiometry. All the samples are grown on glass substrates.

Process I contains two steps. In the first step, Zn powder is evaporated in a low oxygen ambient at 500°C for 2 hrs. The sample from this step is called the "as-grown ZnO". In the second step, the as-grown ZnO film was annealed in dopant gas for an hour to form ZnO:N.

Process II has additional step between the 1st and 2nd step of the process I. In the additional step, a saturated-O_2 is used to anneal the as-grown ZnO sample for an hour. The motivation here was to eliminate O vacancies, which are donors and will compensate acceptors such as N_O.

Hall-effect measurements were performed at room temperature in the van der Pauw configuration, with an Accent 5500 apparatus. Photoluminescence measurements were performed at 4.2 K, and excitation, dispersion, and detection were accomplished, respectively, with a 45-mW HeCd laser, a Spex 1269 1.26-m spectrometer, and a photomultiplier detector. Resolution was better than 0.01 meV in the spectral range important for this study. X-ray diffraction (XRD) was used to characterize the crystal structure of the ZnO films. Scanning electron microscopy (SEM) and energy dispersive X-ray (EDX) were used to obtain the surface morphology and elemental composition of the film, respectively. The acceleration voltage is 5 keV for SEM and 20 kev for EDX.

Results and Discussion

The Hall-effect measurements in Table 1 indicate that annealing ZnO film in 5%NO/N_2 is a feasible method of synthesizing p-type ZnO:N. Each sample was run at least twice, and at different currents, to validate the electrical properties. When p-type conductivity was measured in both runs, it was recorded as p-type. When both runs indicated n-type, it was recorded as n-type. When one run indicated p-type and the other indicated n-type, it was recorded as *ambiguous*-type. The conductivity of the film depended not only on the dopant gas but also on the growth process. For example, under a 350 °C anneal, the ZnO:N film annealed in 5%NO/N_2 mixture exhibited p-type behavior by process I but n-type behavior by process II. When increasing the annealing temperature to 400 °C, both Process I and II produced n-type films for 5%NO/N_2 dopant gas. On the other hand, for the films annealed in pure NO, n-type films were generated at 350 °C and ambiguous-type films at 400 °C, for both processes. This indicates that N_2 plays a critical role during the annealing process. The common belief that pure NO is better than 5%NO/N_2 for producing p-type ZnO is evidently not correct in every situation.

Fig. 1 shows the cross section SEM of the as-grown ZnO. The ZnO film thickness is about 2 μm. The ZnO rods are connected at the substrate. The EDX was taken along the cross section. The EDX has indicated that the as-grown ZnO contains excess Zn existing at the film surface, perhaps mostly at the tips of the nanorods shown in Fig. 2(a). The Zn/O atomic ratio of the as-grown ZnO film is 3:1 at the surface, which proves the existence of excess Zn at the surface of the as-grown film. The excess Zn might be deposited onto the ZnO rods by the Zn vapor during the cooling. These sharp tips shown in Fig. 2(a) disappeared or became dull after annealing in O_2 or in any dopant gas (Fig. 2(b)) and the EDX showed that after annealing in pure O_2, the Zn/O atomic ratio was reduced to 3:4 at the film surface.

Table 1. Hall-effect measurements of ZnO:N

P: Process.   AT: Annealing temperature (°C).

Annealing Gas	P	AT	type	Resis	Mob	Carrier concentration
				Ω·cm	cm^2/V-s	per cm^3
As-grown ZnO	-	-	n	1.08×10^2	0.519	1.11×10^{17}
Pure O_2	II	500	n	4.35×10^3	0.914	1.57×10^{15}
5% NO/N$_2$	I	350	p	1.58×10^1	2.47	1.60×10^{17}
5% NO/N$_2$	II	350	n	4.96×10^1	2.72	4.62×10^{16}
5% NO/N$_2$	I	400	n	3.83×10^1	1.57	1.04×10^{17}
5% NO/N$_2$	II	400	n	2.81×10^2	0.797	2.79×10^{16}
Pure NO	I	350	n	6.57×10^1	1.07	8.85×10^{16}
Pure NO	II	350	n	7.19×10^2	0.827	1.05×10^{16}
Pure NO	I	400	Ambiguous	2.42×10^3	-	-
Pure NO	II	400	Ambiguous	1.15×10^3	-	-

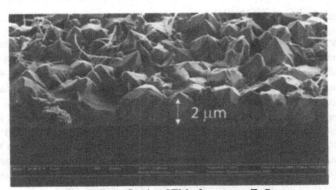

Fig. 1. Cross Section SEM of as-grown ZnO.

Fig. 2 (a) as-grown ZnO (left)   (b) ZnO after annealing in pure O_2 (right)

The possible chemical reactions involved in the growth and annealing processes are presented in Table 2. The p-type conductivity of the films annealed in 5%NO/N$_2$ in Process I is attributed to the formation of Zn$_3$N$_2$ due to the presence of N$_2$ in the 5%NO/N$_2$ mixture by reaction 3 (R3). The formation of Zn$_3$N$_2$ helps to capture the N$_2$ in the 5%NO/N$_2$ and converts N$_2$ to form the Zn-N bond in Zn$_3$N$_2$. Zn$_3$N$_2$ further reacts with NO to form ZnO:N via reaction R4. In other words, reactions R3 and R4 enhance the nitrogen incorporation into the ZnO film.

Table 2. Possible reactions in Processes I and II
R: reaction

	Process I		Process II	
	5% NO/N$_2$	Pure NO	5% NO/N$_2$	Pure NO
Step 1	Zn (s) → Zn (g) (R1) Zn + 1/2O$_2$ → ZnO (R2)			
Step 2	-		Zn + 1/2O$_2$ → ZnO (R2) (Residual Zn reaction with oxygen)	
Step 3	Zn + 1/3N$_2$ → 1/3Zn$_3$N$_2$ (R3) Zn$_3$N$_2$+NO→ZnO:N(R4) ZnO + NO → ZnO:N (R5)	ZnO + NO → ZnO:N (R5)	ZnO + NO → ZnO:N (R5)	

Reaction R4 is affected in two different ways when increasing temperature. At higher temperature the sticking coefficient of the NO in the 5%NO/N$_2$ gas is lower, thus inhibiting R4. On the other hand, all of the chemical reactions, including R4, are temperature dependent and usually are enhanced at higher temperatures. Apparently, 350°C is a critical temperature for the efficacy of 5%NO/N$_2$ in process I. At temperatures lower than 350°C, all of the films annealed in 5%NO/N$_2$ are n-type, evidently because the chemical reactivities of R3 and R4 are lower. At 350°C, the increased chemical reactivities of R3 and R4 produce p-type material. When further increasing the temperature to 400°C, n-type material is obtained due to the poor sticking coefficient of NO, which inhibits R4. When pure NO is used, the sticking coefficient does not play a significant role due to the oversaturation of NO. The film conductivity is thus determined by the temperature dependence of reaction R5. At

350°C, the films are n-type for both processes I and II, but then become of "ambiguous" type (indeterminate) at 400°C. In process II, films grown with 5%NO/N_2 are *n*-type at any temperature. This is because the excess Zn from the 1st step reacts with the pure O_2 in the 2nd step, via reaction R2, before the introduction of the nitrogen dopant gas. Thus, the N_2 in the 5%NO/N_2 mixture does not have any excess Zn with which to react and form Zn_3N_2 and thus creates only n-type behavior at both 350°C and 400°C. Our previous EXFAS study[34] of process II indicated that the N_2 molecules will be trapped in the surface when there is no free Zn to react with and will contribute to the (N-N)$_O$ donor defects.

Table 3 presents some of the photoluminescence properties. The interesting lines can be roughly grouped into seven regions: 1.9 eV (red band); 2.5 eV (green band); 3.31 eV (acceptor band); 3.33 – 3.35 eV (defect lines); 3.360 (Al/Ga lines); 3.362 eV (X lines); and 3.366 eV (surface band). The 3.31-eV band was first seen in *p*-type material [12] and has been observed many times since [35]. Although there is wide disagreement as to the detailed nature of the transition, almost all workers believe that an acceptor is involved. The 3.33 – 3.35-eV lines can have one to five or even more components, depending on the growth process, and since some of these are produced by electron irradiation, they are designated "defect lines" in this work [36]. The sharp lines near 3.360 eV are mainly donor-bound excitons due to Al and Ga, although a defect-related line (at 3.3607 eV) is also known to exist in this region [36]. The 3.362-eV line is also a donor-bound-exciton transition, but its origin is unknown. Finally, a 3.366-eV band, similar to ours, has been shown to be surface-related [37]. From the results listed in Table 3, the relative strengths of the acceptor band (3.31 eV) seem to be correlated with the presence of N, except for the sample annealed only in O_2 at 500 °C. We can explain this result as follows: the as-grown ZnO has excess Zn at the film surface as described earlier. Upon annealing in pure O_2, most of the excess Zn will react with O_2 but some of the Zn atoms will diffuse into the bulk, attaching themselves to various types of acceptors. One such Zn_I-acceptor complex which has been proposed is Zn_I-N_O [36], and even though the sample in question has not been annealed in N, still it likely contains N, a very common background impurity in ZnO.

Conclusions

P-type, nitrogen-doped, ZnO films were formed, for the first time, by thermal evaporation. It was shown that the film conductivity depends not only on the dopant gas but also on the growth and annealing process. The formation of Zn_2N_3 and its further reaction with NO are thought to be important elements in this process. If so, the use of Zn_2N_3 as a starting material might further enhance nitrogen incorporation into the films. Correlations were obtained between electrical and optical properties, so that the latter can serve as useful tools in the development of better p-type ZnO.

Table 3. Photoluminence of ZnO:N

P: Process; AT: Annealing temperature (°C); CT: Conductivity type

Annealing Gas	P	AT	CT	Rank of NBE intensities (1: Strongest)			3.33–3.35 eV (defect lines)	3.31 eV (acceptor band)
				3.360eV	3.362eV	3.366eV		
As-grown ZnO	-	-	n	3	1	2	Very weak	Very weak
Pure O$_2$	II	500	n	3	2	1	Strong	Strong
5% NO/N$_2$	I	350	p	3	2	1	Very weak	Strong
5% NO/N$_2$	II	350	n	3	2	1	Very weak	Strong
5% NO/N$_2$	I	400	n	3	1	2	Very weak	Strong
5% NO/N$_2$	II	400	n	3	2	1	Medium	Strong
Pure NO	I	350	n	3	1	2	Weak	Strong
Pure NO	II	350	N	3	2	1	Strong	Strong
Pure NO	I	400	Amb-iguou	3	1	2	Weak	Strong
Pure NO	II	400	Amb-iguou	3	1	2	medium	Strong

Acknowledgement

The authors are grateful for funding under DOE BES Grant DE-FG02-07ER46389 administrated by Dr. Refik Kortan, and would like to thank T.A. Cooper and W. Rice for the Hall-effect and PL measurements, respectively.

Reference

1. D.C. Look, Mater. Sci. Eng. B **80**, 383 (2001).
2. D.P. Norton, Y.W. Heo, Mater. Today **34–40(7)**, 34 (2004)
3. T. Makino, C.H. Chia, T.T. Nguen, Y. Segawa, Appl. Phys. Lett. **77**, 1632 (2000)
4. A. Tsukazaki, A. Ohtomo, T. Onuma, M. Ohtani, T. Makino, M. Sumiya, K. Ohtani, S.F. Chichibu, S. Fuke, Y. Segawa, H. Ohno, H. Koinuma, M. Kawasaki, Nat. Mater. **4**, 42 (2005)
5. Z.G. Yu, P. Wu, H. Gong, Appl. Phys. Lett. **88**, 132114 (2006)
6. V. Vaithianathan, B.T. Lee, S.S. Kim, Appl. Phys. Lett. **86,** 062101 (2005)
7. F.X. Xiu, Z. Yang, L.J. Mandalapu, D.T. Zhao, J.L. Liu, Appl. Phys. Lett. **87**, 152101 (2005)
8. G.D. Yuan, Z.Z. Ye, L.P. Zhu, Q. Qian, B.H. Zhao, R.X. Fan, L. Graig, S.B. Perkins, Zhang, Appl. Phys. Lett. **86**, 202106 (2005)
9. L.L. Chen, Z.Z. Ye, J.G. Lu, K.P. Chu, Appl. Phys. Lett. **89**, 252113 (2006)
10. J.D. Ye, S.L. Gu, F. Li, S.M. Zhu, R. Zhang, Y. Shi, Y.D. Zhang, X.W. Sun, G.Q. Lo, D.L. Kwong, Appl. Phys. Lett. **90**, 152108 (2007)
11. M. Joseph, H. Tabata, T. Kawai, Jpn. J. Appl. Phys. Part 2 **38**, L1205 (1999)
12. D.C. Look, D.C. Reynolds, C.W. Litton, R.L. Jones, D.B. Eason, G. Cantwell, Appl. Phys. Lett. **81**, 1830 (2002)

13. A.B.M.A. Ashrafi, I. Suemune, H. Kumano, S. Tanaka, Jpn. J. Appl. Phys. Part 2 **41**, L1281 (2002)

14. K. Minegishi, Y. Koiwai, Y. Kikuchi, K. Yano, M. Kasuga, A. Shimizu, Jpn. J. Appl. Phys. Part 2 **36**, L1453 (1997)

15. X.L. Guo, H. Tabata, T. Kawai, J. Cryst. Growth **237–239**, 544 (2002)

16. Z.Z. Ye, J.G. Lu, H.H. Chen, Y.Z. Zhang, L. Wang, B.H. Zhao, J.Y. Huang, J. Cryst. Growth **253**, 258 (2003)

17. C.C. Lin, S.Y. Chen, S.Y. Cheng, H.Y. Lee, Appl. Phys. Lett. **84**, 5040 (2004)

18. J.M. Bian, X.M. Li, X.D. Gao, W.D. Yu, L.D. Chen, Appl. Phys.Lett. **84**, 541 (2004)

19 X. Li, Y. Yan, T.A. Gessert, C.L. Perkins, D. Young, C. Dehart, M. Young, T. J. Coutts, J. Vac. Sci. Technol. A **4**, 1342 (2003)

20 Z.Q. Fang, B. Claflin, D.C. Look, L.L. Kerr, X. Li, J. Appl. Phys. **102**, 023714 (2007)

21 A. Dadgar, A. Krtschil, F. Bertram, S. Giemsch, T. Hempel, P. Veit, A. Diez, N. Oleynik, R. Clos, J. Christen, A. Krost, Superlattice Microstruct. **38**, 245 (2005)

22 W. Liu, S.L. Gu, J.D. Ye, S.M. Zhu, Y.X. Wu, Z.P. Shan, R. Zhang, Y.D. Zheng, S.F. Choy, G.Q. Lo, X.W. Sun, J. Cryst. Growth **310**, 3448 (2008)

23 C.L. Perkins, S.H. Lee, X. Li, S.E. Asher, T.J. Coutts, J. Appl. Phys. **3**, 034907 (2005)

24 M. Tu, Y. Su, C. Ma, J. Appl. Phys. **100**, 053705 (2006)

25 J. Wang, V. Sallet, F. Jomard, A.M. Botelho do Rego, E. Elamurugu, R. Martins, E. Fortunato, Thin Solid Films **24**, 8785 (2007)

26 V. Tvarozek, K. Shtereva, I. Novotny, J. Kovac, P. Sutta, R. Srnanek, A. Vincze, Vac. **2**, 166 (2007)

27 K.S. Ahn, Y.F. Yan, M. Al-Jassim, J. Vac. Sci. Technol. B **4**, L23 (2007)

28 J.G. Lu, Y.Z. Zhang, Z.Z. Ye, L.P. Zhu, L. Wang, B.H. Zhao, Q.L. Liang, Appl. Phys. Lett. **88**, 222114 (2006)

29. H.W. Liang, Y.M. Lu, D.Z. Shen, Y.C. Liu, J.F. Yan, C.X. Shan, B.H. Li, Z.Z. Zhang, J.Y. Zhang, X.W. Fan, Phys. Status Solidi. A **6**, 1060 (2005)

30. Yanfa Yan, and S.B. Zhang, Phys. Rev. Lett. B, Vol. 66, 073202, (2002).

31. W. Xu, Z. Ye, T. Zhou, B. Zhao, L. Zhu, and J. Huang, J. of Crystal Growth, Vol. 265, pp. 133-136, (2004).

32. H.S. Kim, E.S. Jung, W.J. Lee, J.H. Kim,S.O. Ryu, S.Y. Choi, Ceram. Int. **34**, 1097 (2008)

33. Y. Natsume, H. Sakata, Mater. Chem. Phys. **78**, 170 (2002)

34. W. Mu, L. Kerr, N. Leyarovska, Chem. Phys. Lett. **469**, 318 (2009)

35. D.C. Look, B. Claflin, Ya. I. Alivov, and S.J. Park, phys. Stat. sol. (a) **201**, 2203 (2004)

36. D.C. Look, G.C. Farlow, P. Reunchan, S. Limpijumnong, S.B. Zhang, and K. Nordlund, Phys. Rev. Lett. **95**, 225502 (2005)

37. L. Wischmeier, T. Voss, I. Rückmann, J. Gutowski, A.C. Mofor, A. Bakin, and A. Waag, Phys. Rev. B **74**, 195333 (2006)

Electronic and Optical Properties

Mater. Res. Soc. Symp. Proc. Vol. 1201 © 2010 Materials Research Society   1201-H03-03

In Situ Surface Photovoltage Spectroscopy of ZnO Nanopowders Processed by Remote Plasma

Raul M. Peters[1], Stephen P. Glancy[2], J. Antonio Paramo[1], and Yuri M. Strzhemechny[1]
[1]Department of Physics and Astronomy, Texas Christian University, Fort Worth, TX, USA
[2]Department of Engineering Science, Trinity University, San Antonio, TX, USA

ABSTRACT

In many instances the quality of the surface in ZnO nanoscale systems is a key performance-defining parameter. The surface itself could be a very significant source of lattice defects as well as contaminating impurities, and this influence may extend into the sub-surface vicinity. In our work, key element of the surface analysis is the surface photovoltage (SPV) spectroscopy known for its advantages, such as: identification of conduction vs. valence band nature of the defect-related transitions and the defect level positions within the band gap, ability to measure relatively low densities of surface defects as well as their cross sections. Additional information can be obtained from the SPV transient measurements. In our system, SPV characterization is run in high vacuum, complemented by *in situ* remote plasma treatment. This combination of surface-sensitive and surface-specific tools is well-suited for studying surface properties with a high degree of reliability since there is no exposure to common air contaminants between processing and characterization cycles. We employed O/He remote plasma treatments of ZnO nanocrystalline surfaces. In situ SPV spectra and transient measurements of the as-received and processed samples revealed, on the one hand, a number of common spectral features in different ZnO nanopowder specimens, and, on the other hand, a noticeable plasma-driven changes in the surface defect properties, as well as in the overall electronic and optical surface characteristics.

INTRODUCTION

Market-driven technological miniaturization imposes new demands on the device geometry and performance: as the volume of the crystals is shrinking, the relative significance of the crystal surfaces and their properties is increasing. In nanostructures therefore one deals with the domination of surface-enhanced phenomena, many of which are mediated by defects. Thus, understanding and control of the specifically surface as well as near-surface defects comes to the forefront.

Major thrust in the studies of defects in bulk ZnO crystals has been recently focused primarily on overcoming the difficulties with producing a high-quality reliably *p*-type material [1]. Such material is sought because of great potential for spintronic and optoelectronic applications. As of today, efforts to obtain a device-grade *p*-type ZnO have not resulted in a material suitable for mass production. Remarkably, nanoscale ZnO structures are offering numerous applications that do not require resolving the challenge of *p*-type conductivity. Recent reports demonstrated that nanosize ZnO could be employed in such systems as nanotransducers [2], random lasers [3], dye-sensitized solar cells [4], nanosensors [5], etc. Obviously, surface defects influence key performance properties of those promising zinc oxide nanosystems. There is no agreement on the nature of these surface states as well as their relation to bulk defects. Moreover, not only understanding, but the entailing modification and control of the surface defects in nanocrystals are required for performance improvement of ZnO-based photovoltaics, optoelectronic systems, sensors, etc. It is necessary to significantly advance our understanding and ability to manipulate surface defect properties in ZnO nanostructures.

To address this we employed in our studies an *in vacuo* combination of such surface-sensitive and surface-specific tools as surface photovoltage (SPV) spectroscopy and remote plasma (RP) treatment. SPV is known [6] for its ability to detect surface states and distinguish their charge sates and donor- vs. acceptor-like nature. Notable advantages of this technique are:identification of conductivity type, conduction vs. valence band nature of the deep level transitions and the deep level positions within the band gap, ability to measure surface defect densities less than 10^{10} cm^{-2} as well as their cross sections. Additional information can be deduced from the SPV transient measurements. On the other hand, RP (neutral and chemically-active) yields a double benefit. On the one hand, it is an efficient tool for surface cleaning, and, on the other hand, it allows one to tailor surface properties. Chemically active RP may activate such processes as ionization, neutralization, de-excitation, and dissociation within the plasma penetration range and then beyond due to diffusion and local stresses. Our *in situ* arrangement of processing and characterization is uniquely suited for studying surface properties with a high degree of reliability since in between processing and characterization cycles there is no exposure to common air contaminants. Although ZnO nanostructures are realized in many morphologies and dimensionalities, we chose, as a natural starting point, to run experiments on a number of commercially available ZnO nanopowders with different average grain sizes. They are valuable objects for research, firstly, because of their bulk commercial availability and due to their numerous applications.

One has to bear in mind that SPV experiments on nanocrystalline samples pose a challenge. In macroscopic crystals, the efficient separation of the electron-hole pairs, essential for meaningful SPV response, is driven by the built-in field in the space-charge region as well as the Dember effect (light-induced concentration gradient) [7]. When the crystal size becomes comparable to the thickness of the space-charge region and the characteristic diffusion length, detection of the SPV signal may become problematic because of the lower field and faster diffusion [8]. For the smallest particles the additional complication arises from the disappearance of a well-defined band-gap.

In recent years there were only a few published SPV studies of nanoscale ZnO [8-13]. Most of them employed a capacitive SPV geometry, and the spectra did not reveal well-defined surface defect transitions, primarily because of a relatively low signal-to-noise ratio. Moreover, these experiments were performed, as a rule, in ambient air. In our work we implemented the Kelvin-probe geometry of the SPV setup *in vacuo* and *in situ* with RP. We were able to obtain clear spectral signatures of a number of defect states and observe significant plasma-induced evolution of the surface.

EXPERIMENTAL DETAILS

Nanopowder ZnO samples were obtained from four different vendors: Alpha Aesar (AA), Sigma Aldrich (SA), Zochem (Z), and American Elements (AE25 and AE31). The latter supplied two grades with different average grain sizes. Crystal sizes and size distributions were determined from the SEM/TEM images with 30 to 50 nanocrystals per sample being analyzed. The average sizes were found to be as follows: AA ~ 60 nm, SA ~ 90 nm, Z ~ 200 nm, AE31 ~ 80 nm, and AE25 < 25nm. The powders were deposited with a spatula on a surface of a vacuum-compatible conductive carbon tape. In addition, the as-received powders were compressed at room temperature into circular pellets ~ 2 mm thick and ~ 1.3 cm in diameter using a Carver hydraulic column press with the nominal force of 7 metric tons. The pellets were rather brittle pointing to a relatively weak bonding between the nanocrystals. We add the 'PLT' suffix to distinguish the pelletized specimens.

For the RP treatment, a high-grade oxygen/helium gas mixture (flow rate ratios ~ 1:2) was supplied into the vacuum chamber through a needle leak-valve and a Pyrex tube inductively coupled with a Dressler Cesar 133 radio-frequency generator. The plasma flame inside the Pyrex tube was controlled by the flow rate and pressure of the gas as well as the power supplied by the generator. We employed 150 – 200 mTorr of gas pressure and 30-40 W of power of the generator. The samples were placed ~ 15 cm downstream from the plasma-generating coil. The RP exposure time varied between 15 and 40 min. We add the 'RP' suffix to distinguish the RP-treated powders. *In situ* SPV characterization was performed before and after RP processing of the samples in an adjacent vacuum chamber. To minimize the vibration during SPV experiments the vacuum in the chamber was maintained by the ion pump only. The signal was collected by a vibrating Besocke Kelvin Probe S positioned near the surface of the sample of interest and Kelvin Control 07. SPV response was excited by a light of a variable frequency (photon energy range 1.13 eV – 3.65 eV) supplied via the fiber optic bundle coupled through an optical feedthrough with an *ex vacuo* optical train: 250 W QTH lamp as a white light source, a pair of fused silica lenses, band-pass filters, and the Oriel Cornerstone grating monochromator. The light was chopped with the frequency of 100 Hz. The spectral ranges were chosen in combination with the band-pass filters to avoid possible higher order diffraction interferences. Prior to each spectral sequence, transient response curves were obtained. The samples were illuminated directly with white light until the SPV saturation was achieved. Then the experiment continued in the dark until surface state equilibrium was achieved.

RESULTS AND DISCUSSION

As we mentioned above, both the carbon-tape deposited powders and the pelletized samples were characterized as-received and following RP treatments. For all these samples n-type super-bandgap transitions were observed at ~ 3.3. eV. The space-charge regions for all as-received and RP-treated surfaces were assumed to be predominantly depleted. Analysis of the sub-bandgap SPV spectra of the studied samples revealed a variety of transitions, most of which were reproducible in the spectra of different powders. However, we observed substantial vendor-to-vendor discrepancies in both the distribution of the spectral signatures as well as their relative intensities. It must be pointed out that such variations were not unexpected in view of our recent reports on the photoluminescence and positron annihilation experiments involving the same nanopowders, where significant spectral deviations were observed as well [14,15]. Table I lists the synopsis of the observed SPV spectral transitions in the studied samples. Several levels are located in the close proximity of each other, especially in the mid-gap vicinity, where the complementary components can be observed. Recent theoretical calculations for single native defects in bulk ZnO provide reasonable explanation for such ambiguity in assignment, resulting from a multitude of possible transitions with similar locations in the gap and different charge states [16]. Besides, the transitions observed in our samples may be associated not only with individual native defects but also with most common impurities, extended defects, and defect clusters. We may speculate about the origins of the observed levels based on the theoretical predictions [16] for the bulk, yet these assignments may not reflect a possible change in energetics in the presence of the surface. To the best of our knowledge, there were no systematic theoretical studies of the energetics of essentially surface states in nanocrocrystalline ZnO.

Table I. SPV spectral transitions observed in the studied samples.

As-received samples	$E_c - E$, eV	$E - E_v$, eV	RP-processed samples	$E_c - E$, eV	$E - E_v$, eV
Z	1.3	-	Z-RP	1.65, 2.9	-
AA-PLT	1.65	-	AA-PLT-RP	1.65	2.2, 2.4, 2.85
AE25	-	-	AE25-RP	1.5, 1.65	2.2, 2.4, 2.85
AE31	-	-	AE31-RP	1.3, 1.5, 1.65	2.2
AE31-PLT	1.3, 1.5	2.85	AE31-PLT-RP	1.5, 2.9	2.85
SA	-	-	SA-RP	1.5	2.2
SA-PLT	1.3	2.85	SA-PLT-RP	1.3, 1.5, 1.65	2.2, 2.85

For example, transitions shown in Fig. 1 could be tentatively attributed to oxygen vacancies, those shown in Fig. 2 – to Zn vacancies, and in Fig. 3 – to O interstitials.

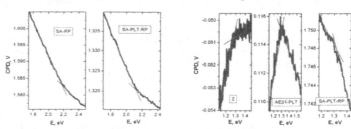

Figure 1. SPV transitions for surface states ~ 2.2 eV above the top of the valence band (left) and ~ 1.3 eV below the bottom of the conduction band (right).

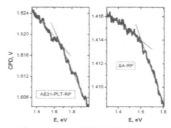

Figure 2. SPV transitions for surface states ~ 1.65 eV below the bottom of the conduction band.

SPV experiments revealed substantial plasma-induced changes for practically all the samples. A greater variety of transitions was observed (see Table I), with defect energetics strongly dependent on a specific sample, as well as plasma parameters. In the plasma-treated samples we generally observed two common effects – a much better signal-to-noise ratio and much faster light-dark equilibration times, both of which point to a substantial increase of surface charge density. Analysis of the plasma-induced changes in the transient characteristics (such as shown in Fig. 4) indicates the increase of the majority carrier concentration at the surface by one to three orders of magnitude. Alternatively, similar to [12] one can argue that oxygen plasma species may be absorbed by the relatively oxygen-deficient surface and thus increase the induced field in the space-charge region, which in turn will enhances the electron-hole pair separation.

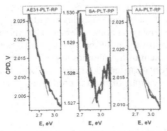

Figure 3. SPV transitions for surface states ~ 2.85 eV above the top of the valence band.

We found that compression of the powders into pellets bears a consistent and visible improvement in the SPV signal-to-noise ratio and spectral stability, a robust evidence of the contribution to enhanced electron-hole pair separation from the reduction of the inter-crystalline space and the increased interaction between grains' surfaces.

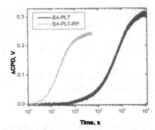

Figure 4. RP-induced changes in the transient SPV properties indicating the increase of the majority carrier concentration at the surface by three orders of magnitude

Comparison of the SPV results with our previously reported photoluminescence and positron lifetime measurements of the same ZnO nanopowders [14,15] revealed only a limited correlation. This may be a good indication that the SPV spectral features have primarily surface origin, whereas luminescence and positron experiments are probing mostly the bulk defect properties. Furthermore, SPV can identify non-radiative gap states undetectable by PL.

Importantly, for several RP-treated samples we observed time evolution of the SPV spectral features, when those samples remained in the characterization chamber for extended periods of time and the spectra were collected on the same spot with an interval of one to several days. We omit detailed description of this spectral evolution due to the space limitations of the manuscript and we will defer such discussion to our further reports. Nevertheless, we have to emphasize in this regard the importance of running our experiments in vacuum, when the plasma-modified surface properties are practically unaffected by the common air contaminants.

CONCLUSIONS

For nanopowder ZnO samples we performed SPV measurements in tandem with RP treatments to determine surface state energetics. Our experiments identified a number of defect signatures present in the spectra of different samples. Assignments of these transitions to native point defects are tentative and require further elucidation and adequate surface-specific theoretical calculations. It was shown that both pelletization of the powders and oxygen plasma treatment lead to a significant improvement of the SPV signal, as well as overall evolution of the spectra. Moreover, oxygen treatment of the samples leads to a substantial increase in the near-surface concentration of the majority carriers. Our observations also provide a good example of plasma serving as a supplementary probe of surface properties.

REFERENCES

1. Ü. Özgür, Ya. I. Alivov, C. Liu, A. Teke, M. A. Reshchikov, S. Doğan, V. Avrutin, S.-J. Cho, and H. Morkoç, *J. Appl. Phys.* **98**, 041301 (2005), and references therein.
2. P. X. Gao, J. Song, J. Liu, and Z. L. Wang, *Adv. Mater.* **19**, 67–72 (2007).
3. H. Noh, M. Scharrer, M. A. Anderson, R. P. H. Chang, and H. Cao, *Phys. Rev. B* **77**, 115136 (2008).
4. Q. Zhang, T. P. Chou, B. Russo, S. A. Jenekhe, and G. Cao, *Angew. Chem. Int. Ed.* **47**, 2402–2406 (2008).
5. Z. L. Wang, *MRS Builletin* **32**, 109-116 (2007).
6. L. Kronik and Y. Shapira, *Surface Science Reports* **37**, 1 (1999), and references therein.
7. H. Dember, *Phys. Z.* **32**, 554 (1931).
8. L. Jing, X. Sun, J. Shang, W. Cai, Z. Xu, Y. Du, and H. Fu, *Solar Energy Materials & Solar Cells* **79**, 133–151 (2003).
9. L. Jing, Z. Xu, X. Sun, J. Shang, and W. Cai, *Applied Surface Science* **180**, 308 (2001).
10. Y. Lin, D. Wang, Q. Zhao, M. Yang, and Q. Zhang, *J. Phys. Chem. B* **108**, 3202 (2004).
11. L. Jing, B. Wang, B. Xin, S. Li, K. Shi, and W. Cai, H. Fu, *Journal of Solid State Chemistry* **177**, 4221 (2004).
12. Y. Lin, D. Wang, Q. Zhao, Z. Li, Y. Ma, and M. Yang, *Nanotechnology* **17**, 2110 (2006).
13. Q. Zhao, T. Xie, L. Peng, Y. Lin, P. Wang, L. Peng, and D. Wang, *J. Phys. Chem. C*, **111**, 17136 (2007).
14. R. M. Peters, J. A. Paramo, C. A. Quarles, and Y. M. Strzhemechny, "Correlation between optoelectronic and positron lifetime properties in as-received and plasma-treated ZnO nanopowders", *Application of Accelerators in Research and Industry*, ed. F. D. McDaniel and B. L. Doyle (AIP, 2009) pp. 965-969.
15. J. A. Paramo, R. M. Peters, C. A. Quarles, H. Vallejo, and Y. M. Strzhemechny, *IOP Conf. Series: Materials Science and Engineering* **6**, 012030 (2009).
16. P. Erhart, K. Albe, and A. Klein, *Phys. Rev. B* **73** 205203 (2006).

Mater. Res. Soc. Symp. Proc. Vol. 1201 © 2010 Materials Research Society   1201-H03-08

Bound-exciton recombination in Mg$_x$Zn$_{1-x}$O thin films

Christof P. Dietrich, Alexander Müller, Marko Stölzel, Martin Lange, Gabriele Benndorf, Holger von Wenckstern, and Marius Grundmann
Institut für Experimentelle Physik II, Universität Leipzig, Linnéstr. 5, 04103 Leipzig, Germany

ABSTRACT

Excitons in semiconductor alloys feel a random disorder potential leading to inhomogeneous line broadening and a lack of knowledge about the dominating recombination processes. Nevertheless, we demonstrate competing localization effects due to disorder (random potential fluctuations) and shallow point defects. We were able to spectrally separate donor-bound and quasi-free excitons within the whole wurtzite-type composition range of Mg$_x$Zn$_{1-x}$O ($0 \leq x \leq 0.33$) using spectrally resolved ($x \leq 0.06$) and time-resolved photoluminescence ($x \geq 0.08$). We found out that donor-bound excitons dominate photoluminescence spectra even for Mg-contents up to $x = 0.18$ and still appear for $x = 0.33$.

INTRODUCTION

Semiconductor alloys are of fundamental importance for today's semiconductor industries. They are used for photonic and electronic applications and offer tunable structural, optical and electronical properties. A multitude of semiconductor alloys has been subject to investigations [1]. By mixing ZnO ($E_g = 3.37$ eV) and MgO ($E_g = 7.5$ eV) in a defined ratio the band gap of the ternary alloy Mg$_x$Zn$_{1-x}$O can be tuned to very short wavelengths. Mg$_x$Zn$_{1-x}$O exhibits excitonic recombination at or above room temperature [2].

The substitution of metal atoms on the cation sublattice leads to a disorder-induced broadening of the spectral lines [3]. We emphasize that due to sizable alloy broadening the recombination of bound and free excitons cannot be spectrally resolved in Mg$_x$Zn$_{1-x}$O for $x > 0.04$. This is the typical situation also for other semiconductor alloys when for all temperatures only one (broad) photoluminescence (PL) peak is observed. Another well-known phenomenon in semiconductor alloys is the temperature-dependent S-shape of the emission maximum due to localization of excitons in the alloy disorder potential which is used to calculate the mean disorder potential depth [4].

We demonstrate that exciton localization in Mg$_x$Zn$_{1-x}$O is due to competing localization mechanisms based on disorder and shallow defects and cannot be exclusively attributed to one of these. Furthermore, we prove the presence of donor-bound excitons up to Mg-contents $x = 0.33$.

EXPERIMENT

The investigated Mg$_x$Zn$_{1-x}$O thin films were grown by pulsed-laser deposition (PLD) on (11-20)-oriented sapphire substrate material. A polycrystalline target consisting of ZnO and MgO was ablated with the 248 nm line of a KrF excimer laser onto the sapphire substrates. These were held at a constant temperature of about 700 °C during the growth. The oxygen partial pressure $p(O_2)$ was varied between 1×10^{-1} and 3×10^{-4} mbar as well as the target composition between 0 wt.-% and 10 wt.-% MgO to achieve different Mg-contents in the thin films. The obtained Mg-contents x range in the whole wurzite-type $Mg_xZn_{1-x}O$ regime from 0.005 to 0.33.

PL at 2 K of the deposited thin films was measured in a He bath cryostat. Measurements between 4.5 K and 300 K were carried out in a temperature-variable He coolfinger cryostat. The time-integrated luminescence was excited with the 325 nm line of a continuous-wave HeCd-laser, the time-resolved luminescence (TR-PL) was excited using either the 696 nm or 810 nm line of a pulsed Ti:Sa laser converted to the UV spectral range by second or third harmonic generation, respectively, with a pulse width of approximately 200 fs. The luminescence was spectrally decomposed using a 320 mm focal length grid monochromator with a 2400 lines/mm grating and detected using either a micro-channel plate photomultiplier or a Peltier-cooled GaAs photomultiplier. For more details of TR-PL setup see [5]. The minimum spectral bandwidth is 0.06 nm and the minimum time resolution is about 20 ps.

The Mg-content x of the thin films was determined performing energy-dispersive X-ray spectroscopy for $x < 0.06$ and Rutherford backscattering spectrometry for $x > 0.08$.

RESULTS AND DISCUSSION

Spectrally resolved luminescence

PL spectra at $T = 2$ K of 5 $Mg_xZn_{1-x}O$ samples with Mg-contents $0.005 \leq x \leq 0.09$ are shown in Fig.1(b). The energetic position of the PL maxima depends on the Mg-content in the thin film (Fig.1(a)). The PL maxima shift towards higher energies with increasing Mg-content due to a change of the alloy band gap. A linear fit of the experimental data yields the equations

$$E_{PL}(2 \text{ K}) = 3.3601(7) \text{ eV} + 1.96(2) \text{ eV} \times x,$$
$$E_{PL}(300 \text{ K}) = 3.309(2) \text{ eV} + 1.85(4) \text{ eV} \times x.$$

For $x = 0$, the dominant transition at 2 K with an energy of 3.3601 eV is attributed to neutral Al-donor bound excitons (I_6-line in [6]). The dominant peak for $x = 0$ at 300 K resembles the transition energy of free excitons in binary ZnO.

PL spectra of Fig.1(b) show two maxima labeled M_1 and M_2. Peak M_1 can be observed for all investigated samples at low temperatures in contrast to peak M_2 which is only observable for $x < 0.031$ due to alloy broadening. For $x = 0.005$, the energetic separation of both peaks (15 meV) is equivalent to the localization energy E_{loc} of Al-donor bound excitons in pure ZnO.

Temperature-dependent PL spectra in temperature range 5 - 290 K of $Mg_{0.005}Zn_{0.995}O$ can be seen in Fig.1(c). The peak M_1 related to Al-donor bound excitons is only observable for temperatures below 160 K due to the finite localization energy. The activation energy for the intensity decrease according to [7] is found to be about $E_T = 16$ meV, similar to E_{loc}. Free excitons (maximum M_2) dominate the PL spectra above 100 K up to room temperature which is also typical in pure ZnO.

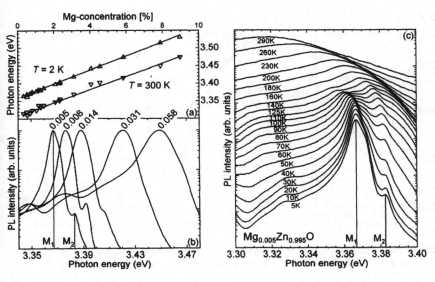

Figure 1: (a) PL maxima vs. Mg-concentration for MgZnO samples at $T = 2$ K (triangles up) and 300 K (triangles down). (b) Low temperature PL spectra of $Mg_xZn_{1-x}O$ samples with $0.005 \leq x \leq 0.058$. The peaks are labeled for $x = 0.005$. (c) Temperature-dependent PL spectra of $Mg_{0.005}Zn_{0.995}O$ sample for temperature range from 5 K to 290 K.

The temperature-dependent spectral positions of maxima M_1 and M_2 for Mg-contents $x = 0.005$, 0.031 and 0.058 are depicted in Fig.2. For $x = 0.005$, the maxima can easily be resolved individually, both peaks follow the temperature-dependent band gap shift [8] of pure ZnO [9] as indicated by the solid lines. For $x = 0.031$, the spectral position of M_1 undergoes a slight S-shape. This indicates a small redistribution of the donor-bound exciton population within the disorder potential [10]. We suppose this situation to be the same for the unbound excitons (M_2). In the following we will treat them as quasi-free. The recombination of quasi-free excitons can only be observed for temperatures above 70 K due to alloy broadening preventing the observation of a similar S-shape. An extrapolation for low temperatures indicates $E_{loc} = 21$ meV, in agreement with the thermal activation energy of donor-bound luminescence, $E_T = 22$ meV.

For $x = 0.058$, alloy broadening is so strong that M_1 and M_2 cannot be resolved individually in the entire temperature range. The spectral positions of the peak undergo a strong S-shape because the peak origin changes from Al-donor bound excitons to quasi-free excitons. Attributing the S-shift to disorder only [11, 12] misses the true nature of exciton localization. Excitons localize on shallow defects as well as in disorder fluctuations whereas localization on shallow defects is dominant in the samples presented so far. Extrapolating the red shift at high temperatures with ZnO band gap shift yields $E_{loc} = 28$ meV and is again as large as the energy for thermal activation, $E_T = 27$ meV. This gives strong evidence for our main statement that the

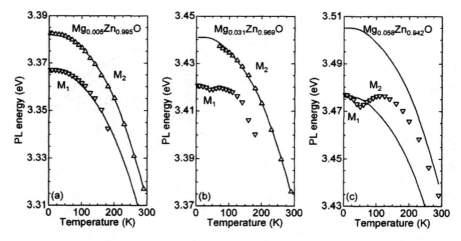

Figure 2: Temperature-dependent spectral positions of $Mg_xZn_{1-x}O$ samples with $x = 0.005$ (a), 0.031 (b) and 0.058 (c). The solid lines are fits of the ZnO band gap shift according to [9].

strong S-shape for $x > 0.04$ is primary due to the delocalization of impurity-bound excitons and not due to the localization of free excitons in the disorder potential.

Time-resolved luminescence

Fig. 3(a) shows typical low temperature transients of $Mg_xZn_{1-x}O$ samples with $x = 0.08, 0.18$ and 0.33 recorded at the PL maximum. All transients behave similar: after the laser excitation pulse a fast intensity decrease can be observed, followed by a slow non-exponential decay which gets significantly slower as the Mg-content increases. The decay time of the fast process is below 1 ns for all investigated samples.

We were not able to obtain a reasonable fit of the data by summing up three mono-exponential functions. Therefore, we used a stretched exponential and a modified power-law as model decay functions. The intensity $I(t)$ of the fast process can be fitted well by a stretched exponential [13] of shape

$$I_{fast}(t) = a_{fast} \exp[-(t/\tau)^\alpha],$$

whereas the intensity of the slow process can be described via the modified power-law [14]

$$I_{slow}(t) = a_{slow} (1 + t/t_0)^{-\beta}.$$

The equations include fit parameters α and β (shape), τ and t_0 (time scale) as well as a_{fast} and a_{slow} (intensity scale), respectively. By fitting the transients with a sum of Eq.3 as well as Eq.4 (as

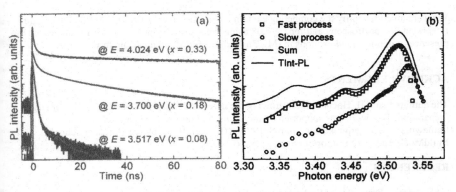

Figure 3: (a) Transients of $Mg_xZn_{1-x}O$ samples with $x = 0.08$, 0.18 and 0.33 recorded at the emission maximum of each sample (solid lines) and fitted by a sum of Eq. 3 and 4 (red dashed lines). (b) Calculated, time-integrated intensities of the decay process present in the transients of $Mg_{0.08}Zn_{0.92}O$, their sum (red solid line) and corresponding spectrum from time-integrated PL (Tint-PL, black solid line).

shown in Fig.3(a)) and integrating over time, we were able to deconvolve the luminescence spectrum into two processes as shown in Fig.3(b). The sum of both processes (red solid line) resembles the measured time-integrated PL spectrum (black solid line) as expected. Again, the maxima of the processes are spectrally separated reflecting the localization energy of donor-bound excitons in ZnO. Therefore, the fast process dominating directly after the laser pulse can be attributed to donor-bound excitons while the slow process is caused by quasi-free, disorder-bound excitons.

As shown in Fig.3(a) the spectral weight of the slow process changes with increasing Mg-content indicating its dominance for Mg-contents above $x = 0.18$. The recombination of donor-bound excitons is overtopped by the recombination of quasi-free excitons but still present for x = 0.33. This is due to the increase of the mean potential barrier within the disorder potential with increasing Mg-content becoming larger than the exciton localization energy on donors.

SUMMARY

In conclusion, we found two competing exciton localization mechanisms in wurtzite-type (Mg,Zn)O alloys. On the one hand, excitons localize on shallow defects, e.g. Al-donors, for low temperatures, but and on the other hand, excitons also localize in the disorder potential fluctuations. Both mechanisms cannot be observed separately for alloy broadening larger than E_{loc} which leads to a more complicated temperature-dependence of the PL energy, so called 'S-shape'.

We were able to identify the competing exciton localization mechanism for $x \leq 0.06$ using time-integrated PL spectroscopy and for $x \geq 0.08$ using time-resolved PL spectroscopy and proved the presence of donor-bound excitons in (Mg,Zn)O alloys for Mg-contents up to $x = 0.33$.

ACKNOWLEDGMENTS

The authors thank H. Hochmuth for growing the investigated samples and G. Ramm for target preparation. This work was supported by the Deutsche Forschungsgemeinschaft in the framework of Schwerpunktprogramm SPP1136 (Gr 1101/10-3), by the Graduate School BuildMoNa and by the European Social Fund (ESF).

REFERENCES

[1] S. Adachi, Properties of Semiconductor Alloys (Wiley, Chichester, 2009).
[2] D.G. Thomas, J. Phys. Chem. Sol. **15**, 86 (1960).
[3] R. Zimmermann, J. Cryst. Growth **101**, 346 (1990).
[4] J. Christen and D. Bimberg, Phys. Rev. B **42**, 7213 (1990).
[5] A. Müller, M. Stölzel, C.P. Dietrich, G. Benndorf, M. Lorenz and M. Grundmann, J. Appl. Phys. **107**, 013704 (2010).
[6] B.K. Meyer et al., phys. stat. sol. B **241**, 231 (2004).
[7] D. Bimberg, M. Sondergelb and E. Grobe, Phys. Rev. B **4**, 3451 (1971).
[8] L. Viña, S. Logothetidis and M. Cardona, Phys. Rev. B **30**, 1979 (1984).
[9] M. Grundmann and C.P. Dietrich, J. Appl. Phys. **106**, 123521 (2009).
[10] Q. Li, S. J. Xu, W. C. Cheng, M. H. Xie, S. Y. Tong, C. M. Che, and H. Yang, Appl. Phys. Lett. **79**, 1810 (2001).
[11] S. Heitsch et al., J. Appl. Phys. **101**, 083521 (2007).
[12] T.A. Wassner, B. Laumer, S. Maier, A. Laufer, B.K. Meyer, M. Stutzmann and M. Eickhoff, J. Appl. Phys. **105**, 023505 (2009).
[13] M. Berberan-Santos, E. Bodunov and B. Valeur, Chem. Phys. **315**, 171 (2005).
[14] M. Berberan-Santos, E. Bodunov and B. Valeur, Chem. Phys. **317**, 57 (2005).

Electric and Magnetic Properties

Mater. Res. Soc. Symp. Proc. Vol. 1201 © 2010 Materials Research Society   1201-H04-02

Light beam induced current measurements on ZnO Schottky diodes and MESFETs

H. von Wenckstern, Z. P. Zhang, M. Lorenz, C. Czekalla, H. Frenzel, A. Lajn and
M. Grundmann

Universität Leipzig, Fakultät für Physik und Geowissenschaften, Institut für Experimentelle
Physik II, Linnéstrasse 5, 04103 Leipzig, Germany

ABSTRACT

The homogeneity of the Schottky barrier potential of reactively sputtered PdO_y/ZnO
Schottky contacts has been investigated by light beam-induced current measurements on the
micrometer scale. It is found that a metallic capping layer, acting as an equipotential surface, is
not necessary for PdO_y/ZnO Schottky contacts in contrast to Ag_xO/ZnO Schottky diodes.
Further, we probed the generated photocurrent of a ZnO-based metal-semiconductor field-effect
transistor for a closed and open channel, respectively. The photocurrent is, in general, one order
of magnitude larger for closed channel conditions. The position of maximum photocurrent
generation shifts towards the drain for higher source drain voltages for closed channel
conditions, whereas it is nearly independent of the source drain potential for an open channel.

INTRODUCTION

Properties of Schottky barrier contacts on ZnO have improved considerably in the past
years either due to refined pretreatment of the ZnO surface [1] or due to the usage of non-
stoichiometric metal oxides as Schottky contact instead of the pure metal [2, 3]. It has been
shown that in the case of a non-stoichiometric silver oxide Schottky contact a metallic capping
layer is necessary to provide an equipotential surface on top of the metal oxide [3]. This ensures
that the electrically active contact area equals the geometric contact area (the contact area is
input to most formulas used to calculate the Schottky contact parameters). Further in, e.g.
Ag_xO/ZnO photodiodes, photo-generated carriers are not efficiently collected if a metallic
capping is omitted [3]. However, comparisons of the spatially resolved photocurrent between
capped and uncapped metal oxide/ZnO Schottky contacts are only published for silver oxide so
far. We investigated the homogeneity of non-stoichiometric palladium oxide Schottky diodes on
ZnO thin films by light beam-induced current (LBIC) measurements on the micrometer scale.
Further we used this method to study the spatial dependence of the generated photocurrent
between the source and drain contacts of a ZnO-based metal-semiconductor field-effect
transistor (MESFET) having a PdO_y Schottky contact as gate.

EXPERIMENT

The ZnO thin films used for the Schottky diodes were grown by pulsed-laser deposition
(PLD) on a-plane sapphire substrates. First a metallic conducting ZnO:Al layer of about 400 nm
thickness was deposited serving as Ohmic back contact of the Schottky diodes [4]. On top of the
ZnO:Al, a nominally undoped ZnO layer of about 2 μm thickness was grown. The net doping
density of this layer is in the mid 10^{16} cm^{-3} range. The growth temperature was about 650°C, the
oxygen partial pressure was 0.016 mbar. The circular Schottky contacts were defined by a

shadow mask. The contacts were realized without any surface pretreatment by reactive dc sputtering of Pd; subsequently a Pd capping layer was sputtered in pure Ar for some of the diodes [3]. The contacts were Au wire bonded; the wires were fixed by a conductive epoxy resin. The MESFETs were grown on quartz glass substrates by PLD. The thickness of the channel is about 30 nm; the net doping density is about 10^{18} cm^{-3}. For MESFET samples the growth temperature was 650°C and the oxygen partial pressure was 3×10^{-4} mbar. The processing of the devices is given in Ref. 5. In this study, however, we used PdO$_y$ as gate contacts.

Properties of PdO$_y$/ZnO Schottky contacts

We have previously shown that reactively sputtered Ag, Pd and Pt contacts are partially oxidized and demonstrated that a capping layer is necessary to provide an equipotential surface indispensible for quantitative analysis for the case of Ag [3]. In this Letter we compare capped and uncapped PdO$_y$/ZnO Schottky diodes. The current-voltage characteristics of such diodes are compared in Fig 1a). Noteworthy, the ideality factor n of the capped diode is with 1.58 considerably smaller than $n = 2.35$ of the uncapped diode. This is clearly evident from the steeper slope of the forward characteristic of the capped diode. A further difference is that the uncapped diode exhibits a much stronger voltage dependence of the reverse current indicating higher lateral potential fluctuations of the Schottky barrier. The higher ideality factor combined with the stronger voltage dependence of the reverse current of the uncapped diode is due to a higher standard deviation of the mean barrier height and its larger voltage coefficient ρ_3 (cf. Ref. 6).

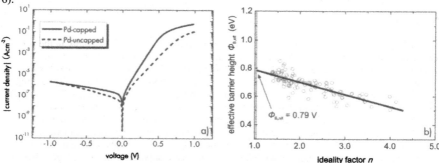

Fig. 1: a) Current density of uncapped and capped reactively sputtered PdO$_y$/ZnO contacts. b) Room temperature effective barrier height of 108 capped reactively sputtered PdO$_y$/ZnO contacts versus the ideality factor. The line represents a linear fit to the data.

For a laterally inhomogeneous barrier potential the effective barrier height $\Phi_{B,\text{eff}}$ (determined from the forward current considering thermionic emission only) should increase for decreasing ideality factor [7,8]. Within this work we investigated 108 Pd capped PdO$_y$/ZnO Schottky diodes. For each diode the ideality factor n and the effective barrier height $\Phi_{B,\text{eff}}$ were determined and are plotted in Fig. 1b). Most of the diodes exhibit ideality factors lower than 2, however, ideality factors up to 4 were observed. A linear relationship between the ideality factor and the effective barrier height is obvious from the figure; a linear regression of the data (solid line in the figure) yields the barrier height $\Phi_{B,\text{nif}}$ at the ideality factor determined by image force lowering (n_{if} = 1.014) of 0.79 V. The homogeneous barrier height $\Phi_{B,hom}$ of a Schottky contact can now be calculated by using $\Phi_{B,hom} = \Phi_{B,nif} + \delta\Phi_{if,0}$, where $\delta\Phi_{if,0}$ is the zero-bias image force lowering [8] to be 0.84 V. The homogeneous or mean barrier heights of another set of PdO$_y$/ZnO Schottky diodes published by Lajn et al. [3] are significantly higher. They were determined by using capacitance-voltage as well as temperature-dependent current-voltage measurements for an individual diode. We attribute this to variation of properties due to different degrees of oxidation of PdO$_y$. This issue is subject to further systematic investigation.

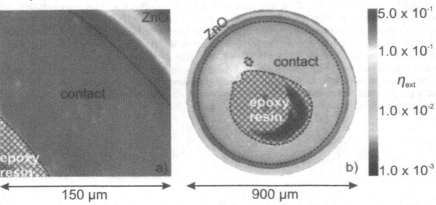

Fig. 2: Map of the external quantum efficiency η_{ext} of a) an uncapped PdO$_y$/ZnO and b) a Pd-capped PdO$_y$/ZnO Schottky contact. The conductive epoxy resin used for bonding absorbs incident light and reduces η_{ext}.

We investigated uncapped and capped PdO$_y$/ZnO Schottky diodes by LBIC measurements. The 325 nm line of a HeCd laser was used for excitation (100 kW/cm^2); the spot size is about 1 µm^2 and determines the resolution of the setup. The LBIC measurement of an uncapped PdO$_y$/ZnO Schottky diode is depicted in Fig. 2a). The current is homogeneously generated except for several points showing lower photocurrent. For uncapped Ag$_x$O/ZnO Schottky diodes the photocurrent is generated very inhomogeneously which was explained by the low conductivity of the Ag$_x$O-layer [3]. Here, the conductivity of the PdO$_y$-layer is sufficient to collect the photocurrent also from the contact edges, in contrast to Ag$_x$O/ZnO. With that it is shown that the electrically active contact area equals the geometrical contact area for PdO$_y$/ZnO. Spots of low photocurrent generation are caused by droplets on the contact as scanning electron microscopy images (not shown) revealed.

The LBIC image of a capped PdO$_y$/ZnO contact is shown in Fig. 2b). Here, the photocurrent increases towards the contact edge which is explained by the decrease of the Pd contact thickness and subsequent increase of transparency for the excitation light towards the edge. For the capped contact there are spots showing considerably lower photocurrents, too. Similar to the uncapped PdO$_y$/ZnO contacts the photocurrent is lower in regions with droplets on the surface.

LBIC measurements on a ZnO-based MESFET

ZnO-based MESFETs on glass substrate with Ag_xO gate contacts are normally-off devices [5]. The MESFET on quartz glass discussed here is normally-on; its threshold voltage is -1 V. The difference in the threshold voltage is most likely due to the fact that Ag, which is an acceptor in ZnO [9], diffuses from the gate contact into the channel [10] which leads to higher compensation and with that to higher threshold voltages. The LBIC measurements were conducted at gate voltages V_G of -1 V (off-state of the transistor) and at 1 V (channel is open) for source-drain voltages V_D ranging between 0 V and 2.4 V. Line scans are depicted for $0.2\ V \leq V_D \leq 0.8\ V$ in fig.3. In general, the photocurrent is about one order of magnitude larger if the transistor is in the off-state (fig. 3a). This is due to the much larger depletion width for $V_G = -1$ V compared to the on-state (fig. 3b) for which the width of the depletion layer should be negligible. For closed channel conditions the device acts as a phototransistor, within the depletion region the generated electron-hole pairs are effectively separated. If the transistor is in the on-state, the device behaves like a photoconductor and the electron-hole pairs are not necessarily separated by the comparatively weak field between source and drain.

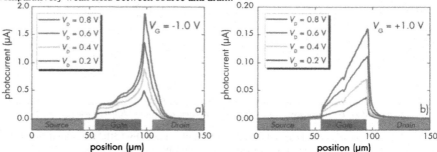

Fig. 3: Photocurrent line scans along the channel of a ZnO-based MESFET with a) $V_G = -1$ V and b) $V_G = 1$ V for different source-drain voltages.

Further, it is evident that in the off-state the largest photocurrent is generated between the gate and the drain contact. The spatial dependence of the photocurrent is caused by the varying depletion layer width in the transistor which is maximal between gate and drain (pinch-off point). For the constant gate voltage of -1 V this pinch-off point is located closer to the drain contact the higher V_D. The peak position of the photocurrent is depicted in fig. 4 in dependence on V_D for $V_G = -1$ V and $V_G = 1$ V, respectively. For $V_G = 1$ V the position of maximal photocurrent generation is determined only by the position of the maximal electric field $E_{max}(x)$. For $V_D > 1$ V, $E_{max}(x)$ is located between gate and drain [11, 12]; within the experimental resolution of about 1 µm its position is independent of V_D for $V_D > 1$ V. Nevertheless, $E_{max}(x)$ increases for higher V_D; thus the maximal photocurrent increases likewise.

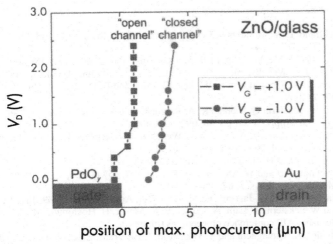

Fig. 4: Position of the maximal photocurrent generation for a ZnO-based MESFET on glass with a gate bias of -1 V and 1 V, respectively, in dependence on the source-drain voltage. The gate consists of a Pd capped PdO_y Schottky contact.

CONCLUSIONS

We have demonstrated that PdO_y/ZnO can be reproducibly and easily fabricated without any pretreatment of the ZnO surface. In contrast to Ag_xO/ZnO Schottky diodes the conductivity of the PdO_y is sufficient to collect the photocurrent of an entire Schottky diode; in principle a metal capping, as used for Ag_xO contacts, is not necessary. However, the current-voltage characteristic improves for Pd-capped PdO_y/ZnO Schottky contacts; the ideality factor decreases considerably. LBIC measurements were further used to study the generation of photocurrent of ZnO-based MESFETs on the micrometer scale. We showed that for closed channel conditions the photocurrent is a measure of the depletion layer width below the gate. Biasing the gate such that the channel is completely opened, the photocurrent signal is proportional to the electric field between source and drain; in particular, the position of the maximum photocurrent corresponds to the position of the maximal electric field between source and drain.

ACKNOWLEDGMENTS

We gratefully thank G. Biehne, H. Hochmuth and Dr. M. Lorenz for their technical support. The work was funded by the Deutsche Forschungsgemeinschaft in the framework of Sonderforschungsbereich 762 "Functionality of Oxide Interfaces" (SFB 762/1) and the Graduate School "Leipzig School of Natural Science - BuildMoNa" (185/1). AL and HvW are grateful for support by the European Social Fund. AL is also supported by the Studienstiftung des deutschen Volkes.

REFERENCES

1. R. Schifano, E. V. Monakhov, U. Grossner, and B. G. Svensson, Appl. Phys. Lett. **91**, 193507 (2007).
2. M.W. Allen, S.M. Durbin and J.B. Metson, Appl. Phys. Lett. **91**, 053512 (2007).
3. A. Lajn, H. v. Wenckstern, Z. Zhang, C. Czekalla, G. Biehne, J. Lenzner, H. Hochmuth, M. Lorenz, M. Grundmann, S. Wickert, C. Vogt, and R. Denecke, J. Vac. Sci. Technol. B **27**, 1769 (2009).
4. H. von Wenckstern, G. Biehne, R. A. Rahman, H. Hochmuth, M. Lorenz, and M. Grundmann, Appl. Phys. Lett. **88**, 092102 (2006).
5. H. Frenzel, M. Lorenz, A. Lajn, H. v. Wenckstern, G. Biehne, H. Hochmuth, and M. Grundmann, Appl. Phys. Lett. **95**, 153503 (2009).
6. J. H. Werner and H. H. Güttler, J. Appl. Phys. **69**, 1522 (1991).
7. R.F. Schmitsdorf and W. Mönch, Eur. Phys. J. B **7**, 457 (1999).
8. W. Mönch, Appl. Phys. A **87**, 359 (2007).
9. Yanfa Yan, M. M. Al-Jassim, and Su-Huai Wei, Appl. Phys. Lett. **89**, 181912 (2006).
10. H. von Wenckstern, A. Lajn, A. Laufer, B. K. Meyer, H. Hochmuth, M. Lorenz and M. Grundmann, AIP conference proceedings (in press).
11. H. Mizuta, K. Yamaguchi and S. Takahashi, IEEE Trans. Electron. Devices **ED-34**, 2027 (1987).
12. Y. Hori, M. Kuzuhara, Y. Ando, and M. Mizuta, J. Appl. Phys. **87**, 3483 (2000).

Mater. Res. Soc. Symp. Proc. Vol. 1201 © 2010 Materials Research Society   1201-H04-05

Influences of polarization effects in the electrical properties of polycrystalline MgZnO/ZnO heterostructure

Huai-An Chin[1], Chih-I Huang[1], Yuh-Renn Wu[1], I-Chun Cheng[1], Jian Z. Chen[2], Kuo-Chuang Chiu[3] and Tzer-Shen Lin[3]

[1] Graduate Institute of Photonics and Optoelectronics, National Taiwan University, Taipei, 10617 Taiwan (R.O.C.)
[2] Institute of Applied Mechanics, National Taiwan University, Taipei, 10617 Taiwan (R.O.C.)
[3] Material and Chemical Research Laboratories, Industrial Technology Research Institute, Hsinchu, 31040 Taiwan (R.O.C.)

ABSTRACT

ZnO has shown great promise for the application in optoelectronic devices. Since the modulation of conductivity is one of the key issues in device performances, we have applied the Monte Carlo method to analyze the mobility of poly-crystalline MgZnO/ZnO heterostructure thin film layer in this paper. The effects of the grain boundary scattering, ionized impurity scattering, as well as phonon scattering are considered. Our study shows that with a design of modulation doping by including the effects of spontaneous and piezoelectric polarization, the grain boundary potential can be suppressed to improve the mobility of the ZnO layer by order(s) of magnitude. Simulation results are also confirmed by our experimental works that polarization effects play an important role to attract carriers and to increase the mobility.

INTRODUCTION

ZnO is a generic n-type transparent wide bandgap semiconductor material, commonly used for transparent thin film transistors. ZnO-based alloys, on the other side, have attracted much attention due to their wide bandgap properties and thus show great potential in ultra-violet optoelectronic and high frequency electronic applications [1]. However, the mobility of ZnO film is a critical issue to application and most as-deposited ZnO layers are amorphous or polycrystalline, where the typical mobility is only around 1-20 cm^2/Vs [2][3]. One of the major cause is the grain boundary scattering. An attempt to screen the grain boundary defects by a high doping level, on the other side, may as well increase the ionized impurity scattering. Thus, we make use of the strong polarization effect of wurtzite ZnO to design a multi-layer MgZnO/ZnO heterostructure to suppress the grain boundary potential in the ZnO layer, while keeping the dopants separated from the transport carriers. This design requires a material system with small lattice mismatch reducing interface influences the carrier transport. The small lattice mismatch of MgZnO/ZnO makes it a promising candidate to practically realize this design [3][4].

In this study, we developed the Monte Carlo program [5] to study the grain boundary scattering effect on the mobility of the polycrystalline MgZnO/ZnO multilayer. It is noted that the mobility is not in the range of 1 or 20 cm^2/Vs and Monte Carlo is reasonable for transport in

73

extended states. If the mobility is very low, conduction may occur by hopping conduction which is not described by Monte Carlo. Also, in order to analyze the effect of modulation doping concentration and the polarization charge to the grain boundary potential, we apply our 2D finite element method (FEM) Poisson and drift-diffusion solver [6] to study the grain boundary potential change due to the defect trap at the grain boundary. We also present experimental results to verify the simulation.

THEORY AND EXPERIMENT

Simulation

In our study, we apply the 2D finite element method (FEM) Poisson and drift-diffusion solver to study the potential change due to the defect trap at the grain boundary. After that, we developed the Monte Carlo program to calculate the mobility of the polycrystalline MgZnO/ZnO multilayer. The material parameters we used for studying ZnO can be found in ref. [1][7], and the grain boundary scattering model proposed by [8] is applied here to study the grain boundary scattering. In our calculations, grain boundaries are treated as a series of parallel planes perpendicular to the applied field [8] as shown in Fig. 1(b). The scattering rate of grain boundary can be expressed as

$$\langle S(k) \rangle = \frac{S^2 m_x}{2\pi^2 \hbar^3 |k_x| d} \times \left[\frac{1 - \exp(-4k_x^2 \sigma^2)}{1 + \exp(-4k_x^2 \sigma^2) - 2\exp(-2k_x^2 \sigma^2)\cos(2k_x d)} \right] \quad (1)$$

where σ is the standard deviation associated with Gaussian distribution which can be expressed as

$$f(x_1, x_2, \ldots, x_n) = \frac{\exp\left\{ -\sum_{i=1}^{N-1} \left[(x_{i+1} - x_i - d)^2 / 2\sigma^2 \right] \right\}}{(2\pi\sigma^2)^{(N-1)/2}} \quad (2)$$

d is the average separation between adjacent grain boundaries, and S corresponds to a product of the energy perturbation and its spatial dimension. A delta function is used to model grain boundary potential located at position x_n with scattering potential strength S. The Hamiltonian $V(x)$ is given by a product of the energy perturbation and its spatial dimension, where the grain boundary potential can be expressed as

$$V(x) = S \sum_{n=1}^{n=N} \delta(x - x_n) \quad (3)$$

The detail derivation can be found in ref. [8]. As shown in Eqn. (1), the grain boundary scattering rate is mainly decided by the strength S and the boundary separation size d. In this study, we choose d in a reasonable ranges from 1 μ m to 50 nm. 2D FEM Poisson and drift-diffusion solver [6] is used here to analyze the change of grain boundary potential. The Poisson equation can be expressed as

$$\nabla^2 V = \frac{n - p + N_A - N_D + N_T}{\varepsilon} \tag{4}$$

where V is the band potential of the device. N_T is the defect trap density. n, p, N_A and N_D are the electron, hole, acceptor, and donor density, respectively. With our 2D FEM program, we can study the grain boundary potential straightforwardly.

Experiment

A 200nm-thick ZnO thin film is first deposited on a Corning Eagle-2000 glass substrate and then annealed at 600°C to reinforce the grain formation. Afterwards, titanium metal is deposited and patterned to form the contact pads [9][10], followed by the deposition of a $Mg_xZn_{1-x}O$ (x = 0.2, 0.3, 0.4) thin film with various thickness to obtain the heterostructure. Prior to the formation of heterostructures, the crystallinity of ZnO thin films is investigated by X-ray diffraction measurements (XRD) and (200) peak is obtained [11][12].The resistance is determined via two-point I-V measurement.

RESULTS AND DISCUSSION

In our model, as shown in Fig. 1(a), the MgZnO layer is grown on top of the ZnO. A 5 nm modulation doping layer is assumed to be 5 nm away from the MgZnO/ZnO interface. Fig. 1(b) shows the calculated shape of grain boundary potential by our 2D Poisson solver. A 5 nm defect traps are assumed at the middle of the device, and the defect density is 1×10^{19} cm^{-3} by defaults.

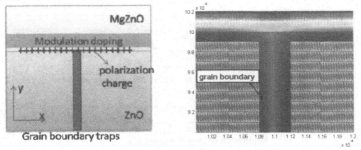

Fig 1. (a) A schematic of device, the grain boundary is perpendicular to x-axis. (b) The calculated grain boundary potential with the 2D Poisson and Drift-diffusion solver.

In order to obtain the scattering strength S, we need to extract the grain boundary potential at the conducting channel, i.e. at MgZnO/ZnO interface. Fig. 2(a) and 2(b) show the cross section view of the grain boundary potential along the MgZnO/ZnO interface without and with polarization effects, respectively.

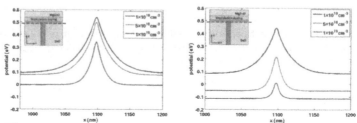

Fig 2. (a) The cross section grain boundary potential without considering the spontaneous and piezoelectric polarization view along x-direction (the position of red dash line in the small figure). (b) The cross section grain boundary potential with considering the spontaneous and piezoelectric polarization view along x-direction (the position of red dash line in the small figure).

In both cases, the suppression of grain boundary potential near the MgZnO/ZnO interface becomes significant when the modulation doping density is higher than 1×10^{19} cm^{-3}. As shown in Fig. 1(b), if the amount of 2.95×10^{12} cm^{-2} polarization charge at the Mg$_{0.15}$Zn$_{0.85}$O/ZnO interface [13][14] is considered, the positive polar charge will further enhance the band bending and suppress the grain boundary potential. The relationship between mobility and modulation doping density is also obtained from simulation, as shown in Fig. 3. Results indicate that as modulation doping density increases, the screening effect becomes stronger and largely reduces grain boundary potential and thus enhances the carrier mobility in ZnO. Improvement in mobility saturates at higher modulation doping density, which might be attributed to the ion scattering.

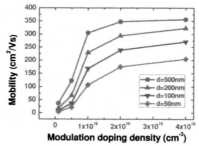

Fig 3. ZnO mobility versus modulation doping density at different grain sizes.

The layer thickness of MgZnO is also very important for the optimization of the channel charge density. The carrier concentration increases with the depth, and tends to become saturated as shown in Fig. 4(a). Therefore, the optimal depth can be obtained from the calculated critical depth of the saturated carrier concentration.

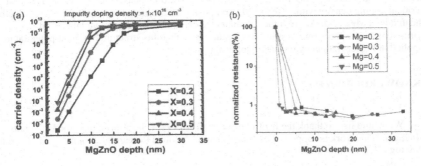

Fig 4. (a) The carrier concentration in the undoped ZnO layer changed with different depth of the $Mg_xZn_{1-x}O$, the ionized impurity is 1×10^{16} cm^{-3}. (b) The measured normalized resistance of polycrystalline $Mg_xZn_{1-x}O/ZnO$ heterostructure versus $Mg_xZn_{1-x}O$ thickness (x = 0.2, 0.3, 0.4, 0.5).

The critical saturated depth is reduced by increasing the Mg composition from 20 to 50 %. When the Mg composition is higher, it contributes larger polarization charge difference between the MgZnO and ZnO layer [14]. A larger MgZnO depth contributes higher carrier concentration in the ZnO layer, screening more grain boundary potential and improving the electron mobility. Therefore, with the increase of the polarization charge, the saturated depth of the carrier concentration can be shortened. In order to verify our simulation result, we deposit polycrystalline MgZnO/ZnO heterostructure with various MgZnO layer thicknesses to evaluate the effect of the polarization charges on the conductivity. Fig. 4(b) shows the relationship between the MgZnO thickness and the resulting normalized resistance of MgZnO/ZnO heterostructure. As the MgZnO thickness increases, the resistance of heterostructure decreases rapidly and then saturates. The enhancement of the interfacial polarization effect is stronger as the Mg content of the $Mg_xZn_{1-x}O$ layer increases; this leads to the decrease of both the saturated resistance of $Mg_xZn_{1-x}O/ZnO$ and the saturated thickness for the $Mg_xZn_{1-x}O$. The saturated electrical conductance is improved by more than two orders of magnitude for all cases, which confirms our simulation result.

CONCLUSIONS

In conclusion, we have presented calculations of the electron mobility characteristics in MgZnO/ZnO multi-layers with the grain boundary scattering and the screening effect of grain boundary potential by polarization field and modulated doping. Monte Carlo program with combination of grain boundary scattering effect, 2D FEM Poisson and drift-diffusion solver is developed to study the polycrystalline device. Our results show that modulation doping will suppress the impurity potential as well as the grain boundary potential, which makes it possible to achieve a high mobility ZnO with even with a polycrystalline phase. Our results also suggest the critical doping density with different Mg compositions. Besides, we present the critical depth of the undoped MgZnO layer from the calculated 2d carrier concentration, and the trends are confirmed by our experimental results. Our results show that this structure has great potential to be applied in higher frequency transistor applications.

ACKNOWLEDGMENTS

We would like to thank the fruitful discussion with Prof. Jasprit Singh in the University of Michigan. This work is supported by Material and Chemical Research Laboratories, Industrial Technology Research Institute of Taiwan.

REFERENCES

1. S. C. Su, Y. M. Lu, Z. Z. Zhang, C. X. Shan, B. H. Li, D. Z. Shen,B. Yao, J. Y. Zhang, D. X. Zhao, and X. W. Fan, Appl. Phys. Lett., vol. 93, p. 082108, 2008.
2. D. J. Cohen, K. C. Ruthe, and S. A. Barnett, J. Appl. Phys., vol. 96, p. 459,2004
3. J. Sun, D. A. Mourey, D. Zhao, S. K. Park, S. F. Nelson, D. H. Levy,D. Freeman, P. C. Corvan, L. Tutt, and T. N. Jackson, IEEE Electron Dev. Lett., vol. 29, p. 721, 2008.
4. H. Tampo, K. Matsubara, A. Yamada, H. Shibata, P. Fons, M. Yamagata,H. Kanie, and S. Niki, Journal of Crystal Growth, vol. 301, p. 358, 2007.
5. Yuh-Renn Wu and Madhusudan Singh and Jasprit Singh, IEEE Trans. Electron Dev., vol. 53, p. 588, 2006.
6. Yuh-Renn Wu and Jasprit Singh, J. Appl. Phys., vol. 101, p. 113712, 2007.
7. B. Guo, U. Ravaioli, and M. Staedele, Computer Physics Communications, vol. 175, p. 482, 2006.
8. R. P. Joshi and A. Srivastava, Appl. Phys. Lett., vol. 69, p. 1786, 1996.
9. S. Kim, B. S. Kang, F. Ren, Y. W. Heo, K. Ip, D. P. Norton, and S. J. Pearton, Appl. Phys. Lett., vol. 84, no. 11,p. 1904, 2004.
10. Y. Shimura, K. Nomura, H. Yanagi, T. Kamiya, M. Hirano, and H. Hosono, Thin Solid Films, vol. 516, p. 5899, 2008.
11. T. Hiramatsu, M. Furuta, H. Furuta, T. Matsuda, C. Li, and T. Hirao, Journal of Crystal Growth, vol. 311, no. 2,p. 282, 2009.
12. P. Nunes, E. Fortunato, and R. Martins, Thin Solid Films, vol. 383, p. 277, 2001.
13. H. Tampo, H. Shibata, K. Maejima, A. Yamada, K. Matsubara, P. Fons,S. Kashiwaya, S. Niki, Y. Chiba, T. Wakamatsu, and H. Kanie, Appl. Phys. Lett., vol. 93, p. 202104, 2008.
14. A. Malashevich and D. Vanderbilt, Appl. Phys. Lett., vol. 93, p. 045106, 2008.

Poster Session I

Mater. Res. Soc. Symp. Proc. Vol. 1201 © 2010 Materials Research Society   1201-H05-03

Residual Stresses in Sputtered ZnO Films on (100) Si Substrates by XRD

F. Conchon[1], P.O. Renault[1], P. Goudeau[1], E. Le Bourhis[1], E. Sondergard[2], E. Barthel[2], S. Grachev[2], E. Gouardes[3], V. Rondeau[3], R. Gy[3], R. Lazzari[4], J. Jupille[4], N. Brun[5]

[1] Laboratoire de Physique des Matériaux (PHYMAT) UMR 6630 - Université de Poitiers, France
[2] Laboratoire de Surface du Verre et Interfaces (SVI), UMR 125, Aubervilliers, France
[3] Laboratoire de Recherche de Saint-Gobain (SGR), Aubervilliers, France
[4] Institut des Nanosciences de Paris (INSP), UMR 7588, Paris, France
[5] Laboratoire de Physique des Solides (LPS), UMR 8502, Orsay, France

ABSTRACT

Residual stresses in sputtered ZnO films on Si are investigated and discussed. By means of X-ray diffraction, we show that as-deposited ZnO films encapsulated or not by Si_3N_4 protective coatings are highly compressively stressed. Moreover, a transition of stress is observed as a function of the post-deposition annealing temperature. After a heat treatment at 800°C, ZnO films are tensily stressed while ZnO films encapsulated by Si_3N_4 are stress-free. With the aid of in-situ X-ray diffraction, we argue that this thermally-activated stress relaxation can be attributed to a variation of the chemical composition of the ZnO films.

INTRODUCTION

ZnO is one of the most interesting II-VI semiconductor with a wide direct band gap of 3.35 eV [1]. ZnO is an optically transparent material of technological importance for its practical and potential applications for short wavelength optoelectronic devices and transparent conductive oxide films, such as blue to UV light-emitting diodes or solar cells electrodes [2]. Various deposition techniques have been applied for ZnO films elaboration such as rf and dc magnetron sputtering [3], metal-organic chemical vapor deposition (MOCVD) [4], pulsed laser deposition (PLD) [5], or filtered cathodic vacuum arc technique [6]. The advantage of the magnetron sputtering technique is the achievement of polycrystalline ZnO films deposition on large area without intentional substrate heating. However, it results in residual stresses which can be detrimental when occurrence of spontaneous delamination, or under scratch during processing or in service. Of the many articles about the preparation of zinc oxide films, only a handful report about stresses in these films whereas many studies have been carried out on the electronic and optical properties of the ZnO bulk, films, and nanostructures [7,8,9,10,11,12]. Stresses being correlated to thin films physical properties, their control hence requires an in depth comprehension of stresses formation and relaxation mechanisms. In the present study, we propose with the aid of X-ray diffraction (XRD) to study residual stresses in sputtered ZnO films deposited on Si. In particular, we will evidence a thermally-induced stress relaxation observed by in-situ XRD, the origin of which can be attributed to the presence of point defects in the ZnO films.

EXPERIMENTS

ZnO films (100 nm thick) were deposited by magnetron sputtering (100) Si substrates. Films were sputtered from a zinc target under a pressure of 1.5×10^{-6} bar, with a power of 1500 W. The content of oxygen in the plasma was equal to 48%. The deposition was conducted without intentional substrate heating. Finally, two kinds of films were analyzed: ZnO films (100 nm thick) directly deposited on Si and ZnO films (100 nm thick) encapsulated on both sides by Si_3N_4 films (40 nm thick) preventing the active film from aging or scratching.

The crystalline quality and residual stress analysis were investigated by XRD with a four-circle Seifert 3003 diffractometer equipped with a Cu X-ray source (1×1 mm^2 point focus) and a Ni filter in the direct beam path to absorb the CuK$_\beta$ radiation. The macro-stresses were evaluated by the $\sin^2 \psi$ method [13,14]. Whereas the curvature method (based on the Stoney's concept [15]) leads to the macroscopic residual stresses implying the whole film volume (crystalline and non-crystalline), the $\sin^2 \psi$ method only concerns the diffracting regions of the film.

In the present study, much attention is paid to the in-situ XRD measurements and their interpretation in terms of stress relaxation mechanisms. For that purpose, a four-circle Bruker D8 diffractometer is used. It is characterized by a Cu X-ray source producing a linear focus, Soller slits determining the divergence ($1°$) of the incident beam. Post-deposition heat-treatments were carried out on the encapsulated and non-encapsulated ZnO films in an hemispheric furnace mounted on the Bruker's diffractometer. This furnace is equipped with a graphite dome and allows to perform annealing under controlled atmosphere. In this study, films were annealed under air up to 800°C with a heating/cooling rate of 1°C/second.

The XRD stress analysis relies on the evaluation of the stress state in the material by the measurement of d-spacings in different directions of reciprocal space. This can be achieved by performing longitudinal scans (the so-called θ-2θ scans) for 14 tilt angles ψ of the sample (ψ being the angle between the normal to the surface and the normal to the diffracting planes). For the general case of a triaxial state of stress, the strain $\varepsilon_{\varphi\psi}^{hkl}$ experienced by a set of lattice planes {hkl} in a direction defined by the rotation angle φ and the tilt angle ψ depends on the components of the stress tensor, σ_{ij} ($i, j = 1$ to 3). For sub-surface regions like thin films, the strain-stress equation can be considerably simplified due to the free surface condition ($\sigma_{i3} = 0$). This state of stress is defined as a planar (or biaxial) stress state.

Furthermore, if a rotationally symmetric stress state is present then $\sigma_{11} = \sigma_{22} = \sigma_\varphi = \sigma$:

$$\varepsilon_{\varphi\psi}^{hkl} = \frac{1}{2} S_2^{hkl} \sigma \sin^2 \psi + 2 S_1^{hkl} \sigma \qquad (1$$

where S_1^{hkl} and $1/2 S_2^{hkl}$ are the so-called X-ray elastic constants [13].

Then, for a macroscopically isotropic material, the strain vs $\sin^2 \psi$ plots exhibit a linear behavior. ZnO presents an hexagonal structure. Using the following values for the stiffnesses C_{ij} ($\times 10^5$ MPa): $C_{11} = 2.097$, $C_{12} = 1.211$, $C_{13} = 1.052$, $C_{33} = 2.109$, $C_{55} = 0.425$ [16], the Young's modulus of ZnO single crystal can be calculated. Here, whatever the crystallographic direction, the Young's modulus value is almost the same, i.e. about 124 GPa with a variation of 3%. Thus, ZnO can be assumed as macroscopically isotropic. Residual stresses are finally derived from the slope of the $\sin^2 \psi$ plot expressed by equation (2) using the Voigt approximation [13].

82

RESULTS AND DISCUSSIONS

Crystalline quality

Firstly, the crystalline quality has been investigated. The first three peaks, (10-10) (0002) and (10-11), of the hexagonal phase (space group P63mc) are clearly identified (figure 1) for all the samples on the θ-2θ scans. Moreover, ZnO films are partially (0002) textured.

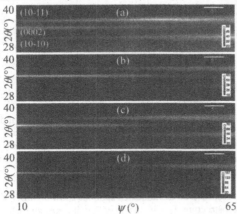

Figure 1. (2θ, ψ) maps of the (10-10) (0002) and (10-11) reflections for the 100 nm thick ZnO film encapsulated by two 40 nm thick Si_3N_4 films (a and b) and for the non-encapsulated 100 nm thick ZnO films (c and d). (a) and (c) are recorded before annealing, (b) and (d) after an annealing at 800°C.

This point is supported by the presence on the (2θ, ψ) maps (figure 1) of an intense diffraction peak for small ψ values for the (0002) reflection and by an increase of the diffracted intensity around $\psi = 60°$ for the (10-11) reflection (60° being the value of the angle between (0002) and (10-11) planes). After an annealing at 800°C of the samples, the (0002) texture is enhanced (figure 1. b and d) and more particularly for the non-encapsulated ZnO film. Indeed, on figure 1.d, the (0002) reflection is more intense and sharpened while the other reflections are almost absent until $\psi = 30°$. This first result highlights the importance of Si_3N_4 coatings during the post-deposition annealing process.

Residual stresses and post-deposition annealing under ambient atmosphere

In order to extract the peak position of the (10-10) (0002) and (10-11) reflections, θ-2θ scans are simulated with Voigt functions whose parameters have been optimized using a standard Levenberg-Marquardt minimization procedure. Then $\ln(1/\sin\theta)$ is plotted vs $\sin^2\psi$ with the corresponding linear fit (continuous lines in figure 2) that is realized by taking into account weights attributed to each experimental point. For the sake of clarity, we show here the $\sin^2\psi$ plots only for the (10-11) reflection but similar results are observed for (10-10) and (0002) reflections. This residual stress analysis reveals that ZnO films are compressively strained as attested by the negative slope observed on the $\sin^2\psi$ plots in figure 2.a and b. Indeed, we obtain

σ = -0.70 (±0.1) GPa for ZnO/Si and σ = -0.83 (±0.03) GPa for ZnO encapsulated by Si_3N_4 on Si which is in good agreement with the values reported by Hinze et al. [17].

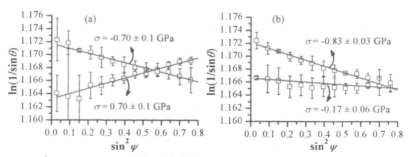

Figure 2. $\sin^2\psi$ plots of the (10-11) reflexion. The 800°C heat treatment induces a change of the slope on the $\sin^2\psi$ plots from negative (untreated: compressive stresses σ = -0.7 GPa) to positive (annealed: tensile stresses σ = 0.7 GPa) for ZnO/Si (a) while the slope only tends to zero after the heat treatment for ZnO encapsulated by Si_3N_4 (untreated: compressive stresses σ = -0.83 GPa, annealed: compressive stresses σ = -0.17 GPa) (b).

XRD measurements recorded on the annealed samples show significant changes in terms of residual stresses and microstructure. Firstly, the annealing induces an increase of the grain size (derived from the width of the XRD profiles using the Scherrer formula) from 10 nm to 36 nm for non-encapsulated ZnO and from 14 nm to 24 nm for the encapsulated one. Concomitantly to this grain growth, an enhancement of the (0002) texture is observed. Furthermore, annealing gives rise to a transition of stress from compressive to tensile for the non-encapsulated film and only an annihilation of the residual stresses for the encapsulated one. These two points are evidenced on figure 3.a and b by the remarkable change of slope on the $\sin^2\psi$ plots after the heat treatment. We obtain after annealing σ = 0.70 (±0.1) GPa for ZnO/Si and σ = -0.17 (±0.06) GPa for ZnO encapsulated by Si_3N_4. These results show that Si_3N_4 layers play an important role in the mechanism responsible for the relaxation of residual stresses during heat treatment. The identification of this mechanism appears then to be a crucial point at this stage to control the amount of residual stresses in these films. For instance, an important parameter comes into play when deposited thin oxide films: the stoichiometry. For sputtered ZnO films, many studies concern the evolution of the stress and the microstructure as a function of the sputtering pressure or the oxygen partial pressure [3,12,17,18,19]. These two parameters strongly influence residual stresses in sputtered ZnO films. Moreover ZnO is well-known to present several types of point defects in large concentration. Hence, thermally-induced changes of the chemical composition is highly expected in the ZnO films. In order to understand the influence of the film stoichiometry on the stress relaxation, in-situ XRD measurements have been carried out.

For each sample, θ-2θ scans are recorded (for ψ = 0) at each temperature plateau (for T = 30°C, 200°C, 400°C, 600°C, 800°C and after cooling at 30°C). Results are shown in figure 3 where is reported the evolution of the (0002) peak position (this means the out-of-plane strain) as a function of temperature. Experimental data presented here have been previously corrected by the different dilatation contributions of the sample holder and the sample.

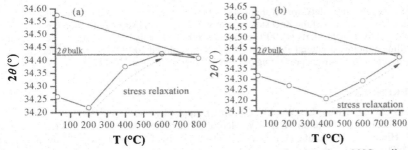

Figure 3. (0002) peak position as a function of temperature upon heating (from 30°C until 800°C) and after cooling (at 30°C) for ZnO/Si (a) and for ZnO encapsulated by Si_3N_4 (b), continuous lines are guides for the eyes. The stress free 2θ (bulk) value of ZnO is also shown. These measurements were recorded under air.

Firstly, for ZnO/Si (figure 3.a), a small decrease of 2θ is observed in the temperature range 30°C-200°C. This can be attributed to compressive thermal stresses (about -0.035 GPa) arising in the ZnO film because of the difference of linear thermal expansion coefficients between the Si substrate ($\alpha_a = 2.6 \times 10^{-6}$ °C^{-1} at room temperature and 4.3×10^{-6} °C^{-1} at 800°C) [20] and the ZnO film ($\alpha_a = 7.1 \times 10^{-6}$ °C^{-1} from RT to 800°C) [21]. The film presenting a larger thermal expansion coefficient than the substrate, is then subjected to in-plane compressive stress upon heating giving rise to a tensile strain in the out-of-plane direction. From 200°C to 600°C we observe a stress relaxation, i.e. 2θ increases tending to its bulk value and then stabilizes up to 800°C. During cooling, high tensile thermal stresses appear (about 0.5 GPa) again because of the difference of linear thermal expansion coefficients. For the encapsulated ZnO film (figure 4 b), the appearance of tensile stresses is the predominant phenomenon up to 400°C. Consequently, in this case the stress relaxation process is shifted toward high temperatures and is then observed in the range 400°C-800°C instead of 200°C-600°C (for ZnO/Si). We can then conclude that the presence of Si_3N_4 protective coatings limit the thermally-induced stress relaxation in the ZnO films. Hence, it can be expected from these results that the stress relaxation is mainly driven by a "chemical" mechanism, this means by a change of the ZnO stoichiometry during annealing. This kind of relaxation mechanism is now frequently encountered in oxide films [22]. In addition, it is well-known that ZnO films present oxygen vacancies V_O [23,24,25] regardless of the deposition process. If V_O is the dominant defect, it may induce large distortion in the ZnO wurtzite lattice. In fact, the large size of the oxygen ions helps to maintain the crystal structure in ZnO and absence of these induces large inward relaxation of the neighboring atoms and then high stresses [26]. As a consequence, we postulate that a variation of the oxygen vacancy concentration during heating (or aging under air) involves stress relaxation.

CONCLUSIONS

In summary, we investigated residual stresses in sputtered ZnO films deposited onto Si substrates with the use of XRD. First, we showed that ZnO films are highly compressively stressed. Then, a transition from compressive to tensile stresses has been evidenced as consequence of a post-deposition annealing at 800°C for non-encapsulated ZnO films. Thanks to in-situ XRD measurements, we highlighted that this change of sign of the residual stresses can be attributed to a variation of the ZnO stoichiometry. This chemical stress relaxation mechanism is thermally activated and starts at 200°C to finish at 600°C. Finally, Si_3N_4 coatings encapsulating ZnO films limit this structural evolution (which is then shifted 200°C higher) and then constitute efficient protective barriers preventing oxygen permeation through ZnO films.

REFERENCES

1. C.S Chen, C.T Kuo, T.B. Wu, I.N. Lin, Jpn. J. Appl. Phys. 36, 1169 (1997)
2. S.J. Pearton, D.P. Norton, K. Ip, Y.W. Heo, T. Steiner, Prog. Mat. Sci. 50, 293 (2005)
3. T. Hiramatsu, M. Furuta, H. Furuta, T. Matsuda, C. Li, T. Hirao, J. Cryst. Growth, 311, 282 (2009)
4. K. Haga, T. Suzuki, Y. Kashiwaba, H. Watanabe, B.P. Zhang, Y. Segawa, Thin Solid Films, 433, 131 (2003)
5. X.W. Sun, H.S. Kwok, J. Appl. Phys. 86, 408 (1999)
6. H.X. Lee, S.P. Lau, Y.G. Wang, B.K. Tay, H.H. Hng, Thin Solid Films, 458, 15 (2004)
7. S.B. Zhang, S.-H. Wei, A. Zunger, Phys. Rev. B 63, 075205 (2000)
8. C-S.Hsiao, S-Y. Chen, W-Li Kuo, C-C. Lin, S-Y. Cheng, Nanotechnology 19, 405608 (2008)
9. S J. Kang, Y H. Joung, Appl. Surf. Sci. 253, 7330 (2007)
10. B. Hwang, K. Park, H-S. Chun, C-H. An, H. Kim, H-J. Lee, Appl. Phys. Lett. 93, 222104 (2008)
11. S.J. Chen, Y.C. Liu, C.L. Shao, C.S. Xu, Y.X. Liu, L. Wang, B.B. Liu, G.T. Zou, J. Appl. Phys. 99, 066102 (2006)
12. M.K. Puchert, P.Y. Timbrell, R.N. Lamb, J. Vac. Sci. Technol. A 14, 2220 (1996)
13. V. Hauk, " Structural and residual stress analysis by nondestructive methods ", Elsevier (1997)
14. J. Tranchant, P.Y. Tessier, J.P. Landesman, M.A. Djouadi, B. Angleraud, P.O. Renault, B. Giraud, P. Goudeau, Surf. Coat. Technol. 202, 2247 (2008)
15. G.G. Stoney, Proc. Soc. Lond. A82, 172 (1909)
16. G. Simmons, H. Wang, " Single elastic constants and calculated aggregate properties: a handbook ", second edition, Cambridge Massachusetts and London England (1971)
17. J. Hinze, K. Ellmer, J. Appl. Phys. 88, 2443 (2000)
18. O. Kappertz, R. Drese, M. Wuttig, J. Vac. Sci. Technol. A, 20, 2084 (2002)
19. R. Drese, M. Wuttig, J. Appl. Phys. 98, 073514 (2005)
20. Y. Okada, Y. Tokumaru, J. Appl. Phys. 56, 314 (1984)
21. K. Ellmer, A. Klein, B. Rech, " Transparent conductive zinc oxide: basics and applications in thin film solar cell ", Springer series in materials science, 104 (2008)
22. F. Conchon, A. Boulle, R. Guinebretière, E. Dooryhée, J-L. Hodeau, C. Girardot, S. Pignard, J. Kreisel, F. Weiss, L. Libralesso, T. L. Lee, J. Appl. Phys. 103, 123501 (2008)
23. S. Dutta, S. Chattopadhyay, A. Sarkar, M. Chakrabarti, D. Sanyal, D. Jana, Prog. Mater. Sci. 54, 89 (2009)
24. J. Zhao, L. Hu, W. Liu, Z. Wang, Appl. Surf. Sci. 253 (2007)
25. K. Wang, Z. Ding, S. Yao, H. Zhang, S. Tan, F. Xiong, P. Zhang, Mater. Res. Bull. 43, 3327 (2008)
26. A. Janotti, C.G. Van de Walle, Phys. Rev. B 76, 165202 (2007)

Mater. Res. Soc. Symp. Proc. Vol. 1201 © 2010 Materials Research Society

A Luminescence Study of Electron-Irradiated ZnO Crystals

M. Avella[1], O. Martínez[1], J. Jiménez[1], B. Wang[2], P. Drevinsky[3], and D. Bliss[3]

[1] GdS Optronlab. Univ. de Valladolid, Paseo de Belén 1, 47011 Valladolid, Spain

[2] Solid State Scientific Corp. 27-2 Wright Rd., Hollis, NH 03049

[3] Air Force Research Laboratory, Sensors Directorate, Hanscom AFB, MA 01731

ABSTRACT

ZnO crystals were irradiated with high energy electrons (1MeV). The main defects created were analyzed by cathodoluminescence (CL) spectroscopy. The main effects of irradiation on the CL spectrum were the partial quenching of the emission, the shift of the visible luminescence to the yellow, and the observation of an additional band and its phonon replicas at ≈3.32 eV. Zinc vacancy related defects are postulated as responsible for the changes induced in the spectrum.

INTRODUCTION

The understanding of point defects is essential for the management of the electronic and optoelectronic applications of semiconductors. ZnO is a very promising semiconductor for UV optoelectronics, because of its large bandgap (3.3 eV), and its large free exciton binding energy (60 meV). On the other hand, ZnO substrates are available to produce homoepitaxial layers suitable for devices. However, reliable p-type material is not yet available, which prevents the application of ZnO for advanced devices. To achieve successful p-type doping, one needs to improve the knowledge about the native defects, which can complex with impurities and also form compensating levels precluding effective p-type conversion [1]. The role of the native point defects in ZnO is still controversial. High energy e-irradiation is an efficient method to create vacancies and interstitial defects, while luminescence techniques are very useful to study the signature of the defects [2]. However, the luminescence spectrum of e-irradiated ZnO has not reported dramatic spectral changes, which is consistent with the strong radiation hardness of ZnO. Positron annihilation studies have revealed the presence of Zn vacancies under high energy electron irradiation, as the main defects generated by the irradiation [3,4]. Zinc vacancies are acceptors which can compensate the n-type nature of as grown ZnO. We present herein a cathodoluminescence (CL) analysis of ZnO crystals irradiated with high energy electrons. The e-irradiation seems mainly generating Zn sublattice primary defects, which interact with the previously existing defects and impurities in the crystal; therefore, the final defects shall depend on the growth conditions and the history of the samples.

EXPERIMENTAL DETAILS

ZnO wafers from different vendors were irradiated at room temperature with high energy electrons (1 MeV) at fluences of 10^{17} cm^{-2}. The wafers from vendor 1 (labelled as M) were annealed at 650°C in 1.4 Torr O_2 atmosphere during 0.5 h. Then they were irradiated. Samples from vendor 2 (labelled as TD) were irradiated without previous annealing.

The CL measurements were carried out at 80 K with a XiClone System from Gatan (Gatan UK) attached to a field emission scanning electron microscope (FESEM) LEO 1530 (Carl Zeiss, Germany). The acceleration voltage of the e-beam was varied between 2 and 30 kV, in order to modify the penetration depth of the probe beam. Some Vickers indentations were done on the samples, in order to compare the effects of the e-irradiation with the mechanically generated defects. Hall effect measurements were carried out on sample TD after irradiation.

RESULTS AND DISCUSSION

The luminescence spectrum of ZnO crystals consists of three main spectral regions, the near band edge (NBE) emission with energies above 3.33 eV, the bands with energies below 3.33 eV, usually associated with donor acceptor pair (DAP) transitions and/or phonon replicas of the different excitonic bands, and the visible luminescence in the green–red spectral range. The relative importance of the different bands depends on the sample growth method and history. The CL spectra of the two samples studied herein, before irradiation, are shown in Fig.1. One observes significant difference between both samples, which reveals different defect compositions. First, the CL emission is significantly higher for sample TD, over one order of magnitude, which suggests a lower defect concentration in that sample with respect to the M sample, which would present a much higher concentration of non radiative recombination centers. The nature of the non radiative recombination centers in ZnO is a matter of controversy; they have been related to extended defects and /or to native defects. In particular, V_{Zn} related complexes were associated with non radiative recombination centers [5]. The CL spectrum of sample TD is characterized by a strong contribution of the phonon replicas, which can be even more intense than the excitonic bands, while sample M presents lower intensity phonon replicas; note that the NBE spectrum consists of several close lying free, FX, and bound, BX, excitonic transitions. This result is also compatible with a better crystalline quality of TD sample. In a previous paper we established a correlation between the relative intensity of the phonon replicas with respect to the NBE band in terms of the crystal quality [6].

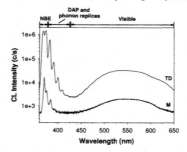

Figure1. CL spectra of both TD and M samples without irradiation (Eb = 10 kV).

The spectral region below 3.33 eV consists of several bands separated each other by ≈70 meV, which is roughly the energy of the LO phonon (72 meV) in ZnO. There is a main band at ≈3.31 eV and several phonon replicas at 3.24, 3.17 and 3.1 eV. The nature of the 3.31 eV band is a matter of controversy. It has been associated with a donor-acceptor pair (DAP) transition [7], but the nature of the defects involved was not revealed. It is also identified as the first phonon replica of the free exciton transition, FX-1LO band [8]. Also, a band related to structural defects was reported around such energy [8-10]. On the other hand, the broad visible band consists of at least three contributions, green (GL), yellow (YL), and orange-red (RL), the origin of which is still a matter of controversy. GL is usually associated with defects related to oxygen deficiency (V_O and Zn_i), while the YL and RL bands have been rather associated with Zn deficiency (V_{Zn}, O_{Zn}, O_i) [11].

The CL spectra before and after irradiation are shown in Figures 2a and b. Sample M does not present significant changes in the CL spectrum after e-irradiation of the annealed sample, Figure 2a; the main effect is the relative quenching of the overall luminescence emission, because of the generation of non radiative recombination centers, but the spectral shape is very similar before and after irradiation; a slight increase in the relative intensity of the visible luminescence with respect to the NBE emission could also be considered, but it is not very relevant. Note that the previous annealing of this sample shifted the visible band to the yellow, and the further irradiation did not shift anymore the band.

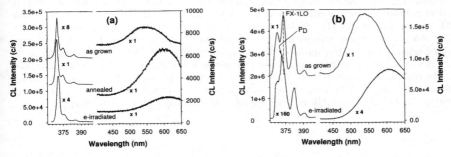

Figure 2.a) CL spectra of sample M before irradiation, annealed and e-irradiated (Eb = 15 kV); b) CL spectra of sample TD before and after e-irradiation (Eb = 30 kV).

The situation is different for sample TD, for which the CL spectrum presents significant changes after irradiation, Figure 2b. The overall CL intensity is drastically reduced, due to the generation of non radiative recombination centers. The NBE spectral region is also modified, one observes changes in the bound exciton bands. The region of the phonon replicas presents noteworthy changes; a new band appears clearly differentiated from the FX-1LO band, the new band labelled P_D is observed in the high energy side of the FX-1LO band. The existence of a defect related band around 3.32 eV appearing as a shoulder of the FX-1LO band was previously reported [12], usually in the neighborhood of crystal defects; in particular, we observed it in TD samples in the regions surrounding the dislocations generated by Vickers indentation [6,12]. A similar band (3.32 eV ($\approx$373 nm)) was seen to be enhanced in e-irradiated ZnO [13].

The visible band shifts from the green to the yellow-red under electron irradiation; furthermore, its relative intensity with respect to the NBE emission is strongly enhanced. That is, under irradiation the GL band is quenched, while the YL band is enhanced, becoming dominant. This behaviour suggests a strong generation of the deep levels responsible for the YL. The visible bands have been reported to undergo changes under the effect of e-irradiation [2] and plastic deformation [14].

On the other hand, Hall effect measurements showed a conversion from n-type to p-type after e-irradiation. Electric isolation was reported under O^+ implantation and it was associated with the formation of Zn vacancies [15]. The conversion to p-type suggests than in addition to the deep levels, the electron irradiation generates acceptor levels, shallow enough to supply holes to the valence band at room temperature. The defects responsible for the P_D band can be adequate candidates. The P_D band presents well defined phonon replicas, which supports the acceptor nature of the defects responsible for that emission.

One of the advantages of the excitation with an e-beam in the CL measurements is the large penetration depth range compared to photoluminescence (PL), which is excited with an UV laser beam. The conventionally used He-Cd laser (325 nm) penetrates around 100 nm, while the penetration of the e-beam can be varied from a few nanometers to 3 μm [16] using the 2 to 30 kV range. This allows observing the effect of the e-irradiation far from the surface where surface defects could interact with the primary defects generated by the e-irradiation. Note that 1MeV electrons cross the sample; therefore the generation of defects by the e-irradiation is a bulk effect, not merely a surface effect. In fact, the CL signature depends on penetration depth, see the CL spectra of irradiated TD sample for two different e-beam acceleration voltages in Figure 3. The two spectra present different features. The structure of the visible band is different for the two e-beam voltages: while for 30 kV the GL is quenched and the YL is strongly enhanced, in the spectrum acquired at 10 kV, the visible band presents a relatively lower intensity and appears as the convolution of both GL and YL bands. On the other hand, the P_D peak is observed in both cases but with different contribution. Note that at 10 kV the subsurface damage is still influencing the spectrum, while at 30 kV the subsurface damage does not play a major role.

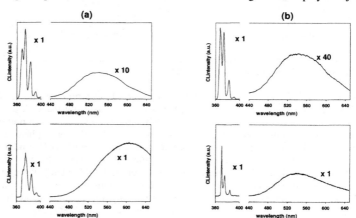

Figure 3. CL spectra of sample TD before (top) and after (bottom) e-irradiation. a) Eb = 30 kV; b) Eb = 10 kV.

The primary defects produced by electron irradiation are vacancies and interstitials. The atomic displacement rate can be estimated from the Macinley-Feshbach formula [17], using the displacement energy, E_d. The E_d values for ZnO, 41.4 eV for O and 18.5 eV for Zn, were estimated by Van Vechten [18]. According to this, the displacement rate is higher for Zn; therefore, Zn sublattice defects are expected in a higher concentration compared to O sublattice defects. Tuomisto *et al.* [3] estimated a very low displacement rate of 0.03 cm^{-1}, which means that the defects should recombine at low temperature [2-4]. Frenkel pairs are formed as primary defects, but due to the high mobility of the interstitials they recombine at relatively low temperature, therefore, the concentration of defects at room temperature must be low. The recombination of the Frenkel pairs on one side, and the complex formed by the primary defects on the other side, must be influenced by the background defects and impurities present in the samples.

The P_D band observed under electron irradiation at 3.32 eV seems to be similar to the one induced by plastic deformation, Figure 4. It is probably associated with V_{Zn} defects. A relation between V_{Zn} and a luminescence band at 3.34 eV was reported [19]; note that that measurement was carried out at 10 K; therefore, it can be related to our band at 3.32 eV, which was measured at 80 K.

The presence of dominant Zn sublattice defects induced by the e-irradiation is consistent with our CL data, which reveal a prevalence of V_{Zn} related defects; in particular, the formation of a shallow acceptor related to V_{Zn} and responsible for the P_D band, the enhancement of the YL, which is usually associated with V_{Zn} complexes [11], and the formation of non radiative recombination centers, also associated with V_{Zn} complexes [5]. It is the ability of V_{Zn} to form different complexes that could explain the differences observed among the different samples and the depth dependence of the e-irradiation effect. At the present time, one cannot account for such observations as the quenching of the GL band under e-irradiation. Tentatively, one can associate the GL band to complexes involving V_O, which might be restructured under the e-irradiation. The quenching of this band under O_2 annealing suggests that the V_O defects are involved in the GL emission.

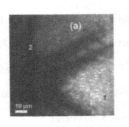 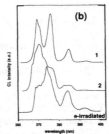

Figure 4. a) CL image of an indentation in as-grown TD sample; b) CL spectra in a defect free area (1), in an indented area (2) and after e-irradiation (3) (Eb = 30 kV).

CONCLUSIONS

Electron irradiation of ZnO crystals generates defects, for which the signature was revealed by spectrally resolved CL. The main consequences of the e-irradiation depend on the interaction of the primary defects generated by the irradiation with the as-grown defects. The spectral changes induced by the electron irradiation were interpreted in terms of the formation of zinc vacancy complexes.

ACKNOWLEDGMENTS

The Spanish group was funded by European Office for Aerospace Research and Development (EOARD) (Grant FA-8655-09-1-3073-3) and by the Spanish Government (through research projects MAT 2007-63617 and MAT2007-66741-C02-02).

REFERENCES

1. M. D. McCluskey and S. J. Jokela, *J. Appl. Phys.* **106**, 071101 (2009).
2. L. S. Vlasenko and G. D. Watkins, *Phys. Rev. B* **71**, 125210 (2005).
3. F. Tuomisto, V. Ranki, K. Saarinen, and D. C. Look, *Phys. Rev. Lett.* **91**, 205502 (2003).
4. C. Koscum, D. C. Look, G. C. Farlow, and J. R. Sizelove, *Semicond. Sci. Technol.* **19**, 752 (2004).
5. S. F. Chichibu, T. Onuma, M. Kubota, A. Uedono, T. Sota, A. Tsukazaki, A. Ohtomo, and M. Kawasaki, *J. Appl. Phys.* **99**, 093505 (2006).
6. J. Mass, M. Avella, J. Jiménez, A. Rodríguez, T. Rodríguez, M. Callahan, D. Bliss, and Buguo Wang, *J. Cryst. Growth* **310**, 1000 (2008).
7. T. B. Hur, G. S. Jeen, Y. H. Hwang, and H. K. Kim; *J. Appl. Phys.* **94**, 5787 (2003).
8. B. K. Meyer, H. Alves, D. M. Hofmann, W. Kriegseis, D. Forster, F. Bertram, J. Christen, A. Hoffmann, M. Strasburg, M. Dworzak, U. Haboeck, and A. V. Rodina, *Phys. Stat. Sol. (b)* **241**, 231 (2004).
9. M. Schirra, R. Schneider, A. Reiser, G. M. Prinz, M. Feneberg, J. Biskupek, U. Kaiser, C. E. Krill, R. Sauer, and K. Thonke, *Physica B* **401-402**, 362 (2007).
10. Y. Ohno, H. Koizumi, T. Taishi, I. Yonenaga, K. Fujii, H. Goto, and T. Yao, *Appl. Phys. Lett.* **92**, 011922 (2008).
11. H. C. Ong and G. T. Du, *J. Cryst. Growth* **265**, 471(2004).
12. J. Mass, M. Avella, J. Jiménez, M. Callahan, D. Bliss, and Buguo Wang, *J. Mater. Res.* **22**, 3526 (2007).
13. M. A. Hernández-Fenollosa, L. C. Damonte, and B. Marí, *Superlatt. and Microstr.* **38**, 336 (2005).
14. V. Coleman, J. E. Bradby, C. Jagadish, and M. R. Phillips, *Appl. Phys. Lett.* **89**, 082102 (2006).
15. A. Zubiaga, F. Tuomisto, V. A. Coleman, H. H. Tan, C. Jagadish, K. Koike, S. Sasa, M. Inoue, and M. Yano, *Phys. Rev.B* **78**, 035125 (2008).
16. K.Kanaya, and S.Okayama; *J. Phys. D Appl. Phys.* **5**, 43 (1972).
17. F. Agulló-López et al, *Point defects in materials*, Academic Press, New York, 1988.
18. J. A. Van Vechten, in *Handbook on semiconductors*, ed. by T.S.Moss and S.P.Keller (North Holland, Amsterdam 1988) ch.1.
19. A. Zubiaga, J. A. García, F. Plazaola, F. Tuomisto, K. Saarinen, J. Zúñiga-Pérez, and V. Muñoz Sanjosé, *J. Appl. Phys.* **99**, 05316 (2006).

Mater. Res. Soc. Symp. Proc. Vol. 1201 © 2010 Materials Research Society   1201-H05-11

Electrical and thermal stress analysis of In$_2$O$_3$-Ga$_2$O$_3$-ZnO thin-film transistors

Mami Fujii[1], Tomoki Maruyama[1], Masahiro Horita[1], Kiyoshi Uchiyama[1], Ji Sim Jung[2], Jang Yeon Kwon[2], Yukiharu Uraoka[1,3]

[1] Nara Institute of Science and Technology, 8916-5 Takayama, Ikoma, Nara 630-0192, Japan
[2] Samsung Advanced Institute of Technology,
Mt. 14-1, Nongseo-Dong, Giheung-Gu, Yongin-Si, Gyeonggi-Do, 446-712, Korea
[3] CREST, Japan Science and Technology Agency, Honcho, Kawaguchi, Saitama 332-0012, Japan

ABSTRACT

Degradation of In$_2$O$_3$-Ga$_2$O$_3$-ZnO (IGZO) thin-film transistors (TFTs)) was studied. We evaluated degradation caused by applying gate voltage and drain voltage stress. A parallel shift of the transfer curve was observed under gate voltage stress. Joule heating caused by the drain current was observed. We tried to reproduce this degradation of the transfer curve change by device simulation. When we assumed the trap level as the density of state (DOS) model and increased two kinds of trap density, we obtained properties that show the same trends as the experimental results. We concluded that two degradation mechanisms occur under gate and drain voltage stress conditions. And then, we tried to improve the TFT characteristics using high pressure water vapor (HPV) annealing. We also found that the cooling conditions after HPV annealing affect the IGZO TFT characteristics.

INTRODUCTION

Oxide semiconductor TFTs are promising devices for driving circuits of next-generation displays[1]. However, the properties of the oxide semiconductor channel and the interface between the oxide semiconductor and the gate insulator have not been entirely clarified, which help the interpretation of degradation mechanism. Recently, hydrogen effect on the electrical properties of oxide semiconductor TFTs has attracted much attention. Nomura et al reported that the hydrogen has a favorable influence[2] on the IGZO TFT characteristics. On the other hand, Cho et al reported the hydrogen has an unfavorable influence[3]. In this study, we discuss the degradation mechanism by utilizing device simulation and the improved characteristics by HPV annealing, which improves the characteristics of Si TFTs with silicon dioxide film as gate insulator due to the passivation effect of oxygen and hydrogen.[4]

THEORY

Electrical and thermal analysis

We fabricated bottom gate top contact IGZO TFTs, and process conditions are follows: Mo metal was deposited on a glass substrate by sputtering. After the patterning of the gate metal by wet etching, a SiNx ($\varepsilon = 7.0$) film with the thickness of 400 nm was deposited as a gate insulator by plasma-enhanced chemical vapor deposition (PECVD) at 300°C. Then an IGZO film with the thickness of 70 nm was deposited in a mixture of argon and oxygen gases. After patterning the IGZO channel layer by wet etching, source and drain electrodes were formed by the lift-off process. The IGZO channel layer and the source/drain electrodes were fabricated by rf sputtering. Finally, the fabricated TFT was annealed for 1 h at 350°C in N_2 ambient.

Various gate and drain voltage stresses were applied to the IGZO TFTs and changes in electrical properties were examined to evaluate reliability at room temperature. To analyze the degradation mechanism, thermal analysis was performed using a thermal infrared imaging system (Infra Scope II). We measured the surface temperature of the sample during operation. InSb charge-coupled device (CCD) was used as the infrared detector, where the temperature sensitivity is 0.1K.

Device simulation

We tried to reproduce changes in the electrical properties observed experimentally using the device simulation software ATLAS. In this case, we assumed a trap density in the band gap of IGZO film shown in the Fig 1. The user-specifiable parameters are the shape of states (Gaussian-state or tail-state) and the trap type of states (donor-like or acceptor-like). We can also set the characteristic decay energy for the tail-state, and the half bandwidth and the energy level for the Gaussian-state within the energy gap. The a-IGZO structural disorder is expressed as tail state. Tail states are attributable to the distribution of bonding angles or lengths. The oxygen vacancies of a-IGZO were generally defined as the Gaussian distributed donor-like trap density located near the conduction band edge[5]. In this study, we proposed Gaussian-distributed acceptor-like trap density located near the valence band edge as shown in Fig. 1. The subscripts of T/G and A/D stand for tail/Gaussian states and acceptor-/donor-like states, respectively.

High Pressure water Vapor annealing

The HPV treatment was performed at 260°C for 1 hour in the pressure of, 1.41 MPa. We examined cooling conditions after HPV annealing as follows; (1) rapid cooling down, where samples were exposed to the atmosphere at the end of annealing, and (2) slow cooling down, where samples were cooled in wet ambient. We measured electrical properties of IGZO TFTs before and after the HPV annealing. And we observed depth profiles of each element in IGZO thin films by Secondary Ion Mass Spectroscopy (SIMS).

RESULTS AND DISCUSSION

The gate voltage dependence of TFT degradation was investigated. The threshold voltage shifted parallel to positive direction with increasing gate stress voltage. After stress imposition, threshold voltage recovered without any annealing. The details are discussed in Ref.6. To examine the degradation of TFTs during operation in peripheral circuits, the effects of drain bias are very important. We measured the changes in electrical properties under Vg=Vd=20 V stress. The amount of the shifts in the threshold voltage was not changed or decreased compared with the case of gate only stress. In addition, the S value changed after 10000 s of bias stress. Then, electrons were trapped not at the interface but at spatially deep places in the channel layer of a-IGZO thin films.

Transfer curves shifted to positive directions and the S value was increased for 10000 s of bias stress. This degradation mode under gate and drain bias stress was different from the case of gate only stress. For example, the amount of the shift in the transfer curve decreased under gate and drain bias stress compared with that for gate only stress, implying that degradation occurred in both the positive and negative voltage directions. These results suggest that there is a possibility of the transfer curve shifting parallel toward a positive voltage by gate voltage stress and toward a negative voltage by drain voltage stress at the same time under gate and drain voltage stress conditions. Therefore, we estimated two degradation modes which shift positively and negatively. We analyzed thermal distributions in TFTs during gate and drain voltage stresses. The thermal distributions under various drain voltage stresses (0, 10, 20 V) at a fixed gate voltage stress (20 V) are shown in Fig. 2. Here, the substrate temperature was fixed to 60 °C.

Since a marked increase in the maximum temperature was observed with an increase of the drain voltage stress under the constant gate voltage stress, we confirmed that the Joule heating was due to the drain current on the TFT channel layer. We thought that this thermal heating caused the threshold voltage shift when the drain voltage stress was applied. In the degradation of Si TFTs, the degradation of the threshold voltage shift was accelerated by thermal heating in the TFT channel layer. Unlike the case of Si TFTs, the amount of the threshold voltage shift decreased under the drain voltage stress as we above explained. Therefore, the transfer curve shifted parallel toward a negative voltage caused by the drain voltage stress.

We assumed that two degradation modes occurred under the conditions of applying gate and drain voltage stresses.
1. Threshold voltage shift toward a positive voltage due to the positive gate voltage stress.
2. Threshold voltage shift toward a negative voltage due to the drain voltage stress.
So far, several DOS models of IGZO-TFTs have been proposed.[7,8,9] In our previous model, transfer curves shifted parallel in the positive direction with decreasing N_{GD} close to the conduction band. However, it is difficult to explain why the trap density decreased with increasing stress time. Therefore, we proposed a new DOS model including an N_{GA} in the deep level in addition to the previous DOS model. The energy level of the N_{GA} was set to 0.1 eV, and N_{GD} was set to 2.7 eV from the valence band edge. The energy level of N_{GD} was the measured value reported in Ref. 10.

Between the off-region and the threshold voltage region, the Fermi level is on the 0.4 eV below the conduction band edge.[11] By applying gate voltage, the Fermi level moves up to N_{GD} trap level until reachreaching to the threshold voltage; then, Then, it moves through the N_{TA} and goes to the conduction band in the ON region. Therefore, N_{GD} increase imposes an S increase,

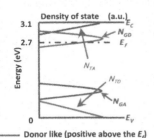

Donor like (positive above the E_f)
Acceptor like (negative below the E_f)

Fig.1 Density of state (DOS) model used in this study.

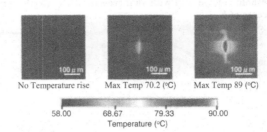

No Temperature rise   Max Temp 70.2 (ºC)   Max Temp 89 (ºC)

58.00   68.67   79.33   90.00
Temperature (ºC)

Fig.2 Thermal images obtained by thermal infrared imaging system. Vg=20V and various frain voltage was applied. Gate size were W/L=150/8 μ m. Substrate temperature is 60°C

and N_{TA} increase imposes field effect mobility and on current decrease. The field effect mobility and the on current were not changed because the density of N_{TA} was constant.

In the simulation using the new model, we obtained the transfer curve dependence on the N_{GA} as shown in Fig. 3 (a). In this simulation, the value of N_{GD} was 3×10^{17} cm^{-3}eV^{-1}. We succeeded in reproducing the parallel shift of the transfer curve toward a positive voltage by increasing only the amount of N_{GA}. When the other parameters were changed, the S values and the on current or the saturation current were varied from the experimental results. Next, we reproduced the experimental results of the S value change. When the N_{GD}, close to the conduction band edge, and the N_{GA} increased, the transfer curve shifted toward a positive voltage without any changes in the on current or the saturation current. However, in the condition of high N_{GD} and N_{GA} trap density, the amount of the shift was decreased and the S value changed compared with the condition of the low trap density, which were similar to the case of 10000 s bias stress in the experimental results. N_{GD} trap density is presumably due to the oxygen vacancies (VO) and it seems that VO increase caused by weak bond breaking. In general, VO makes a donor-like trap in IGZO.[5,12,13] We above explained the relationship between the Joule heating and the drain voltage stress. Therefore, the Joule heating by the drain current is one of the causes of the increase of the N_{GD} trap density (VO). It would appear that N_{GD} and N_{GA} have different origins, and so it is absolutely essential to investigate the origin of N_{GA}. The amount of the shift depended on the trap density at both N_{GA} and N_{GD}. In these simulations, transfer curves shifted parallel toward a positive voltage as N_{GA} increased and toward a negative voltage as N_{GD} (VO) increased.

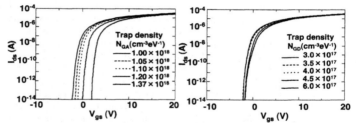

Fig. 3. Simulation results of transfer curve dependence on trap density; (a) parallel threshold voltage shift with an increase of N_{GA}, (b) the S value increase increaswith an increase of N_{GD}.

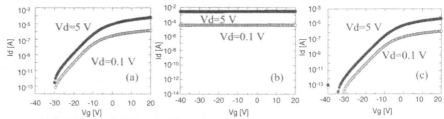

Fig. 4. IGZO TFT characteristics before and after HPV annealing; (a) before annealing, (b) after HPV annealing with rapid cooling in the high pressure vapor conditions, (c) after HPV annealing with slow cooling in the atmosphere.

Next, we investigated the effect of hydrogen and oxygen on the IGZO TFT characteristics. Fig. 4 shows the transfer curve of IGZO TFTs with the structure above mentioned before and after HPV annealing. After HPV annealing followed by the rapid cooling in the atmosphere, the sample did not show a transistor operation and IGZO channel films became of high conductivity. In the case of the slow cooling in the HPV conditions, the TFT characteristics were not changed.

Figure 5 shows depth profiles of hydrogen and oxygen in IGZO/SiO2/Si structures before and after HPV annealing with various take-out temperature. Compared with the before annealing, hydrogen and oxygen ion counts increased after annealing. With a decrease of the take-out temperature of the sample, the ion counts of hydrogen and oxygen were decreased in the IGZO thin film. We understand that these results caused by the decomposed H$^+$ and OH$^-$ from H$_2$O vapor under high pressure ambient. At the beginning of HPV annealing, reoxidization reactions undergo into the oxide thin films as shown in the equation (1) caused by decomposed H$^+$ and OH$^-$ taken into the film effectively. And next, expected to the reductive reaction develops in the oxide thin films by oxygen elimination reactions as the equation (2) in the cooling condition.

We concluded that the oxygen vacancies were compensated by OH$^-$ taken into the IGZO thin films effectively during the HPV annealing. Then, IGZO channel has high conductivity as shown in the TFT characteristics in Fig 4 (b) even though oxygen ion counts increased. Therefore, it would appear that the hydrogen is concerned in the carrier generation. Next, in the case of the rapid cooling, it is speculated that the reductive reaction (Equation (2)) did not occurred. These results speculated that the IGZO thin film conductivity depending on the amount of the hydrogen. And the cooling condition after annealing is important for controlling of TFT characteristics.

$$M + V_O^{\bullet\bullet} + M\text{-}O + H^{\bullet} + OH' + 2e' \rightarrow M\text{-}OH + M\text{-}OH \qquad (1)$$
$$M\text{-}OH + M\text{-}OH \rightarrow M + V_O^{\bullet\bullet} + M\text{-}O + H_2O(gas) + 2e' \qquad (2)$$

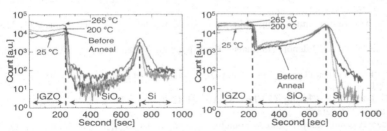

Fig. 5. Depth profiles of hydrogen and oxygen in IGZO/SiO$_2$/Si structures before and after HPV annealing with various taken-out temperature. (a) Hydrogen (b) Oxygen

CONCLUSIONS

We investigated the degradation of In$_2$O$_3$- Ga$_2$O$_3$-ZnO (IGZO) thin-film transistors under DC stress conditions. We examined the drain voltage dependence in addition to the gate voltage dependence. The amount of the threshold voltage shift under gate and drain voltage stress is smaller than in the case of only gate voltage stress. We confirmed Joule heating effect due to the drain current in the TFT channel layer. We analyzed the degradation mechanisms by device simulation. N_{GD} trap density increased with increasing Joule heating caused by the drain voltage. It would appear that N_{GD} and N_{GA} have different origins, and so it is absolutely essential to investigate the origin of N_{GA}. In these simulations, transfer curves shifted parallel toward a positive voltage as N_{GA} increased and toward a negative voltage as N_{GD} increased. Therefore, we could reproduce the transfer curve shifts observed in the experimental results when we assumed the Gaussian-shaped accepter-like states.

We tried to reform the IGZO thin film by reducing the defect states with high pressure vapor annealing. IGZO thin films became high conductive after HPV annealing, however, this electrical properties was changed with the cooling conditions. Therefore, HPV treatment is possibly the effective to improving the oxide-semiconductor thin film.

ACKNOWLEDGMENTS

The authors thank Professor T. Kamiya of Tokyo University of Technology, Professor. M. Furuta of Kochi University of Technology, and Professor M. Kimura of Ryukoku University for useful advice.

REFERENCES

1. K. Nomura, H. Ohta, A. Takagi, T. Kamiya, M. Hirano, and H. Hosono: Nature 432, 488 (2004).
2. Kenji Nomura, Toshio Kamiya, Hiromichi Ohta, Masahiro Hirano, and Hideo Hosono, Applied Physics Letters 93, 192107 (2008).
3. Doo-Hee Cho, Sang-Hee Ko Park, Shinhyuk Yang, Chunwon Byun, Kyoung Ik Cho, Minki Ryu, Sung Mook Chung, Woo-Seok Cheong, Sung Min Yoon, and Chi-Sun Hwang, IMID 2009 Tech. Dig, 26-1, 318-321, (2009)
4. Toshiyuki Sameshima, Kenji Sakamoto, Mitsuru Satoh, Thin Solid Films 335, 138-141 (1998).
5. Toshio Kamiya, Hidenori Hiramatsu, Kenji Nomura, Hideo Hosono, J. Electroceram, 17 : 267-275 (2006)
6. Mami Fujii, Yukiharu Uraoka, Takashi Fuyuki, Ji Sim Jung, and Jang Yeon Kwon, Jpn. J. Appl. Phys, 48, 04C091 (2009)
7. T. C. Fung, C. S. Chuang, C. Chen, K. Abe, H. Kumomi and J. Kanicki: AM-FPD '08, Dig. Tech. Pap., p.251, (2008)
8. Hideo Hosono, Kenji Nomura, Youichi Ogo, Tomoya Uruga, Toshio Kamiya, J. Of Non-Cryst. Solids, 354, 2796-2800 (2008)
9. Hsing-Hung Hsieh, Toshio Kamiya, Kenji Nomura, Hideo Hosono, and Chung-Chih Wu, Appl. Phys. Lett., 92, 133503 (2008)
10. M. Kimura, T. Nakanishi, K. Nomura, T. Kamiya, and H. Hideo: Appl. Phys. Lett. 92, 133512 (2008).
11. Kazushige Takechi, Mitsuru Nakata, Toshimasa Eguchi, Hirotaka Yamaguchi, and Setsuo Kaneko, Jpn. J. Appl. Phys. 48, 011301 (2009)
12. A. F. Kohan, G. Ceder, and D. Morgan, Chris G. Van de Walle, Phys. Rev. B 61, 15 019, (2000)
13. Byung Du Ahn, Hyun Soo Shin, Hyun Jae Kim, Jin-Seong Park, and Jae Kyeong Jeong, Appl. Phys. Lett., 93, 203506, (2008)

Mater. Res. Soc. Symp. Proc. Vol. 1201 © 2010 Materials Research Society   1201-H05-14

Fabrication, processing and characterization of thin film ZnO for integrated optical gas sensors

Eliana Kamińska[1], Anna Piotrowska[1], Iwona Pasternak[1], Michał A. Borysiewicz[1],
Marek Ekielski[1], Krystyna Gołaszewska[1], Witold Rzodkiewicz[1], Tomasz Wojciechowski[2],
Elżbieta Dynowska[2], Przemysław Struk[3], Tadeusz Pustelny[3]

[1]Institute of Electron Technology, Al. Lotników 32/46, 02-668 Warsaw, Poland
[2]Institute of Physics, Polish Academy of Sciences, Al. Lotników 32/46, 02-668 Warsaw, Poland
[3]Silesian University of Technology, ul. Akademicka 2A, 44-100 Gliwice, Poland

ABSTRACT

Zinc oxide layers deposited on quartz substrates by means of RF reactive magnetron sputtering with subsequent RTP annealing in a nitrogen flow at 400°C and in an oxygen flow at 500°C have been investigated in applications to waveguide structures. The ZnO films reveal a highly c-oriented columnar structure with a surface roughness of 4.3 nm. Annealing causes a significant increase of the lattice constant to the value of 5.210±0.001 Å suggesting the relaxation of the stress in the film. The annealing process causes a significant improvement of propagation properties of the fabricated waveguide structures in comparison to structures using as-deposited ZnO films. The minimal attenuation coefficient of the 630 nm thick films was found to be 2.8 and 3.0 dB/cm for TE0 and TM0 modes respectively.

INTRODUCTION

Recently, optical gas sensors built of an optical waveguide with a sensing layer on top of it, integrated with input and output couplers attract considerable interest [1-3]. In this respect, ZnO with its unique combination of optical properties and sensing capabilities offers an opportunity to use it for both the waveguide and sensing elements. However, there are still important technological steps to be mastered in order to fabricate practical devices. The growth of non-doped material presents a special challenge. It is commonly recognized that intrinsic defects such as vacancies, interstitials and extended structural defects as well as residual impurities contribute to background n-type conductivity and lattice constraints in nominally undoped ZnO. The structural properties strongly determine the optical properties of ZnO thin layers and thus influence the feasibility of the use as waveguides [4-7].

The fabrication of unstressed, high quality non-doped ZnO material with well defined microstructure and small surface roughness for use in integrated optical sensors is the focus of this paper.

EXPERIMENTAL

Zinc oxide layers were deposited on quartz substrates by means of RF reactive magnetron sputtering. A ceramic stoichiometric ZnO target (99.999% purity) was used in this study. The sputtering processes were conducted in an oxygen-argon mixture. The working pressure p_{Ar+O2} was equal 1×10^{-2} mbar, partial O_2 pressure $p_{O2} = 3 \times 10^{-3}$ mbar. The deposition rate was 24 nm/min. The as grown ZnO films were annealed in a nitrogen flow at 400°C for 10 min. and subsequently in an oxygen flow at 500°C for 10 min. Rapid Thermal Annealing (RTA) was used for all the thermal processes.

Thermal stress evolution during heating was determined using the wafer curvature method (WCM). The samples were placed inside the Tencor FLX 2320 chamber and heated up to 500°C and cooled down to room temperature in a N_2 flow. The surface roughness was measured with atomic force microscopy and optical properties by means of spectroscopic ellipsometry. The images of cross-sections of the films were obtained by field-emission scanning electron microscopy. Waveguiding properties were studied using a setup composed of a semiconductor laser with wavelength of $\lambda = 677$ nm, polarizer, photodiode, rotating goniometric table with the angular resolution of 0.01 deg, and homodyne nanovoltmeter. The propagation loss, recorded by a digital camera, was calculated in LabView software. The $Bi_{12}GeO_{20}$ prism with a refractive index of 2.5407 or Bragg gratings with periods of 1.6 and 2.0 μm were used as optical couplers.

Bragg grating couplers were etched in ZnO films by inductive coupled plasma technique with the source operated at 13.56 MHz. Positive $UV5^{0.8}$ photoresist was applied as the etch mask. The etching was carried out using BCl_3/Ar chemistry at 700 W ICP power, 20 W RF power, and total gas follow rate of 60 sccm. The chamber pressure was kept at 10 mTorr.

RESULTS AND DISCUSSION

Film structure

XRD spectrum of as deposited and annealed ZnO films is show in figure 1. The (00.2) peak presented at 43.36° indicates that ZnO is highly textured with the c-axis oriented perpendicularly to the substrate surface. For as deposited ZnO film the c-axis lattice constant c = 5.215±0.001 Å is lower than the bulk value (c = 5.2066 Å [8]). After annealing subsequently in nitrogen and oxygen, a shift of the peak in the spectrum is observed ($2\theta = 43.65°$), signifying increase of the lattice constant to the value of 5.210±0.001 Å suggesting the relaxation of the stress in the film.

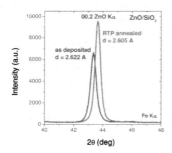

Figure 1. XRD spectra of the as deposited and annealed ZnO films deposited on quartz substrates.

Further evidence of the stress relaxation after annealing can be yielded from thermal stress measurements shown in figure 2. For the as-deposited sample, the stress characteristics show a significant increase of the compressive stress for the region between 300 and 400°C after which it decreases which may indicate a change in structure. A measurement of an annealed film shows no such behaviour which leads to a conclusion that there has been a relaxation after annealing.

a)   b)

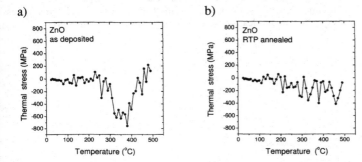

Figure 2. Thermal stress change in sputter-deposited ZnO thin films a) as deposited, b) after annealing.

The highly textured, columnar growth of the film can be seen in the SEM cross-section image below. The film structure consists of columnar grains of nearly uniform size perpendicular to the surface of the substrate. Post-deposition annealing causes a small growth of grains, to which stress relaxation is related. The root mean square roughness values of the surface of ZnO layers measured by AFM are 4.2 nm for the as-deposited layer and 4.3 nm for the annealed film.

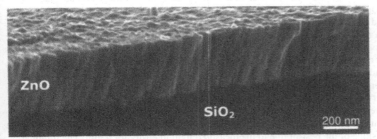

Figure 3. Cross-sectional SEM images of the of ZnO film deposited on quartz.

Optical properties

The formation of a planar waveguide requires a guiding layer with a refractive index higher than those of the substrate and cladding. On it as well as on the thickness of the guiding layer depends the number of the guiding mode. Ellipsometric measurements showed close correlation between lattice constants and refractive indices of the ZnO films (see figure 4). From these measurements it can be seen that the refractive index of the ZnO film after annealing is lower than for the as-deposited film. This effect, which can be observed through the whole spectral range, is most pronounced in the visible and near infra-red ranges. Particularly, for as-deposited and annealed films for the 677 nm light used in the waveguide experiment the refractive indices were 1.9776 and 1.9614 respectively. This change can be attributed to the relaxation of compressive stress in the film through the increase in the film thickness (by 25 nm). This causes a decrease in layer density which in turn leads to a decrease in the refractive index [9].

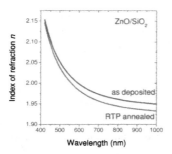

Figure 4. Refractive index as a function of wavelength for the as-deposited and annealed ZnO films.

The planar waveguide structure investigated in this communication consisted of a thin annealed ZnO layer having a refractive index $n = 1.9614$ deposited on a quartz substrate ($n = 1.456$) and air was the cladding layer ($n = 1$). The as-deposited ZnO film exhibited great light attenuation (> 20 dB/cm) and will be not discussed here. The results of numerical analysis carried out by OptiFDTD software demonstrate that the minimal thickness of the waveguiding ZnO layer could not be lower than 200 nm (see figure 5a). For layer thicknesses in the range of 200 – 300 nm a monomode propagation could be observed. For layers thicker than 400 nm multimode propagation occurs.

The modal characteristics of ZnO waveguides with thickness of 330 and 630 nm are shown in figures 5b and 5c. In accordance with theoretical calculations, both TE and TM modes were excited in presented planar ZnO waveguides. Photographs of the streaks of guided light in the ZnO films observed for TE0 and TM0 modes are shown in figure 6.

In order to estimate the attenuation coefficient value, measured light intensity was plotted as a logarithmic function of distance from the coupler (see figure 7). The loss was determined as the negative of the slope of the best-fit line to the measured light intensity for a film. The minimal propagation loss in the 630 nm thick films was found to be 2.8 and 3.0 dB/cm for TE0 and TM0 modes respectively. Comparison of the attenuation coefficients for TE and TM modes is shown in Table I.

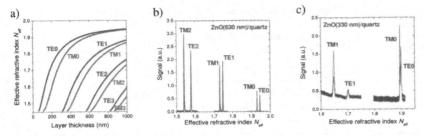

Figure 5. Calculated effective refractive index N_{eff} as a function of ZnO waveguide thickness (a), and modal characteristics of waveguides with thicknesses of 630 nm (b) and 330 nm (c).

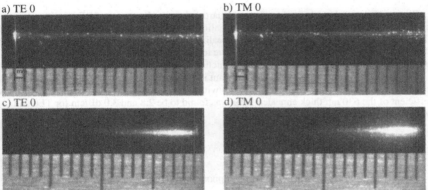

Figure 6. Propagation of light in ZnO waveguides with thickness of 630 nm (a, b) and 330 nm (c, d).

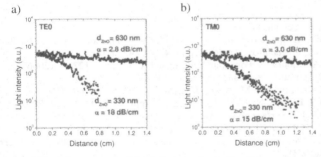

Figure 7. The dependence of the light intensity on the distance from coupler for TE0 and TM0 modes.

Table I. Attenuation coefficients for ZnO waveguides.

ZnO waveguide thickness [nm]	TE Mode	Attenuation coefficient [dB/cm]	TM Mode	Attenuation coefficient [dB/cm]
630	0	2.8	0	3.0
	1	4.8	1	6.1
	2	9.3	2	8.0
330	0	18	0	15
	1	32	1	27

SUMMARY

Zinc oxide layers have been deposited on quartz substrates by means of RF reactive magnetron sputtering with subsequent RTP annealing in a nitrogen flow at 400°C for 10 min. and in an oxygen flow at 500°C for 10 min. The annealing process causes a significant increase of the lattice constant to the value of 5.210±0.001 Å due to the relaxation of the stress in the film. The ZnO film reveals highly c-oriented columnar structure with surface roughness of 4.3 nm.

The annealing process used gives a significant improvement of the propagation properties of the waveguide structures compared to structures with as-deposited films.The minimal attenuation coefficient of the 630 nm thick annealed films was found to be 2.8 and 3.0 dB/cm for TE0 and TM0 modes, respectively.

ACKNOWLEDGMENTS

The presented investigations were partly supported within the grant of Ministry of Science and Higher Education No. N515 025 32/1887.

REFERENCES

1. P.V. Lambeck, Meas. Sci. Technol. **17**, R93 (2006).
2. T. Mazingue, L. Escoubas, L. Spalluto, F. Flory, P. Jacquouton, A. Perrone, E. Kamińska, A. Piotrowska, I. Mihailescu, P. Atanasov, Applied Optics **45**(7), 1425 (2006).
3. J.P. Atanas, R. Asmar, A. Khoury, A. Foucaran, Sensors and Actuators A **127**, 49 (2006).
4. H.W. Kim, N.H. Kim, Materials Science and Engineering B **103**, 297 (2003).
5. N. Mehan, V. Gupta, K. Sreenivas, A. Mansingh, J. Appl. Phys. **96**(6), 3134 (2004).
6. M.K. Ryu, S.H. Lee, M.S. Jang, G. N. Panin, T.W. Kang, J. Appl. Phys. **92**(1), 154 (2002).
7. S.S. Lin, J.L. Huang, D.F. Lii, Surface and Coatings Technology **176**, 173 (2004).
8. T. Hiramatsu, M. Furuta, H. Furuta, T. Matsuda, Jpn. J. Appl. Phys. **46**(6A), 3319 (2007).
9. R. Ondo-Ndon, F. Pascal-Delannoy, A. Boyer, A. Giani, A Foucaran, Materials Science and Engineering B **987**, 68 (2003).

Mater. Res. Soc. Symp. Proc. Vol. 1201 © 2010 Materials Research Society   1201-H05-16

Electrical and optical properties of CVT-grown ZnO crystals

Koji Abe and Masanori Oiwa
Department of Electrical and Electronic Engineering, Nagoya Institute of Technology, Gokiso, Showa, Nagoya 466-8555, Japan

ABSTRACT

Effects of H_2O partial pressure on ZnO crystal growth by chemical vapor transport (CVT) have been investigated. The use of H_2O causes the increase in growth rate of ZnO, indicating that H_2O acts as a dominant oxygen source in ZnO growth by CVT. The use of H_2O also improves structural, electrical, and optical properties of the CVT-grown ZnO crystals. A sharp X-ray rocking curve for the ZnO (0002) reflection was obtained, and the full width at half maximum value was 38 arcsec. Strong near band edge emission was observed in photoluminescence spectra at room temperature. Both carrier concentration and Hall mobility increased with partial pressure of H_2O. The dependence of the carrier concentration on temperature indicates that there exist two donors in the CVT-grown ZnO crystals. The estimated ionization energy for the shallow donor was 35 ± 5 meV and that for the deep donor was 115 ± 5 meV.

INTRODUCTION

Zinc oxide (ZnO) has attracted great attention because of its potential application for near-UV optoelectronic devices. Various crystal growth techniques such as hydrothermal growth, molecular beam epitaxy (MBE), and chemical vapor transport (CVT) have been used to produce ZnO crystals [1–4]. Hydrothermal growth is widely used to produce bulk ZnO crystals. However, hydrothermally grown ZnO crystals contain impurities from a mineralizer solution. Lithium, which is known as a common impurity in hydrothermally grown ZnO crystals, produces carrier compensation. The unintentional lithium doping can be a major obstacle to realize practical devices [1]. CVT has potential for high purity ZnO crystal growth because Zn and O atoms are transported from ZnO powder to seed crystals by the chemical reaction of ZnO with C (ZnO+C↔Zn+CO). Partial pressure in the reaction tube of CVT affects characteristics of ZnO crystals. We have reported that the residual carbon atoms in ZnO crystals decreases with increasing partial pressure of CO_2 in the reaction tube [5]. When H_2O is introduced into the reaction tube, the H_2 partial pressure increases owing to the chemical reaction of H_2O with Zn (H_2O+Zn→ZnO+H_2). In this study, effects of the partial pressure of H_2O on CVT-grown ZnO crystals have been investigated.

EXPERIMENTAL DETAILS

ZnO crystals were grown by CVT. The details of the CVT system are described in a previous report [5]. A graphite crucible was filled with 1.8 g of 5N-purity ZnO powder and placed on the susceptor in a vertical quartz tube. Zn-polar ZnO substrates used as seed crystals were cut from a hydrothermally grown ZnO crystal. The ZnO substrate was fixed on the inside of a crucible lid. Temperatures of the substrate (T_s) and ZnO powder (T_p) were defined as temperatures of the lid and susceptor. During 2-hour crystal growth, T_s and T_p were kept at 840°C and 1020°C, respectively. The quartz tube was filled with a gas mixture to investigate effects of H_2O on ZnO crystals. Sample A was grown in a gas mixture of Ar, CO_2 (230 Torr), and H_2O (25 Torr). Sample B was grown in a gas mixture of Ar and CO_2 (230 Torr). The total pressure at room temperature was adjusted to 610 Torr using Ar.

The structural properties of ZnO crystals were investigated by using X-ray rocking curve measurements. Photoluminescence (PL) measurements were performed using the 325 nm line from a He-Cd laser. Hall measurements were conducted in the temperature range from 79 K to 356 K using the van der Pauw configuration. The ZnO substrate was removed via mechanical polishing before Hall measurements.

RESULTS AND DISCUSSION

The ZnO growth rate for sample A was 100 μm/h and that for sample B was 50 μm/h. The use of H_2O increased the growth rate. However, H_2O partial pressure had little effect on the amount of the ZnO powder which remained in the crucible after 2-hour growth. These results suggest that H_2O plays an important role in oxidizing Zn.

X-ray rocking curves for (0002) reflection of CVT-grown ZnO crystals are shown in figure 1. The Full width at half maximum (FWHM) value of the X-ray rocking curve depended on H_2O partial pressure. The FWHM value of 38 arcsec was obtained for sample A. The FWHM value for sample B was 96 arcsec. It was found that crystallinity of CVT-grown ZnO crystals is improved by using H_2O as oxygen source.

Figure 2 shows PL spectra of CVT-grown ZnO crystals. The intensity of room temperature PL increases with partial pressure of H_2O, as shown in figure 2(a). The intensity of near band edge emission was enhanced more than that of the broad emission at around 2.5 eV. It has been reported that the broad emission is related to oxygen vacancies in ZnO crystals [6]. These results are consistent with the X-ray diffraction data shown in figure 1. Bound exciton lines at 3.353 eV and 3.360 eV are observed in the 7 K PL spectra in figure 2 (b). Sample A had the bound exciton line at 3.360 eV stronger than that of sample B. Lavrov et al. have reported

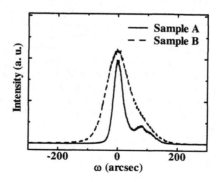

Figure 1. X-ray rocking curves for (0002) reflection of CVT-grown ZnO crystals.

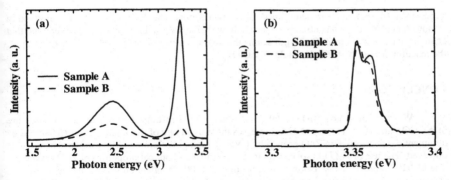

Figure 2. PL spectra of CVT-grown ZnO crystals at (a) room temperature and (b) 7 K.

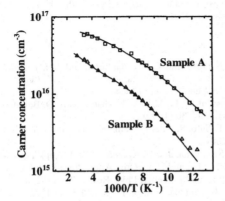

Figure 3. Carrier concentration of CVT-grown ZnO crystals as a function of temperature. The solid lines represent the calculated carrier concentration.

that hydrogen at the bond-centered lattice site in ZnO results in a PL line at 3.3601 eV [7]. It is suggested that the use of H_2O increases hydrogen-related donors.

The carrier concentration of CVT-grown ZnO crystals as a function of temperature is shown in figure 3. The samples exhibited n-type conduction. Although the carrier concentration of sample A was higher than that of sample B, sample A had higher Hall mobility than sample B. The maximum Hall mobility for sample A was 1100 cm^2/Vs at 81 K and that for sample B was 800 cm^2/Vs at 90 K. The use of H_2O results in good crystallinity of CVT-grown ZnO crystals, which improves electrical properties. The carrier concentration as a function of temperature is calculated using the charge neutrality condition. In the calculation, we have assumed that there exist two donors and a compensating acceptor. The solid line shown in figure 3 represents the calculated carrier concentration. The ionization energies of 35±5 meV (shallow donor) and

115±5 meV (deep donor) are determined for the two donors. The deep donor was not observed in sample A because the concentration of the shallow donor was too high. It have been reported that H atoms act as donors in ZnO [7,8]. It seems that hydrogen-doped ZnO crystals are formed by the reaction of H_2O with Zn ($H_2O+Zn \rightarrow ZnO+H_2$).

CONCLUSIONS

We have fabricated ZnO crystals on Zn-polar ZnO substrates by CVT using a gas mixture of Ar, CO_2, and H_2O. The use of H_2O increases the growth rate of ZnO, indicating H_2O acts as a dominant oxygen source in ZnO growth by CVT. A high growth rate of 100 µm/h has been observed. The use of H_2O improves structural, electrical and optical properties of ZnO crystals. The FWHM value of X-ray rocking curve for the ZnO (0002) reflection was 38 arcsec. Both carrier concentration and Hall mobility increased with partial pressure of H_2O. The dependence of the carrier concentration on temperature indicates that there exist two donors in the CVT-grown ZnO crystals. The estimated ionization energy for the shallow donor was 35±5 meV and that for the deep donor was 115±5 meV.

REFERENCES

1. K. Maeda, M. Sato, I. Niikura, and T. Fukuda, Semicond. Sci. Technol. 20, S49 (2005).
2. H. Tampo, A. Yamada, P. Fons, H. Shibata, K. Matsubara, K. Iwata, S. Niki, K. Nakahara, and H. Takasu, Appl. Phys. Lett. 84, 4412 (2004).
3. J.-M. Ntep, S. Said Hassani, A. Lusson, A. Tromson-Carli, D. Ballutaud, G. Didier, R. Triboulet, J. Cryst. Growth 207, 30 (1999).
4. M. Mikami, S. Hong, T. Sato, S. Abe, J. Wang, K. Masumoto, Y. Masa, and M. Issiki, J. Cryst. Growth 304, 37 (2007).
5. K. Abe, Y. Banno, T. Sasayama, and K. Koizumi, Jpn. J. Appl. Phys. 48, 021101 (2009).
6. X. L. Wu, G. G. Siu, C. L. Fu, and H. C. Ong, Appl. Phys. Lett. 78, 2285 (2001).
7. E. V. Lavrov, F. Herklotz, and J. Weber, Phys. Rev. B 79, 165210 (2009).
8. M. D. McCluskey and S. J. Jokela, J. Appl. Phys. 106, 071101 (2009).

Mater. Res. Soc. Symp. Proc. Vol. 1201 © 2010 Materials Research Society   1201-H05-18

Effect of Metal Oxide Nanoparticles on the Mechanical Properties and Tacticity of Poly(Methyl Methacrylate)

Wantinee Viratyaporn and Richard Lehman

Rutgers University, 607 Taylor Road, Piscataway, New Jersey, 08854-8065

Abstract

Nanoparticles were incorporated into poly(methyl methacrylate) matrix by the mean of *in situ* bulk polymerization. Particle chemistry, size, shape, and percent loading were experimental variables in the synthesis and mechanical properties were assessed, particularly impact resistance, which showed improvement at the optimal particle loading. In assessing the mechanisms of this improvement, the elongated shape of zinc oxide particles appears to promote crack deflection processes to introduce a pull-out mechanism similar to that observed in fiber composite systems. Raman spectroscopy was performed to examine the effect of polymer chain conformation and configuration with the addition of nanoparticles.

Introduction

Polymer composites are useful in various applications from household goods to sophisticated electronic materials. In particulate/polymer composite systems, the polymer-particle surface interaction is a critical factor that influences the final properties of the polymer composites system.[1, 2, 3] The quality of the interaction determines to a significant extent the load-transfer efficiency and hence the mechanical properties.[4, 5] Naturally, this factor increases in significance as the size of the particles decreases, and the effect becomes dominant when the particles are at the nanometer scale. Nanoparticles have special effects on polymers since the particle size is of the same order as the polymer chain gyration.[6] Consequently, the nanoparticles can have a strong affect on the configuration and conformation of the surrounding polymer which leads to alterations in the bulk properties of the composite.[7, 8, 9] In this work, we sought to investigate this bound interfacial layer in nanoparticle/polymer composites using Raman spectroscopy and mechanical characterization. Our experimental system was comprised of aluminum oxide and zinc oxide nanoparticles in a PMMA matrix. Various particle sizes and volume fractions were chosen as experimental variables. The polymer nanocomposites were prepared by in situ bulk (radical) polymerization. To minimize the need for in-depth dispersion studies, nanoparticles pre-dispersed in propylene glycol methyl ether acetate were selected. Five combinations of nanoparticles types and sizes were used; two types of aluminum oxides (20 nm and 45 nm) and three types of zinc oxide (20 nm, 35 nm, and 70 nm). The aluminum oxide particles were spherical whereas the zinc oxide particles were acicular.

With regard to mechanical behavior, electron microscopy images show that the nanoparticles in the PMMA lead to the debonding followed by a shear yielding mechanism around the nanoparticles. Moreover, only at a specific particular volume fraction, unusually high (but reproducible) impact strength was observed, apparently resulting from debonding of the matrix and crack deflection arising from the fast loading rate of the 3.5 m/s impact test. This phenomenon was observed only for the zinc oxide particle composites and seems to be due to the acicular shape of these particles. Raman spectroscopy showed changes in the matrix polymer tacticity when nanoparticles were incorporated, an effect that almost certainly arises from the interaction between the polymer chain and nanoparticle surface. The molecular processes by which these conformational changes in the polymer are generated were addressed from a chemical perspective. Metal oxides are known to have hydroxyl layer on their surface. These polar groups have the potential to interact with polar groups on the polymer chains via secondary bonding and reduced the localized Coulombic energy. Such a mechanism, which is essentially one of dipole-dipole interaction, appears to cause the PMMA chains to orient such that the polar acrylic pendant group is closest to the oxide particle surface. Results of these considerations will be presented using qualitative chemical structures of the bonded interfacial region and the association of the polymer/oxide dipole moieties.

EXPERIMENTS

Material and Polymer Nanocomposite Compositions

Poly(methyl methacrylate), PMMA, was purchased from Polysciences, Inc. (Warrington, PA) as micro bead having the size of 200 μm with Mw = 75000. The stabilized methyl methacrylate, MMA, was purchased from Acros Organics (Morris Plains, NJ). Density for both PMMA and MMA are 1.19 g/cm³ and 0.93 g/cm³ respectively. Azobis isobutyronitrile (AIBN) was chosen as the initiator in this work. The pre-dispersed nanoparticle propylene glycol monomethyl ether acetate (PGMEA or PMA), Aluminum oxide (Al_2O_3) and zinc oxide (ZnO) nanoparticles were purchased from Nanophase Technologies, Romeoville, IL. Additional details of the nanoparticles are listed in the Table 1.

Table 1. Purchased Nanoparticles Information

Trade Name	Type	v/o in PGMEA	d*	Morphology
NanoArc® R1130PMA	Al_2O_3	10.35	20	Sphere
NanoDur® X1130PMA	Al_2O_3	21.23	45	Sphere
NanoArc® Q1102PMA	ZnO	10.35	20	Elongated
NanoTek® ZH1102PMA	ZnO	10.35	35	Elongated
NanoTek® Z1102PMA	ZnO	14.76	70	Elongated

*Mean particle size

Processing, Mechanical Testing, and Characterization

The neat PMMA and polymer nanocomposites were synthesized by in situ bulk polymerization. Prior to polymerization, inhibitor was removed from the methyl methacrylate monomer (MMA) by passing the monomer through an aluminum oxide column where the inhibitors were removed by adsorption. In the case of neat PMMA, 1 wt% of AIBN was added and stirred with the magnetic bar for 5 minutes. Then, sheet molds, prepared from two glass plates sealed with window spacing tape was filled with the degassed mixture and placed into a water bath at 50°C for 24 h to polymerize the composite and to develop initial strength. A final cure was conducted at 95°C for 1 h. For the polymer nanocomposites, the polymerization steps were the same except the nanoparticles were sonicated in the PMMA syrup, the mixture of 15 vol% PMMA micro-bead and 85 vol% purified-MMA, for 20 min at 350 Watts, for a total energy application of 420 kJ. Subsequently, the 1 wt% of AIBN initiator was introduced to the mixture and stirred with a magnetic bar for 5 minutes. Then, the mixture was put under a mechanical vacuum for 5 minutes before filling into the sheet mold. Three relative concentrations were chosen, 0.5, 1.0, and 1.5, for each polymer nanocomposite system.

The polymerized sheet was machined cut into small sample bars (63 x 12.5 x 3.1 mm) for both tensile and impact testing. The notched Izod impact was performed on the instrumented pendulum impact machine (Instron POE2000) according to the ASTM D256A. While, tensile testing was performed according to ASTM D638-03 with Universal Testing Machine from MTS (QTest/25). To increase accuracy in the measurement, the extensometer was used. The test was performed with speed of 1 mm/min at room temperature. Both tests were performed at room temperature. Scanning electron microscopy was used to observed the fractured surface of tensile and impact test sample. The fractured samples were mounted on an SEM stud and vacuum degassed overnight followed by gold coated. The SEM images were obtained by Leo-Zeiss Gemini 982 FESEM at 4keV. Additionally, Raman shift spectra were collected with an inVia Raman microscope in the wavenumber range 200-3200 cm^{-1}. The excitation wavelength was 785 nm and magnification was 50x resulting approximately 1 µm of beam diameter on the sample surface. A grating beam path of 1200 l/mm was used. Each spectrum accumulation was replicated 16 times and averaged to increase accuracy. Subsequently, data were analyzed with Wire2 (SP9), software provided by Renishaw. The symmetric stretching of CC_4 (813 cm^{-1}) was chosen as the basis for data normalization due to its constant intensity and position.

DISCUSSION

Addition of nanoparticle affects the stiffness of neat PMMA as shown in the Figure 1a. The tensile modulus of the polymer nanocomposites decreases with increasing nanoparticle content, result anticipated from the plasticizing effect of PGMEA and the weak interactions between polymer chain and nanoparticle surface via secondary interaction, such as hydrogen bonding. Such weak interactions permit increased chain mobility around the nanoparticle surfaces as

111

illustrated in the micrograph (Figure 2a) where the PMMA polymer matrix is stretched from the particle surface to create microvoids. Overall, in the case of PMMA/Al$_2$O$_3$ systems, the tensile modulus decreases by 19% and 27% for 45 nm and 20 nm, respectively. Similarly, the modulus for PMMA/ZnO systems shows the reduction of 15%, 29%, and 30% as the particle size decrease from 70 nm to 35 nm and 20 nm.

The effect of nanoparticle on the impact resistance is shown in the Figure 1b, which shows a significant increase in impact resistance for the zinc oxide composites at one volume percent filler. On the other hand, the addition of aluminum oxide nanoparticles did not enhance the impact properties of PMMA. A major contributor to the zinc oxide particle effect is the acicular shape of the zinc oxide particle, a physical feature that promotes the crack deflection and introduces a high pull-out energy failure mechanism during impact. Such pullout mechanisms are documented and common in fiber reinforced composites.[10] Interestingly, the smallest particle size (20 nm) provides higher impact improvement, as would be expected from the perspective of particle physics. At 1 vol% of zinc oxide addition, the impact resistance increases 324%, 298%, and 292% as the particle size increase from 20 nm to 35 nm and 70 nm, as shown in Figure 1b.

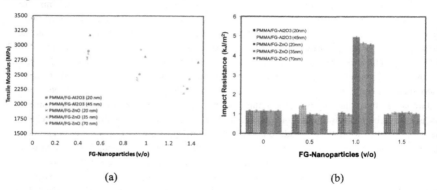

(a)   (b)

Figure 1: Mechanical properties of PMMA nanocomposites of various polymer nanocomposite systems at three relative volume fractions: (a) tensile modulus and (b) impact resistance.

Another qualitative measure of the role of the nanoparticles during failure can be inferred from fracture surface texture as various loading rates (Figure 2). In this figure, both the effect of nanoparticles and loading rate, tensile compared to impact, can be observed. Clearly the addition of the nanoparticles greatly alters the glass-like fracture of the neat PMMA. Furthermore, the slow loading rate of the tensile testing allows for greater polymer relaxation and stretching around the nanoparticles as compared with the fast fracture impact test.

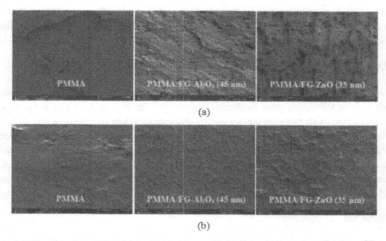

(a)

(b)

Figure 2: Micrographs of PMMA and its nanocomposites at relative concentration of 1 vol% at 5000x magnification: (a) tensile fracture surfaces and (b) impact fracture surfaces.

The addition of nano-size particles into the polymer matrix can perturb conformation as well as configuration of polymer chains. The analysis of original Raman spectra, as shown in Figure 3, indicates that the addition of nanoparticles leads to a change of polymer chain configuration toward syndiotactic sequences as seen in Figure 3a. In addition, the presence of surface hydroxyl group on metal oxide surface along with moiety in the system proceeds de-esterification of ester group in polymer chain.[11]

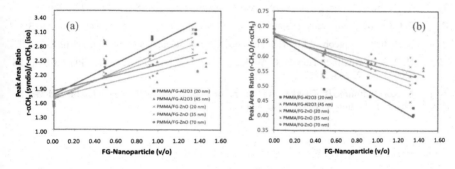

Figure 3: Ratios of deconvoluted peaks from Raman spectra represent relative content of (a) syndiotactic/isotactic in composites and (b) hydrolysis reaction

As a result, the vibration of ester group decreases with relative to α-CH$_3$ group (Figure 3b). This effect is observed to be increase as the surface area of nanoparticles becomes larger in smaller particle size.

SUMMARY AND CONCLUSION

In situ bulk polymerization of polymer nanocomposites results in enhancement of impact resistance without sacrificing other mechanical properties. The high surface area and elongated shape of small zinc oxide particles creates energy absorbing fracture mechanisms that are responsible for the higher impact resistance. Raman results reveal that *in situ* polymerization of polymer nanocomposites increase the syndiotactic sequence in polymer chain as compare to the isotactic one. In addition, some of the ester groups in the PMMA are hydrolyzed by the high dipole character of the oxide surface in general and more specifically the polarity of the adsorbed hydroxyl groups.

REFERENCES

1. K. Liao and S. Li, Appl. Phys. Lett. **79** (2001), 4225-4227.

2. M. Du, B. Guo, Y. Lei, M. Liu, and D. Jia, Polymer **49** (2008), 4871-4876.

3. W. Zhou, B. Wang, Y. Zheng, Y. Zhu, J. Wang, and N. Qi, Chem. Phys. Chem. **9** (2008), 1046 – 1052.

4. A. Durmus, A. Kasgoz, and C. W. Macosko, J. Macromol. Sci. Phys. **47** (2008), 608–619.

5. G. Sui, W. H. Zhongb, X. P. Yang, and Y. H. Yu, Mater. Sci. Eng. A **485** (2008), 524–531.

6. A. J. Crosby and J.-Y. Lee, Polymer Reviews **47** (2007), 217–229.

7. Y. Grohens, M. Brogly, C. Labbe, and J. Schultz, Polymer **38** (1997), 5913-5920.

8. P. Carriere, Y. Grohens, J. Spevacek, and J. Schultz, Langmuir **16** (2000), 5051-5053.

9. D. Ciprari, K. Jacob and R. Tannenbaum, Macromolecules **39** (2006), 6565-6573.

10. H. P. S. A. Khalil, S. Hanida, C. W. Kang, and N. A. N. Fuaad, J. Reinf. Plast. Comp. **26** (2007), no. 2, 203-218.

11. K. Konstadinidis, B. Thakkar, A. Chakraborty, L. W. Potts, R. Tannenbaum, and M. Tirrell, Langmuir **8** (1992), 1307-1317.

Mater. Res. Soc. Symp. Proc. Vol. 1201 © 2010 Materials Research Society   1201-H05-20

Effect of growth conditions on structural and electrical properties of Ga-doped ZnO films grown by plasma-assisted MBE

V. Avrutin[1*],   H.Y. Liu[1], N. Izyumskaya[1], M.A. Reshchikov[2],
Ü. Özgür[1], A.V. Kvit[3], P.M. Voyles[3], and H. Morkoç[1,2]

[1] Department of Electrical and Computer Engineering, Virginia Commonwealth
University, Richmond, VA 23284
[2]Physics Department, Virginia Commonwealth University, Richmond, VA 23284
[3]Deparment of Materials Science and Engineering, University of Wisconsin-Madison,
Madison, WI 53706

Abstract: ZnO has recently attracted a great deal of attention as a material for transparent contacts in light emitters and adsorbers. ZnO films heavily doped with Ga (carrier concentration in the range of 10^{20} - 10^{21} cm^{-3}) were grown on a-plane sapphire substrates by RF plasma-assisted molecular beam epitaxy. Oxygen pressure during growth (i.e. metal (Zn+Ga)–to–oxygen ratio) was found to have a crucial effect on structural, electrical, and optical properties of the ZnO:Ga films. As-grown layers prepared under metal-rich conditions exhibited resistivities below 3×10^{-4} Ω-cm and an optical transparency exceeding 90% in the visible spectral range. In contrast, the films grown under the oxygen-rich conditions required thermal activation and showed inferior structural, electrical, and optical characteristics even after annealing.

INTRODUCTION

Due to the well-established fabrication technology, indium-tin oxide (ITO) is the predominant material for transparent conducting electrodes [1]. However, the scarcity of indium in nature (and associated cost) limits the application of ITO. Zinc oxide heavily doped with aluminum (AZO) or gallium (GZO) is receiving increasing attention as a material with potential to replace ITO for transparent electrode applications in solar cells, light-emitting devices, and transparent thin film transistors [2,3,4,5]. AZO and GZO were demonstrated to have low resistivity and high transparency in the visible spectral range and, in many cases, outperform ITO. For instance, Agura et al. [4] reported a very low resistivity of ~8.5×10^{-5} Ω-cm for AZO, and Park et al. [5] reported even lower resistivity of ~8.1×10^{-5} Ω-cm for GZO, both values are nearly the same as the lowest reported resistivity of ~7.7×10^{-5} Ω-cm for ITO [6]. Ga is an excellent n-type dopant in ZnO with a more compatible covalent bond length (1.92 Å for Ga–O and 1.97 Å for Zn–O) than that of Al (2.7 Å for Al–O) [7].

GZO thin films can be grown by various methods, including ion plating [8], sol-gel [9], magnetron sputtering [3], metal organic chemical vapor deposition [2], pulsed laser deposition [5], and molecular beam epitaxy (MBE) [10,11,12]. The MBE technique, with its precise control over the process parameters, allows one to gain insight into the nature of physical phenomena governing the optical and electrical properties of a material and thus provide valuable information for further advancements. However, the influence of MBE growth conditions on the electrical and optical properties of GZO is not studied sufficiently. Only the effect of a Ga/Zn supply ratio have been reported so far for MBE growth [10,11,12], and to the best of our knowledge only one group [12] has reported MBE-grown GZO films with a sufficiently low resistivity of 3×10^{-4} Ω-cm and transmittance of >85% in the visible range. In this work, we demonstrate the critical effect of oxygen pressure during plasma-assisted MBE growth on electrical, structural, and optical properties of GZO films.

EXPERIMENTAL DETAILS

GZO layers were grown on a-plane sapphire substrates by RF plasma-assisted MBE. First, a ~10-nm-thick ZnO buffer layer was grown at a substrate temperature of 300 °C to provide smooth nucleation surface. Then, GZO films with thickness varying in the range from 100 to 750 nm were deposited at a substrate temperature of 400°C. During growth, the temperatures of Ga and Zn cells were kept constant at 600°C and 350°C, respectively. Reactive oxygen was supplied from an RF plasma source operated at 400 W. The flux of reactive oxygen was controlled by supplying O_2 through a mass-flow controller with corresponding pressure in the growth chamber during growth. Three series of GZO films were grown under three different oxygen pressures $P_{O2} = 4.5 \times 10^{-6}$, 8.0×10^{-6}, and 1.5×10^{-5} Torr, corresponding to metal (Zn+Ga) rich, near-stoichiometric (reactive oxygen to incorporated Zn ratio $\approx$ 1:1), and oxygen-rich conditions, respectively. The reactive oxygen–to–incorporated Zn ratios were assessed from growth rate $vs.$ oxygen pressure dependence. Ga content of the films found from secondary ion mass-spectrometry was ~1 at. %. Rapid thermal annealing (RTA) at a temperature of ~650°C in N_2 atmosphere was performed to activate Ga donors and increase the conductivity of the GZO films. Surface morphology and crystal quality of the GZO films were characterized by reflection high energy electron diffraction (RHEED), x-ray diffraction (XRD), and transmission electron microscopy (TEM). Electrical properties were studied by Hall effect measurement using the van der Pauw configuration, and optical properties were investigated by photoluminescence (PL) and transmittance measurements. For transmittance measurements, after the spectrum of the light from a tungsten lamp passing through the sample was measured, the GZO film was completely etched away using a 10% aqueous HCl solution and the measurement was repeated under exactly the same experimental conditions and on the same region of the sample to obtain the reference transmission through the substrate.

RESULTS AND DISCUSSION

In situ monitoring of RHEED patterns during growth of the GZO films studied revealed a strong effect of the oxygen pressure on the growth mode. Figure 1 shows that, as P_{O2} was increased from 4.5×10^{-6} to 8×10^{-6} Torr, the RHEED pattern of GZO changed from two-dimensional (2D) [Figure 1(a)] to three-dimensional (3D) growth mode [Figure 1(b)]. When P_{O2} was further increased to 1.5×10^{-5} Torr, the RHEED pattern transformed from 3D growth mode into partially poly-crystal growth mode indicated by short arcs superimposed on the spotty diffraction [Figure 1(c)] after growing the first 10-15 nm of GZO. Further increase in P_{O2} resulted in polycrystalline layers. The effect of oxygen pressure on the crystal perfection of the GZO films was revealed by XRD: the full width at half maximum (FWHM) of the (0002) rocking curves measured for the ~300-nm films grown at $P_{O2} = 1.5\times10^{-5}$, 8.0×10^{-6}, and 4.5×10^{-6} Torr are 0.907°, 0.623°, and 0.488°, respectively [Figure 1(d)]. Thus, the films grown under the metal-rich conditions exhibit better crystal quality in terms of FWHM.

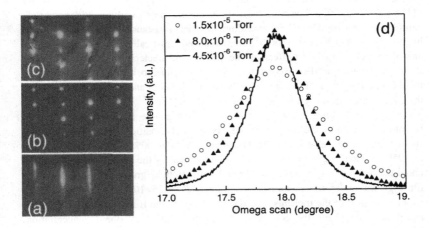

Figure 1. Typical RHEED patterns recorded for the [11-20] azimuth of GZO films grown at $P_{O2} =$ (a) 4.5×10^{-6} (630-nm film), (b) 8.0×10^{-6} (300-nm film), (c) 1.5×10^{-5} Torr (300-nm film), respectively and (d) rocking curves (0002) XRD reflection measured from GZO films grown at different P_{O2} with comparable film thickness of 300 nm.

The effect of oxygen pressure on the crystal perfection of the GZO films was confirmed by TEM. Figure 2 compares cross-sectional TEM images of the samples grown under different oxygen pressures. One can see that the layer deposited under the metal-rich conditions [Figure 2(a)] is grown epitaxially with a columnar structure composed by slightly misoriented domains (smaller than 0.5°). Selective area diffraction (SAD) reveals very strong epitaxial relationship between the sapphire substrate and the GZO layer. Predominant extended defects in this film are dislocations, the concentration of which decreases drastically from interface to the surface of the structure, since most of the dislocations recombine near the interface within the first 100 nm from the substrate. The film grown under the oxygen-rich conditions has a textured structure [Figure 2(b)], as evidenced by Moiré patterns observed in the vicinity of the interface. High-angle boundaries are the predominant

117

extended defects in this film. SAD confirms this observation and reveals the arc structure typical for textured materials.

The GZO films deposited under the metal-rich conditions (P_{O2}=4.5×10^{-6} Torr) exhibited superior electrical characteristics (conductivity, electron mobility, carrier concentration) as compared to those of the layers grown at higher P_{O2} (1.5×10^{-5} and 8.0×10^{-6} Torr). The as-grown films deposited under the metal-rich conditions exhibited the lowest values of sheet resistance, R_S, reducing from 30 to 4.6 Ω/□ with the film thickness increasing from 100 to 630 nm. For example, resistivity, carrier concentration, and Hall mobility of the 630-nm thick GZO layer were 3×10^{-4} Ω-cm, 7x10^{20} cm^{-3}, and 32 cm^{-2} V s^{-1}, respectively. On the other hand, the as-grown films deposited under the oxygen-rich conditions showed very high R_S, typically in the kΩ/□ range, and the layers deposited under the near-stoichiometric conditions exhibited R_S of ~100 Ω/□. Upon annealing, the sheet resistance of the layers grown at higher P_{O2} (1.5×10^{-5} and 8.0×10^{-6} Torr) reduced drastically. To the contrary, the annealing had virtually no effect on the electrical characteristics of the GZO films grown under the metal-rich conditions, apparently due to better crystal quality of these films. The effect of the oxygen pressure on electrical and transport properties of the GZO films is discussed in more detail elsewhere [13].

Figure 3(a) compares transmittance spectra from the as-grown and annealed samples grown under the metal-rich and oxygen-rich conditions. For all the samples studied, transmittance below the ZnO bandgap is close to 90%, with the highest value of ~95% at wavelength longer than 385 nm observed for the annealed sample grown under the metal-rich conditions [see Figure 3(a)]. Both as-grown and annealed GZO films grown under the metal-rich conditions (P_{O2}=4.5×10^{-6} Torr) demonstrate a strong blue shift of the transmission edge attributed to the Burstein-Moss shift of the Fermi level deep into the conduction band, indicating high donor concentration, which is in agreement with the electrical measurements, revealing the highly conductive nature of these samples. According to the electrical data, the layers grown under the oxygen-rich conditions became highly conductive only upon RTA, which agrees well with the optical data indicating that the transmission edge shifts significantly to shorter wavelengths only in the annealed sample (see Figure 3).

PL spectra of GZO films measured at room temperature before and after annealing are presented in Figures 3 (b) and (c), respectively. In both cases, the PL spectra [Figures 3 (b) and (c)] from the film grown under metal-rich conditions (P_{O2}=4.5×10^{-6} Torr) exhibits only strong near band edge (NBE) emission, broadened and shifted to higher energies (maximum at 3.35-3.4 eV), that agrees with the Fermi level position well above the conduction band minimum. To the contrary, the PL spectra of the as-grown GZO films deposited at P_{O2}=8.0×10^{-6} and 1.5×10^{-5} Torr show only weak NBE emission and a strong, broad defect-related emission around 1.75 eV. The nature of the defect-related band needs further studies. After annealing of these two samples, the intensity of the defect-related emission decreases, while the intensity of the NBE emission increases by a factor of ~30 and the peak maximum shifts significantly to shorter wavelengths, indicating improvement of the crystal quality and increase in concentration of free electrons, in a good agreement with the electrical and XRD data. Nevertheless, the NBE line from the sample grown under metal-rich conditions shows the strongest blue shift, indicating higher concentration of free carriers.

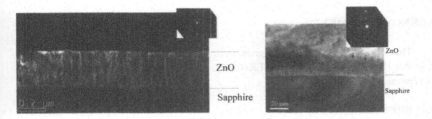

Figure 2 Cross-sectional dark-field TEM image of the GZO films grown under (a) metal-rich (P_{O2}=4.5×10^{-6} Torr) and (b) oxygen-rich (P_{O2}=1.5×10^{-5} Torr) conditions. Arrows indicate positions of moiré contrast due to overlapping of small grains with different orientation.

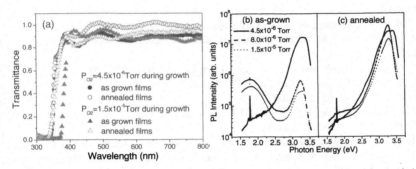

Figure 3. Optical properties of GZO films: (a) transmittance of both as-grown and annealed GZO grown at different oxygen pressures (Note that transmittance higher then 100% is an measurement artifact due to the Fabri-Perrot interference) and PL of as-grown (b) and annealed (c) GZO films.

CONCLUSIONS

Highly conductive and highly transparent GZO films have been achieved by plasma-assisted MBE. A dramatic effect of the metal (Zn+Ga)-to-oxygen ratio on the electrical, structural, and optical properties of the GZO films was revealed. The as-grown layers deposited under metal-rich conditions exhibited a resistivity below 3×10^{-4} Ω-cm and optical transparency exceeding 90% in the visible spectral range, which is accompanied with a large blue shift of the transmission edge. The blue shift is attributed to the Burstein-Moss shift of the Fermi level deep into the conduction band owing to high electron concentration. The results of XRD, TEM, and PL measurements indicate superior crystal quality of the layers grown under metal-rich conditions. In contrast, the films grown under near-stoichiometric and oxygen-rich conditions exhibit inferior crystallinity as well as electrical and optical properties.

ACKNOWLEDGMENTS

This work was supported by a grant from the Department of Energy, Basic Energy Sciences (Project No. DE-FG-02-08ER46547), through a sub-grant from the University of Wisconsin.

REFERENCES

1. S. J. Kim, IEEE Photon. Technol. Lett. 17, 1617 (2005).
2. H. H. Chen, A.D. Pasquier, G. Saraf, J. Zhong, and Y.C. Lu, Semicond. Sci. Technol. 23, 045004 (2008).
3. S.J. Tark, M.G. Kang, S. Park, J. H. Jang, J.C. Lee, W.M. Kim, J.S. Lee, D. Kim, Curr. Appl. Phys., 9, 1318 (2009).
4. H. Agura, H. Suzuki, T. Matsushita, T. Aoki, and M. Okuda, Thin Solid Films, 445, 263 (2003).
5. S.-M. Park, T. Ikegami, K. Ebihara, Thin Solid Films, 513, 90 (2006).
6. H. Ohta, M. Orita, M. Hirano, H. Tanji, H. Kawazoe, and H. Hosono, Appl. Phys. Lett., 76, 2740 (2000).
7. Z. Yang, D. C. Look, and J. L. Liu, Appl. Phys. Lett. 94, 072101 (2009).
8. T. Yamada, A. Miyake, S. Kishimoto, H. Makino, N. Yamamoto, and T. Yamamoto, Appl. Phys. Lett. 91, 051915 (2007).
9. W.J. Park, H.S. Shin, B.D. Ahn, G.H. Kim, S.M. Lee, K.H. Kim, and H.J. Kim, Appl. Phys. Lett. 93, 083508 (2008).
10. H. Kato, M. Sano, K. Miyamoto, T. Yao, J. Cryst. Growth 237–239,538 (2002).
11. H. J. Ko, Y. F. Chen, S. K. Hong, H. Wenisch, T. Yao, and D. C. Look, Appl. Phys. Lett. 77, 3761 (2000).
12. T. Muranaka, A. Nisii, T. Uehara, T. Sakano, Y. Nabetani, T. Akitsu, T. Kato, T. Matsumoto, S. Hagihara, O. Abe, S. Hiraki and Y. Fujikawa, J. Korean Phys. Soc. 53, 2947 (2008).
13. H.Y. Liu, V. Avrutin, N. Izyumskaya, M.A. Reshchikov, Ü. Özgür, and H. Morkoç, submitted to Phys. Stat. Sol. : Rapid Res. Lett.

Mater. Res. Soc. Symp. Proc. Vol. 1201 © 2010 Materials Research Society   1201-H05-24

Effect of Aluminum Nitrate Concentration in Zinc Acetate Precursor on ZnO:Al Thin Films Prepared by Spray Pyrolysis

Samerkhae Jongthammanurak[1], Sirirak Phakkeeree[2], Yot Boontongkong[1] and Chanchana Thanachayanont[1]

[1]National Metal and Materials Technology Center, 114 Thailand Science Park, Pahonyothin Rd., Klong 1, Klong Luang, Pathumthani 12120, Thailand
[2]Department of Materials Engineering, Kasetsart University, Bangkok Thailand

ABSTRACT

Aluminium-doped zinc oxide (ZnO) films have been prepared by spray pyrolysis technique using the mixed solution of zinc acetate dihydrate and aluminium nitrate nonahydrate in methanol. Concentration of aluminum in the solution was varied in a range of 1, 3 and 5 atomic percent. The results from X-ray diffraction showed that the preferred orientation of ZnO films changed to the [002] direction when the concentration of aluminum in the solution exceeded 1 atomic percent. ZnO films deposited from the 3 atomic percent Al containing solution had the largest grains and showed the lowest resistivity of 75 Ω-cm. Addition of aluminum into the precursor solution shifted the absorption edge towards longer wavelengths.

INTRODUCTION

Transparent conductive oxide (TCO) offers advantages of electrical conductivity and transparency to visible light. Indium tin oxide (ITO) thin films have been widely used in making electrodes of photovoltaic and optoelectronics devices [1-4]. However, rising cost of ITO in the last five years necessitates the replacement of ITO with cheaper materials in order to keep the devices in reasonable price. ZnO is an attractive candidate for the substitution because of its abundance. Doping of Ga, In, Sn and Al into ZnO thin films is demonstrated to decrease the electrical resitivity [1, 5-8].

The ZnO thin films can be deposited by magnetron sputtering [8-12], pulse laser deposition [13], sol gel [14] and spray pyrolysis [15-20] techniques. Al-doped ZnO thin films deposited by magnetron sputtering demonstrate the electrical resistivity in the range of 10^{-4} Ω-cm [8-12]. They are used as electrodes in organic solar cells [8] and light emitting devices [9], showing performance comparable to ITO. The effect of Al doping concentration on the resistivity is investigated. The range of Al doping in ZnO films deposited by the spray pyrolysis technique is varied from 0.1 to 5 atomic percent [15-18]. On the other methodology, the effect of film thickness on optical and electrical properties was studied in Al-doped ZnO films with Al doping concentration of 0.15 and 4 atomic percent [19, 20]. In the investigation of the effect of Al doping concentration on ZnO films, aluminium chloride was used as the dopant [15-18, 20].

In this study, we investigated the effect of aluminium concentration on ZnO thin film prepared by spray pyrolysis technique, by using aluminium nitrate as the dopant. Concentration of aluminum in the precursor solution was varied in a range of 1, 3 and 5 atomic percent. Preferred orientation in the [002] direction was observed in ZnO films doped with Al concentration, which exceeded 1 atomic percent. The ZnO films doped with Al concentration of 3 atomic percent had the largest grains and lowest resistivity of 75 Ω-cm.

EXPERIMENT

Undoped and Al-doped ZnO thin films were deposited at 430°C on glass substrates by spray pyrolysis using the solution of 0.05 molar zinc acetate dihydrate in methanol. Aluminium nitrate nonahydrate ($Al(NO_3)_3 \cdot 9H_2O$) of 1, 3 and 5 atomic percent of zinc acetate dihydrate concentration were added into the solution to make the precursors for Al-doped ZnO film deposition. The crystal structures of the ZnO films were studied by using X-ray diffraction (XRD) technique (Rigaku TTRAX III), with Cu Kα radiation operating at 50 kV, 300mA at room temperature. The microstructures were examined using scanning electron microscopy (SEM) (JEOL 6301-F). The resistivity of the films was determined by four-point probe measurement at room temperature. Optical transmittance of the films was measured using Shimadzu UV-VIS-NIR spectrophotometer.

DISCUSSION

Figure 1 shows the XRD patterns of the undoped and Al-doped ZnO films. All films were polycrystalline ZnO with hexagonal wurtzite structure. Peaks of aluminum oxide were not observed. Preferred orientation of undoped ZnO films was in the [101] direction. However, ZnO films with aluminum concentration of more than 1 atomic percent showed preferred crystal orientation in the [002] direction. The preferred orientation in the [002] direction was observed in ZnO thin films deposited using zinc nitrate [21]. Thus the observed preferred orientation in the (002) direction was possibly due to the nature of the aluminum nitrate precursor, which was used in the experiments. The similar preferred orientation of the (002) planes was observed in the Al-doped ZnO films deposited by sol-gel spin coating [14].

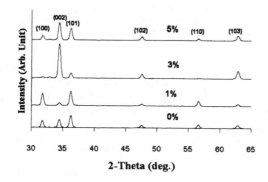

Figure 1. X-Ray diffraction patterns of undoped and Al-doped ZnO films with varying Al concentrations in the precursor solution

SEM images of ZnO films doped with 0, 1, 3 and 5 Al atomic percent are shown in Figure 2 a – d, respectively. The surface morphology of Al-doped ZnO films was changed with the addition of aluminum nitrate nonahydrate in the precursor solution. Undoped ZnO films showed randomly oriented grains with the diameters in a few hundred nanometers (Fig. 2a). ZnO films doped with Al concentration of 1 atomic percent had slightly larger grains compared to the undoped films; the grains looked like agglomeration on the surface (Fig. 2b). ZnO films doped with Al concentration of 3 atomic percent showed different surface morphology with partial hexagonal grains (Fig. 2c). ZnO films doped with Al concentration of 5 atomic percent had grains with varying sizes. Its surface morphology is mixed between the agglomerates and partial hexagonal grains (Fig. 2d).

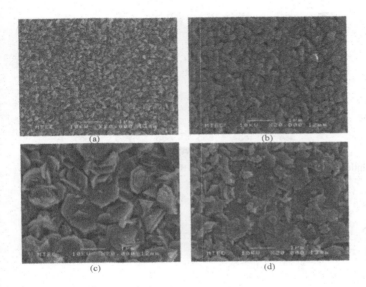

Figure 2. SEM images of (a) undoped ZnO film and ZnO films with Al concentrations of (b) 1, (c) 3 and (d) 5 atomic percent.

We measured the film thickness from the cross-section SEM images. The thickness and electrical resistivity of ZnO films are reported in Table 1. The undoped and ZnO films doped with Al concentrations of 3 and 5 atomic percent had the thickness in a range of approximately 920-1020 nm. Among these films, ZnO films doped with Al concentration of 3 atomic percent had the lowest resistivity of 75 Ω-cm. Undoped ZnO films had the highest resistivity of 2,907 Ω-cm. Addition of 3 atomic percent aluminum nitrate reduced the resistivity by more than an order of magnitude.

Table I. Thickness and resistance of ZnO films with Al concentrations

Al doping (%)	Thickness (nm)	Sheet Resistance [Ω/square]	Resistivity [Ω-cm]
0	1020	2.85×10^7	2,907
1	417	1.88×10^6	78
3	920	8.16×10^5	75
5	960	7.86×10^6	754

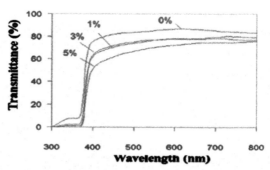

Figure 3. Optical transmittance of ZnO films with different Al concentrations

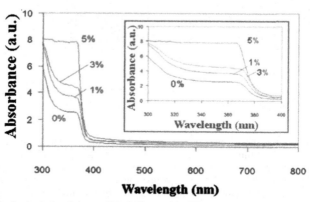

Figure 4. Optical absorbance of ZnO films with different Al concentrations. Addition of aluminum shifted the absorption edge towards the longer wavelength

Transmittance spectra of the ZnO films were characterized as a function of Al concentration. Figure 3a shows the optical transmittance spectra in a range of 300-800 nm wavelengths. The undoped ZnO film was approximately 80% transparent in the 400-800nm range; this was comparable to the transmittance at 550 nm reported in Ref. 15. The ZnO films doped with 3 and 5 Al atomic percent were approximately 70% and 60% transparent, respectively. The addition of Al concentrations into the ZnO films decreased the optical transmittance of the ZnO films in the 400-800nm range. This observation was in agreement with the previous studies [15-17]. Below the wavelength 380 nm, the undoped film was distinguished by its relatively high transmittance of ~ 7%.

Figure 4 shows the optical absorbance of ZnO films with different Al concentrations. Below a wavelength of 400nm located the absorption edge, at which the sharp reduction of the absorbance took place. We noticed that addition of Al shifted the absorption edge towards the longer wavelength (the inset in Figure 4). We proposed that the presence of Al in the films may have changed the energy gap of ZnO. Previous studies by Lee et al show that the increase in carrier concentration shifts the absorption edge towards the shorter wavelengths, which is described as the Burstein-Moss effect [17,18]. Nevertheless, our ZnO films had relatively high resistivity; the carrier concentration in the films was not sufficiently high to cause the increase in the energy gap. Tensile strain effect, if present in our ZnO films, may have caused the reduction in the energy gap and shifted the absorption edge to longer wavelengths. To ascertain our hypothesis, further investigation of strain in the films by using XRD technique will be performed.

CONCLUSIONS

ZnO films with Al concentrations of 0, 1, 3 and 5 atomic percent were deposited on glass substrates at 430°C using spray pyrolysis technique. These films were polycrystalline with hexagonal wurtzite structure. Preferred orientation in the [002] direction was observed in ZnO films doped with Al concentration, which exceeded 1 atomic percent. The ZnO films doped with Al concentration of 3 atomic percent had the largest grains and lowest resistivity of 75 Ω-cm. The addition of Al reduced the optical transmittance in the 400-800 nm range and shifted the absorption edge towards longer wavelength hypothetically due to strain effect.

ACKNOWLEDGMENTS

This work is financially supported by National Metal and Materials Technology Center (MTEC), Thailand via Project no IH-0018/52 and IH-0083/53. The authors would like to thank Dr. Tosawat Seetawan of Sakon Nakhon Rajabhat University, Dr.Weerasak Somkhunthot of Loei Rajabhat University and Dr. Ratchatee Techapiesancharoenkij of Kasetsart University Thailand for insightful discussion.

REFERENCES

1. T. Minami, *Semicond. Sci. Technol.* **20**, S35-43 (2005).
2. H.L. Hartnagel, A.L. Dawar, A.K. Jain and C. Jagadish, *Semiconducting transparent thin films* (Taylor&Francis, 1985).

3. F.L. Zhang, M. Johansson, M.R. Andersson, J.C. Hummelen and O. Inganäs, *Synthetic Metals* **137**, 1401-1402 (2003).
4. D.C. Oertel, M.G. Bawendi, A.C. Arango and V. Bulović, *Appl. Phys. Letts.* **87**, 213505 (2005).
5. K.T. Ramakrishna Reddy, H. Gopalaswamy, P.J. Reddy and R.W. Miles, *J. Crys. Gro.* **210**, 516-520 (2000).
6. M.S. Tokumoto, A. Smith, C.V. Santilli, S.H. Pulcinelli, A.F. Craievich, E. Elkaim, A. Traverse and V. Briois, Thin Solid Films **416**, 284-293 (2002).
7. S.M. Rozati, S. Moradi, S. Golshahi, R. Martins and E. Fortunato, Thin Solid Films **518**, 1279-1282 (2009).
8. G.B. Murdoch, S. Hinds, E.H. Sargent, S.W. Tsang, L. Mordoukhovski and Z.H. Lu, *Appl. Phys. Letts.* **94**, 213301 (2009).
9. D. Xu, Z. Deng, Y. Xu, J. Xiao, C. Liang, Z. Pei and C. Sun, *Phys. Letts. A.* **346**, 148 (2005).
10. M. Chen, Z.L. Pei, C. Sun, J. Gong, R.F. Huang and L.S. Wen, *Mater. Sci. Eng.* **B85**, 212-217 (2001).
11. J. Yoo, J. Lee, S. Kim, K. Yoon, I. Jun Park, S.K. Dhungel, B. Karunagaran, D. Mangalaraj and J. Yi, *Thin Solid Films* **480-481**, 213-217 (2006).
12. Y.M. Chung, C.S. Moon, W.S. Jung and J. G. Han, *Thin Solid Films* **515**, 567-570 (2006).
13. H. Agura, A. Suzuki, T. Matsushita, T. Aoki and M. Okuda, *Thin Solid Films* **445**, 263-267 (2003).
14. S. Mridha and D. Basak, *J. Phys. D: Appl. Phys.* **40**, 6902-6907 (2007).
15. B. Joseph, P.K. Manoj and V.K. Vaidyan, *Ceram. Inter.* **32**, 487-493 (2006).
16. M.T. Mohammad, A.A. Hashim and M.H. Al-Maamory, *Mater. Chem. Phys.* **99**, 382-387 (2006).
17. S.M. Rozati and Sh. Akesteh, Mater. Char. 58, 319-322 (2007).
18. J.-H. Lee and B.-O. Park, *Mater. Sci. and Eng. B* **106**, 242 (2004).
19. M.A. Kaid and A. Ashour, *Appl. Surf. Sci.* **253**, 3029-3033 (2007).
20. W.T. Seebar, M.O. Abou-Helal, S. Barth, D. Beli, T. Höche, H.H. Afify and S.E. Demian, *Mater. Sci. in Sem. Proc.* **2**, 45-55 (1999).
21. E. Bacaksiz, M. Parlak, M. Tomakin, A. Özçelik, M. Karakiz and M. Altunbaş, *J. Alloys and Comps.* **466**, 447 (2008).

Mater. Res. Soc. Symp. Proc. Vol. 1201 © 2010 Materials Research Society   1201-H05-25

High quality p-type ZnO film grown on ZnO substrate by nitrogen and tellurium co-doping

S. H. Park[1,2], T. Minegishi[1], J. S. Park[1], H. J. Lee[1], T. Taish[3], I. Yonenaga[3], D. C. Oh[4], M. N. Jung[5], J. H. Chang[5], S. K. Hong[6], and T. Yao[1]

[1] Center for Interdisciplinary Research, Tohoku University, Sendai, 980-8578, Japan
[2] National Institute for Materials Science, Tsukuba, 305-0044, Japan
[3] Institute for Material Research, Tohoku University, Sendai, 980-8577, Japan
[4] Department of Defense Science and Technology, Hoseo University, 336-795, Korea
[5] Major of Semiconductor Physics, Korea Maritime University, Pusan 606-791, Korea
[6] Nano Information Systems Engineering, Chungnam National Univ., Daejeon, 305-764, Korea

ABSTRACT

Nitrogen and tellurium co-doped ZnO (ZnO:[N+Te]) films have been grown on (0001) ZnO substrate by plasma-assisted molecular beam epitaxy. The electron concentration of tellurium doped ZnO (ZnO:Te) gradually increases, compared with that of undoped ZnO (u-ZnO). On the other hand, conductivity of ZnO:[N+Te] changes from n-type to p-type characteristic with a hole concentration of 4×10^{16} cm^{-3}. However, nitrogen doped ZnO film (ZnO:N) still remains as n-type conductivity with an electron concentration of 2.5×10^{17} cm^{-3}. Secondary ion mass spectroscopy reveals that nitrogen concentration ([N]) of ZnO:[N+Te] film (2×10^{21} cm^{-3}) is relatively higher than that of ZnO:N film (3×10^{20} cm^{-3}). 10 K photoluminescence spectra show that considerable improvement of emission properties of ZnO:[N+Te] with an emergence of narrow acceptor bound exciton ($A^{o}X$, 3.359 eV) and donor-acceptor pair (DAP, 3.217 eV), compared with those of u-ZnO. Consequently, high quality p-type ZnO with high N concentration is realized by using Te and N co-doping technique due to reduction of Madelung energy.

INTRODUCTION

ZnO has a direct band gap of 3.37 eV at room-temperature and has also attracted attention as a useful material for UV optoelectronic applications. The large excitonic binding energy of 60 meV also raises the interesting possibility of utilizing excitonic effects in room-temperature devices [1]. In order to realize ZnO-based semiconductor devices, it is indispensable to fabricate ZnO films with p-type as well as n-type conductivity by controllable extrinsic doping. However, ZnO have naturally only n-type conductivity due to the presence of native defects such as oxygen vacancy and zinc interstitial, which hampers obtaining high-quality p-type ZnO layers required for optoelectronic devices. Although there are many reports on p-type nitrogen doped ZnO film by various growth method, high quality p-type ZnO films have not been achieved in terms of poor electrical and optical properties [2-4]. This classical problem is originated from the low solubility limit of ZnO.

In order to solve solubility limit of ZnO, Yamamoto et al. has proposed firstly the alternative co-doping method (N+III-element) based on *initio* electronic band structure calculations [5]. They have reported that co-doping using reactive III-element enhances the incorporation of N acceptor in p-type co-doped ZnO because n-type doping using III-element species reduces the Madelung energy. Based on these predictions, many researchers have tried to grow p-type ZnO by using co-doping method [6-9]. However, the typical problems such as reduction of emission intensity and/or limits of hole concentration ($\sim 10^{16} cm^{-3}$) still remain, in spite of a high concentration of N atoms ($10^{20} cm^{-3}$) [10].

In the present study, the tellurium as the isoelectronic impurity is selected as a co-dopant. It has been well-known that VI-elements are naturally abundant in ZnTe, while II-elements are rich in ZnO. Therefore, the improvement of crystal quality is expected due to more stoichiometrically ionic balance. Also, it has been already reported that Te doping increase nitrogen solubility in ZnSe [11]. Based on report above, N solubility may increase due to the reduction of Madelung energy in system co-doped with Te atom as an isoelectronic donor and N atom an acceptor. In this paper, N and Te codoped ZnO films are grown on (001) ZnO substrates by plasma-assisted molecular-beam epitaxy (P-MBE) and they are systematically investigated in terms of their N incorporation efficiency, optical properties and electrical properties.

EXPERIMENT

All substrate were supplied from Tokyo Denpa Co. Ltd., which was 5 mm×5 mm×0.5 mm pieces cut from c-plane wafers. To obtain the flat surface of (0001) ZnO substrate, they were thermally annealed at 1040 °C by using open-tube typed furnace under oxygen ambient for 30 minutes. Consequently, substrate with smooth surface (RMS value = 1.0 Å) could be prepared for homoepitaxy. The homoepitaxial ZnO thin films were grown by using plasma-assisted molecular-beam epitaxy (P-MBE). Subsequently, 10-nm-thick Zn-face ZnO buffer layers were grown at 400 °C and they were annealed at 850 °C. Then six kinds of samples (sample A~F), which is described in Table 1, were grown on the buffer layer at growth temperature as low as 500 °C. Zinc flux (J_{Zn}) and O_2/N_2 flows were 1 Å/s and 2/0.5 sccm, respectively. The thickness of grown samples was 500~600 nm.

The quantitative nitrogen concentration was estimated by using secondary ion mass spectroscopy (SIMS) measurement with Cs positive ion. Hall effect measurement were performed in the van der Pauw configuration by using a direct current of 0.5~10 µA, and magnetic field of -3.5 ~ +3.5 kG. Data were complied employing both positive and negative currents and reversing magnetic field, which was measured about 5 times per a sample in order to obtain the accurate result. Photoluminescence (PL) spectra were measured at 10 K using the 325 nm line from a He-Cd laser as an excitation source and a monochromator with a focal length of 32 cm.

DISCUSSION

Figure 1 shows the carrier concentration (n_c) of u-ZnO (open square), ZnO:Te (close square), ZnO:N (open circle) and ZnO:[N+Te] (close circle) with elevating Te fluxes (J_{Te}), respectively. As Te flux is irradiated up to 0.03 Å/s, carrier concentration gradually increases to $2.4×10^{18} cm^{-3}$, in comparison with that of u-ZnO ($4×10^{16} cm^{-3}$). Meanwhile, when J_{Te} become as high as 0.06 Å/s, it eventually decreases down to 9×10^{16} cm^{-3}. All of Te doped ZnO films are proved as n-type conductivity. Generally, the degradation of crystal quality and/or surface may cause the increase of electron concentration by forming the unintentional donor type defects. However, we have already observed that surfaces of ZnO:[N+Te] with low Te flux (J_{Te} = 0.01 Å/s) exhibit smooth (data is not shown here), which have almost same roughness to that of u-ZnO (RMS = ~9 Å). Thus an effect of other extrinsic defects is not insufficient to explain the increasing electron concentration.

Increment of n_c with low Te doping ($J_{Te} \leq 0.03$ Å/s) can be explained more persuasively by using a relationship of electronegativity between Te and O. Note that electronegativity of Te (2.1) is relatively smaller than that of O (3.44). If Te atoms are incorporated into O-atom site, those will be positively charged, which it may act as isoelectronic donor center in ZnO. On the other hand, the slight decrease of free electron density in higher J_{Te} (0.06 Å/s) results from compensation defects trapping electrons in the metastable structure, based on previous report [12].

ZnO:N film grown without Te doping has carrier concentration of 2.5×10^{17} cm^{-3}, which is higher than that of u-ZnO. Unfortunately, it shows n-type conductivity because numerous unintentional defects-complexes formed in ZnO affect on degradation of conductivity. The conductivity of ZnO:[N+Te] eventually changes from n-type to p-type characteristic with hole concentration (p) in range of $1\sim4\times10^{16}$ cm^{-3}. This phenomenon can be understood in terms of Madelung energy of ZnO. If N as acceptor is incorporated into ZnO film with additional Te atom as isoelectronic donor, total energy will be more stabilized by ionic neutralization. There are some reports on such minimization of total energy experimentally and theoretically, when nitrogen acceptors interact donor such as III-group element, causing a realization of p-type conductivity [5, 13, 14].

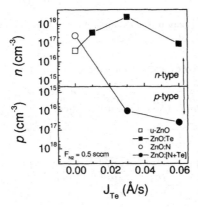

Figure 1 Carrier concentration (n_c) of u-ZnO (open square, sample A), ZnO:Te (close square, sample B, C), ZnO:N (open circle, sample D) and ZnO:[N+Te] (close circle, sample E, F) with elevating Te fluxes (J_{Te}), simultaneously.

To confirm the conductivity behavior, the Hall voltages of both u-ZnO (sample A) and ZnO:[N+Te] (sample E) were measured by changing magnetic field in the range of -3.5 ~+3.5

129

kG, as shown in Fig. 2. The negative slope of Hall voltage is observed in u-ZnO, indicating the n-type conductivity. Meanwhile, ZnO:[N+Te] film shows the positive slope behavior for all four current/ field combinations, which gives confidence that the layer is truly p-type conductivity. Consequently, the p-type ZnO films can be fabricated by using N and Te (N+Te) codoping. Detailed electrical properties and N concentration of all samples are summarized in Table I.

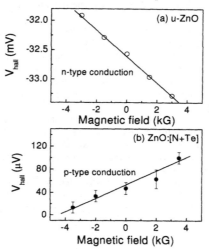

Figure 2 Hall voltages of both u-ZnO (sample A) and ZnO:[N+Te] (sample E) with changing magnetic field in the range of -3.5 ~+3.5 kG.

Table 1. Electrical properties and N concentration of all samples grown at different growth conditions

#		N_2 flow (sccm)	J_{Te} (A/s)	[N] (/cm³)	n_c (/cm³)	Carrier type	μ (cm²/Vs)	ρ (Ω cm)
A	u-ZnO	-	-	-	4×10^{16}	n	94	1.73
B	ZnO:Te	-	0.01	-	3.6×10^{17}	n	69	0.25
C	ZnO:Te	-	0.03	-	2.4×10^{18}	n	11	0.23
D	ZnO:N	0.5	-	3×10^{20}	2.5×10^{17}	n	8	3
E	ZnO:[N+Te]	0.5	0.03	2×10^{21}	5.8×10^{15}	p	21	49
F	ZnO:[N+Te]	0.5	0.06	1×10^{21}	4×10^{16}	p	11	13

Figure 3 shows the depth profile of nitrogen atom for (a) ZnO:[N+Te] (sample E) and (b) ZnO:N (sample D), respectively, which was determined by SIMS measurement. We have used other homoepitaxial ZnO:N film as a reference for the evaluation of [N] in the samples. The [N] of three samples, involving a reference, were measured simultaneously in order to compare the accurate [N]. It is found that ZnO:[N+Te] have relatively higher [N] (2×10^{21} cm⁻³), compared with that of ZnO:N (3×10^{20} cm⁻³). This result indicates that Te doping can enhance N solubility in ZnO. This phenomenon is one of origins on conversion to p-type conductivity. The Madelung energy of ZnO:N is expected to be reduced by codoping Te atom, as discussed previously. Accordingly, incorporation efficiency of nitrogen atom increases in ZnO film doped Te atom because it has more stabilized state by reduction of Madelung energy, which can help to solve the N solubility limit of ZnO.

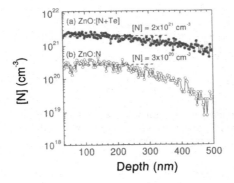

Figure 3 Depth profile for nitrogen atom for (a) ZnO:[N+Te] (sample E) and (b) ZnO:N (sample D), respectively.

Figure 4 shows the normalized photoluminescence spectrum of near band edge for u-ZnO (sample A), ZnO:N (sample D) and ZnO:[N+Te] (sample E), respectively, which were measured at 10 K. The u-ZnO film shows dominant neutral donor bound exciton emission (D^oX, 3.366 eV). On the other hand, ZnO:[N+Te] shows dominant emission line at 3.359 eV. There is obvious difference on emission energy between such specific emission of ZnO:[N+Te] and D^oX of u-ZnO. It can be assigned as acceptor bound excition (A^oX), based on optical studies in previous many reports [15-17]. Surprisingly, it has greatly narrower linewidth (1.2 meV) rather than that of D^oX (3.8 meV) in u-ZnO film, as shown in inset of Fig.4, which indicates emission properties of ZnO:[N+Te] significantly improve. In addition, the donor-acceptor pair recombination (DAP) and their 72 emV- 1 LO phonon replicas are observed at 3.217 eV and 3.145 eV, respectively [18]. On the other hand, ZnO:N have broad DAP emission line at 3.253 eV and free-electron to acceptor (*FA*) at 3.297 eV [19]. However, A^oX emission line is not observed in ZnO:[N+Te], even if a weak D^oX is still identify at 3.364 eV. The PL spectra shown in Fig. 4 quantitatively supports the evaluated nitrogen concentration in the ZnO:[N+Te] films. Normally, unintentional defects formed during N-doping into ZnO affect on the serious degradation of electrical and optical properties. However, considerable degradation in ZnO:[N+Te] is not observed, compared with u-ZnO. This obvious improvement of optical properties in ZnO:[N+Te] strongly indicates that defect complexes are effectively suppressed by codoping N and Te, which is regarded as another origin for reasonable p-type conductivity. Both the improvement of emission properties and enhancement of N solubility are ascribed to decrease of Madelung energy. Consequently, high quality p-type ZnO film is realized by (N+Te) codoping, even though incorporated N concentration is beyond solubility limit.

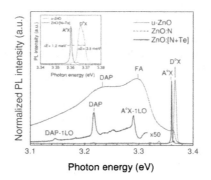

Figure 4 10K photoluminescence spectrum of near band edge for u-ZnO (sample A), ZnO:N (sample D) and ZnO:[N+Te] (sample E). Inset shows the detailed $A^{o}X$ (3.359 eV) and $D^{o}X$ (3.366 eV) in sample E and sample A, respectively.

CONCLUSIONS

In conclusion, ZnO film has been grown on (0001) substrate by using N and Te (N+Te) co-doping technique. Te atoms cause the increase of electron concentration in ZnO, which indicates Te atom acts as isoelectronic donor. ZnO:[N+Te] film grown by homoeptiaxy shows the p-type conductivity with a hole concentration of 4×10^{16} cm^{-3}. The p-type ZnO:[N+Te] film has two critical features; First, its N concentration increases up to 2×10^{21} cm^{-3}, compared with that in ZnO:N (3×10^{20} cm^{-3}). Second, its emission properties are greatly improved, emerging a dominant $A^{o}X$ emission line with narrow line broadening of 1.2 meV. Such specific behaviors of p-type ZnO:[N+Te] are originated from the reduction the Madelung energy. Consequently, (N+Te) co-doping technique helps to realize a high quality p-type ZnO beyond the solubility limit.

REFERENCES

1. D. M. Bagnall, Y. F. Chen, Z. Zhu, M. Y. Shen, T. Goto, and T. Yao, Appl. Phys. Lett. **73**, 1038 (1998)
2. A. Tsukazaki, A. Ohtomo, T. Onuma, M. Ohtani, T. Makino, M. Sumiya, K. Ohtani, S. Chichibu, S. Fuke, Y. Segawa, H. Ohno, H. Koinuma, and M. Kawasaki, Nat. mat. **4**, 42 (2004)
3. D. K. Hwang, S. H. Kang, J. H. Lim, E. J. Yang, J. Y. Oh, J. H. Yang, S. J. Park, Appl. Phys. Lett. **86**, 222101 (2005)
4. D. C. Look, D. C. Reynolds, C. W. Litton, R. L. Jones, D. B. Eason, and G. Cantwell, Appl. Phys. Lett. **81**, 1830 (2002)
5. T. Yamamoto, H. Katayama-Yoshida, Jpn. J. Appl. Phys. **38**, L166 (1999)
6. M. Joseph, H. Tabata, and T. Kawai, Jpn. J. Appl. Phys. 38, L2505 (1999)
7. M. Kumar, T. H. Kim, S. S. Kim, and B. T. Lee, Appl. Phys. Lett. **89**, 112103 (2006)

8. L. L. Chen, J. G. Lu, Z. Z. Ye, Y. M. Lin, B. H. Zhao, Y. M. Ye, J. S. Li, and L. P. Zhu, Appl. Phys. Lett. **87**, 252106 (2005)
9. G. D. Yuan, Z. Z. Ye, L. P. Zhu, Q. Qian, B. H. Zhao, R. X. Fan, C. L. Perkins, S. B. Zhang, Appl. Phys. Lett. **86**, 202106 (2005)
10. K. Nakahara, H. Takasu, P. Fons, A. Yamada, K. Iwata, K. Matsubara, R. Hunger, and S. Niki, J. Cryst. Growth **237**, 503 (2002)
11. Y. Fan, J. Han, R. L. Gunshor, J. Walker, N. M. Hohnson, and A. V. Nurmikko, J. Electron. Mater. **24**, 131 (1995)
12. H. Matsui, H. Saeki, H. Tabata and T. Kawai, Jpn. J. Appl. Phys. **42**, 5494 (2003)
13. S. B. Zhang, S.-H, Wei, Y. Yan, Physica B **302**, 135 (2001)
14. K. S. Ahn, Y. Yan, S. Shet, T. Deutsch, J. Turner, and M. Al-Jassim, Appl. Phys. Lett. **91**, 231909 (2007)
15. I. Suemune, A.B.M. A. Ashrafi, M. Ebihara, M. Kurimoto, H. Humano, T. Y. Seoung, B. J. Kim, Y. W. Ok, Phys. Stat. Sol. (b) **241**, 640 (2004)
16. J. Lu, Q. Liang, Y. Zhang, Z. Ye and S. Fujita, J. Phys. D: Appl. Phys. **40**, 3177 (2007)
17. X. H. Wang, B. Yao, D. Z. Shen, Z. Z. Zhnag, B. H. Li, Z. P. Wei, Y. M. Lu, D. Z. Zhao, J. Y. Zhang, X. W. Fan, L. X. Guan, C. X. Cong, Sol. Stat. Comm. **141**, 200 (2007)
18. L. Wang, N. C. Giles, Appl. Phys. Lett. **84**, 3049 (2004)
19. J. W. Sun, Y. M. Lu, Y. C. Liu, D. Z. Shen, Z. Z. Zhang, B. Yao, H. H. Li, J. Y. Zhang, D. Z. Zhao, and X. W. Fan, J. Appl. Phys. **102**, 043522 (2007)

Mater. Res. Soc. Symp. Proc. Vol. 1201 © 2010 Materials Research Society

Growing Zn$_{0.90}$Co$_{0.10}$O Diluted Magnetic Semiconductors by r. f. Magnetron Sputtering

Musa Mutlu Can[1,2], Tezer Fırat[1] and Şadan Özcan[1]
[1] Hacettepe University, Physics Engineering Department, Beytepe, 06800, ANKARA, Turkey
[2] University of Delaware, Material Science and Engineering Department, 19716, Newark, DE, USA

ABSTRACT

Zn$_{0.90}$Co$_{0.10}$O particles, synthesized by mechanical milling and thermal treatment, were pressed at 25 tons to form a 2" target for a radio frequency (r. f.) magnetron sputtering system. Using this target, thin films were deposited on (0001) oriented sapphire (α-Al$_2$O$_3$) substrates under 30W, 60W and 120W r. f. powers. Structural analyses of these films were done with X-Ray Diffractometer (XRD), Energy Dispersive X-Ray Spectrometry (EDS), X-Ray Photo Spectroscopy (XPS) and Atomic Force Microscopy (AFM). The ZnO films were deposited with (0002) preferred direction, which was coherent to (0001) ordered α-Al$_2$O$_3$. Impurity phases, such as Co clusters, CoO and Co$_3$O$_4$, were not detected with the surface analyses of Zn$_{0.90}$Co$_{0.10}$O thin films. Substituted Co atoms in the host ZnO matrix were identified by the binding energy peak of Co2p$_{3/2}$, 781.3±0.4eV, and the energy difference of ~15.61±0.03eV between Co2p$_{1/2}$ and Co2p$_{3/2}$. These results also proved that there were no Co clusters or Co$_3$O$_4$ phases in the lattice. Homogeneity of Co atoms in the lattice was shown by EDS spectra. It was understood that the higher r. f. power caused the more homogeneous distribution of Co and Zn atoms in thin films. Distributions of Co and Zn on the film surface, deposited under 120W, were found as 8.1±0.1% (normalized atomic ratio) and 91.7±0.7% (normalized atomic ratio), respectively, and the surface roughness of thin film was demonstrated by AFM figures as 14.2±0.1nm.

INTRODUCTION

Oxide semiconductors (zinc oxide (ZnO), indium tin oxide (ITO) and cadmium oxide (CdO)) have gained importance in the last two decades due to their transparency in the visible region and their good electrical conductivity [1]. In addition to these properties, ZnO semiconductors have a wide band gap (3.3eV) at room temperature [2,3], a high reflective index [3] and a high piezo electrical constant [3]. These features make ZnO useful in the fields of photovoltaic cells, gas sensors, chemical sensors, transducers, pressure sensors and anti reflector coatings [2-4].

ZnO is also a well-known oxide semiconductor for research in diluted magnetic semiconductors. The main idea is to dope transition metals (V, Cr, Mn, Fe, Co, Ni and Cu) inside the ZnO lattice [1-5]. Magnetic behavior in the lattice is expected due to s-d and p-d hybridization [3,5] between the s and p levels of ZnO and the d level of transition metals. Molecular beam epitaxy (MBE), radio frequency (r. f.) magnetron sputtering, chemical vapor deposition (CVD), pulsed laser deposition (PLD), sol-gel and solid state reaction techniques have been used for thin film and bulk growing processes [3-6].

This study used a Zn$_{0.90}$Co$_{0.10}$O target which was synthesized by solid state reaction techniques. Thin films were grown from the target directly on sapphire substrates using an r. f. magnetron sputtering system. The quality of a film depends on many parameters such as substrate temperature, growing atmosphere, r. f. power, distance between the target and substrate, post annealing temperature and cooling rate. The purpose of this study is to find the best growing parameters of Zn$_{0.90}$Co$_{0.10}$O thin films with homogeneous Co distribution and low surface roughness by using an r. f. magnetron sputtering system.

EXPERIMENTAL

All depositions were done by one gun with a $Zn_{0.90}Co_{0.10}O$ target, placed vertically with respect to the sapphire (α-Al_2O_3) substrate.

(a)   Preparation of $Zn_{0.90}Co_{0.10}O$ Target

The XRD results (Figure 1) of the synthesized $Zn_{0.90}Co_{0.10}O$ samples showed coherent peaks with ZnO (ICDD card is PDF#00-036-1451) and $ZnWO_4$ (ICDD card is PDF#00-015-0774) structures. Measured tungsten (W) impurities were around 1.5% due to erosion from the WC (tungsten carbide) vials and balls of the mechanical miller, which consist of 94% W and 6% Co. No Co or Co composites in $Zn_{0.90}Co_{0.10}O$ samples have been observed. Synthesized $Zn_{0.90}Co_{0.10}O$ particles were pressed under 25tons to form a 2" diameter target which was used in the r. f. magnetron sputtering system.

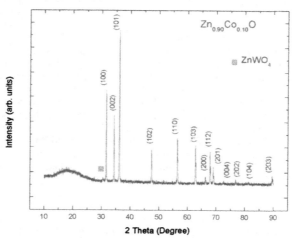

Figure 1. X-ray diffraction pattern of synthesized $Zn_{0.90}Co_{0.10}O$ powders. The indexes are arranged according to ICDD cards of ZnO (PDF#00-036-1451) and $ZnWO_4$ (PDF#00-015-0774).

(b)   Growing Process of $Zn_{0.90}Co_{0.10}O$ Thin Films

$Zn_{0.90}Co_{0.10}O$ thin films were directly grown on 3x3x0.33 mm^3 α-Al_2O_3 substrates, oriented through (0001). Growth steps were prearranged for every film: (1) pressure of the chamber was decreased to 1.2×10^{-5}Torr, (2) gas flow pressure was set to 40mTorr with the partial pressure of ¼ O_2 and ¾ Ar, (3) substrates were heated to 650^0C and held 30min at this temperature for thermal stability, (4) applied r. f. powers were 30W, 60W and 120W, (5) every sample was pre-sputtered for 10min and then deposited at specific time intervals and (6) post annealing was done at 750^0C under 1000Torr O_2 pressure up to 30min.

Structural analyses were performed using x-ray powder diffraction (XRD) patterns, energy dispersed x-ray spectroscopy (EDS), scanning electron microscopy (SEM), x-ray photoelectron spectra (XPS) and atomic force microscopy (AFM).

XRD spectra were taken using a Rigaku model diffractometer with CuK_α radiation. Scanning rates were 0.02°/min from 10-90°. Co atoms inside the ZnO lattice were detected by XPS. XPS measurements were done under 234W power with mono chromate Al K_α radiation in the 0 to 1200eV energy range with 0.10eV steps. Elements in the samples were detected with EDS using a Bruker-Axs XFlash 3001 SDD/EDS spectrometer. Surface roughness was measured by AFM.

RESULTS AND DISCUSSIONS

XRD spectra were demonstrated in Figure 2a and Figure 2b. Three dominant peaks have been observed at around 34.6°, 41.8° and 72.9°. One of them, 41.8°, is from α-Al_2O_3 (0001) and the others are well-matched with ZnO planes of (0002) and (0004). In addition, the diffraction peak from the (1000) ZnO plane was also observed for the thin film deposited for 90min under 120W; however, the peak is not dominant. The dominance of the peak from the (0002) plane showed that the preferred growth was in the ZnO (0002) // α-Al_2O_3 (0001) direction. The lack of detection of Co and Co composites in the XRD patterns can be related to the settlement of Co atoms inside the ZnO lattice.

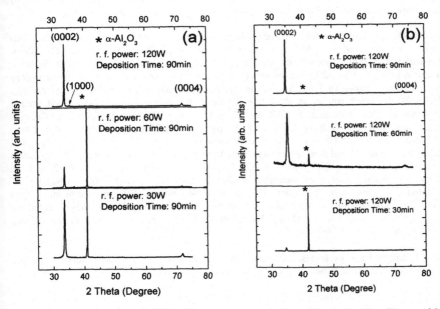

Figure 2. X-ray powder diffraction patterns of $Zn_{0.90}Co_{0.10}O$ thin films which were deposited on sapphire (α-Al_2O_3) (a) under r. f. power of 30, 60 and 120W for 90min and (b) under r. f. power of 120W for 30, 60 and 90min.

R. f. power effect on crystallinity was explored with the XRD patterns in Figure 2a. Dominance of the (0002) planes was seen in all patterns. The peak of α-Al_2O_3 (0001) disappeared with increased r. f. power up to 120W as the result of increasing thickness. Similarly, disappearance was also observed by increasing deposition time under constant r. f. power of 120W as seen in Figure 2b.

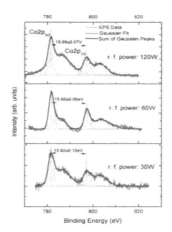

Figure 3. XPS spectra of Co2p of $Zn_{0.90}Co_{0.10}O$ thin films. Deposition times of each film, grown under 30W, 60W and 120W r. f. power, were 90 minutes.

Co substitutions with Zn atoms in the ZnO lattice were discovered with XPS spectra. All spectra were arranged according to the C1s binding energy (285eV) position. XPS values, demonstrated in Table.1, were found by fitting spectra to Gaussian peaks as seen in Figure 3. Similar results, mentioned before in reference [7], were observed for the electronic energy levels of Zn2p, shown in Table 1, and the energy differences between $Zn2p_{1/2}$ and $Zn2p_{3/2}$, 23.10±0.02eV. Furthermore, as shown in Figure 3, the curve of Co2p was fitted with 4 Gaussians which corresponded to Co2p electronic levels as $Co2p_{3/2}$, shake up peak of $Co2p_{3/2}$, $Co2p_{1/2}$ and shake up peak of $Co2p_{1/2}$. XPS spectra of Co2p energy levels of thin films, deposited at 30, 60 and 120W for 90min, were shown in Figure 3. Positions of the $Co2p_{3/2}$ and $Co2p_{1/2}$ energy levels and their differences were used to find the valance state and substitution of Co atoms inside the ZnO lattice [8-10]. A shift of the electronic state of $Co2p_{3/2}$ from 778eV, the binding energy of metallic Co clusters, to a higher energy of 781.62±0.03eV indicated Co^{+2} ions inside the lattice. The 15.64±0.08eV energy difference between $Co2p_{3/2}$ and $Co2p_{1/2}$ also signified Co-O bonds in the samples. Additionally these values revealed Co atoms inside the ZnO lattice and the absence of Co clusters due to the lack of the binding energy peak at around 778eV [11-13].

Table 1. Binding energy levels, Zn2p and Co2p, of thin films which were deposited under 30W, 60W and 120W r. f. power.

r. f. power (W)		30	60	120
Electronic Energy Levels of Zn2p (eV)	$Zn2p_{3/2}$	1022.40±0.01	1022.20±0.01	1021.90±0.01
	$Zn2p_{1/2}$	1045.46±0.02	1045.30±0.01	1044.97±0.01
Electronic Energy Levels of Co2p (eV)	$Co2p_{1/2}$	797.24±0.09	797.37±0.03	797.16±0.05
	$Co2p_{3/2}$	781.62±0.03	781.77±0.03	781.47±0.02

Homogeneity of Co distributions is another important parameter for the quality of thin films. Co atom distributions on the surface were figured out by EDS measurements. Records of the measurements were shown in Table 2. EDS data was taken from 5 different regions of thin films: four of them from the corners and the 5th from the center. In addition, the written data included the errors of standard deviation of five measurements.

Table 2. EDS data of the samples, grown under 30W, 60W and 120W for 90min.

Number of Measurements	r. f. power: 30W Deposition Time: 90min		r. f. power: 60W Deposition Time: 90min		r. f. power: 120W Deposition Time: 90min	
	Normalized Atomic Ratio of Cobalt (Co) %	Normalized Atomic Ratio of Zinc (Zn) %	Normalized Atomic Ratio of Cobalt (Co) %	Normalized Atomic Ratio of Zinc (Zn) %	Normalized Atomic Ratio of Cobalt (Co) %	Normalized Atomic Ratio of Zinc (Zn) %
1th	7.6	92.4	8.1	92.0	8.4	91.6
2nd	5.9	94.1	7.0	93.0	7.7	92.3
3rd	5.6	94.4	4.8	95.2	8.5	91.5
4th	9.6	90.4	9.4	90.6	7.9	92.0
5th	3.0	97.0	5.5	94.5	7.9	92.1
Mean Value	6.3±0.9	93.7±1.6	7.0±0.6	93.0±1.3	8.1±0.1	91.9±0.7

Average percentages of normalized atomic weight of Co atoms were 6.3±0.9%, 7.0±0.6% and 8.1±0.1% for the thin films grown under 30W, 60W and 120W, respectively. The inverse proportion between error and increasing power was evidence of uniformly separated Co atoms in the ZnO lattice. Results showed that Co distribution in the film grown under 120W was more homogeneous than the others.

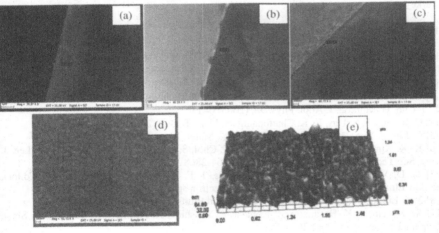

Figure 4. Cross-sectional SEM figures of $Zn_{0.90}Co_{0.10}O$ thin films, grown under 120W r. f. power with deposition times of (a) 30, (b) 60, (c) 90min. (d) SEM and (e) AFM figures of surface of the deposited film under 120W for 90min.

Thicknesses of the thin films depended on deposition time and were determined from cross-sectional SEM pictures as shown in Figure 4. The values were ~100nm, ~400nm and ~1000nm for 30, 60 and 90min deposition times, respectively. In addition to thickness, because of the homogeneous Co distribution, surface roughness was only investigated for film deposited under 120W for 90min. The roughness was determined by SEM and AFM measurements, shown in Figure 4d and Figure 4e, respectively. The roughness was found to be 14.2 ± 0.1nm.

Conclusion

The purpose of this study was to find which parameters were important to form uniformly distributed Co atoms inside a ZnO lattice. A $Zn_{0.90}Co_{0.10}O$ target, which was synthesized by solid state reaction techniques, was used to grow thin films. Preferred growths of the ZnO structure were identified as ZnO (0002) parallel to α-Al_2O_3 (0001). It was found that Co substitution with Zn occurred independently from r. f. power. However, Co homogeneity is dependent on r. f. power. Increased r. f. power formed homogeneously distributed Co atoms. Co clusters or different Co composites have not been detected in the thin films. Also, W impurities, which are present in the target due to corrosion of the vial and from the balls of the mechanical miller, have not been detected on the film surface.

Acknowledgment

Thanks to Dr. Evren Çubukçu and Prof. Dr. Erkan Aydar from the Department of Geology Engineering in Hacettepe University for the EDS measurements. The authors also would like to thank Roy Murray, University of Delaware, Department of Physics, for the help in editing the manuscript. This work is supported by The Higher Education Institution of Turkey (YÖK).

References

[1] H. L. Hartnagel, A. L. Dawar, A. K. Jain, and C. Jagadish, Semiconducting Transparent Thin Films, Institute of Physics Publishing, Bristol and Philadelphia, (1995).
[2] S. Venkataprasad Bhat, F. L. Deepak, Sol. Stat. Comm., 135 (2005) 345-347
[3] Ü. Özgür, Ya. I. Alivov, C. Liu, A. Teke, M. A. Reshchikov, S. Doğan, V. Avrutin, S.-J. Cho, H. Morkoç, J. Appl. Phys. **98**, (2005) 041301
[4] C. K. Ghosh, S. Das, K. K. Chattopadhyay, Phys. B, 399 (2007) 38-46
[5] C. Liu, F. Yun, H. Morkoç, J. Mater. Scien.: Mater. Elec., 16 (2005) 555-597
[6] K.-K. Kim, J.-H. Song, H.-J. Jung, W. – K. Choi, S. – J. Park, J. – H. Song, J. – Y. Lee, J. Vac. Sci. Technol. A 18 (Nov/Dec 2000) 6, 2864-2868
[7] C. D. Wagner, W. M. Riggs, L. E. Davis, J. F. Moulder, G. E. Muilenberg (Editor), Perkin-Elmer Corp. Handbook of x-ray photoelectron spectroscopy (1979), Pg.84-84
[8] X. J. Liu, C. Song, F. Zeng, F. Pan, Thin Solid Films, 516 (2008) 8757-8761
[9] M. Naeem, S. K. Hasanain, M. Kobayashi, Y. Ishida, A. Fujimori, S. Buzby, S. I. Shah, Nanotechnology, 17 (2006) 2675-2680
[10] J. Hays, K. M. Reddy, N. Y. Graces, M. H. Engelhard, V. Shutthanandan, M. Luo, C. Xu, N. C. Giles, C. Wang, S. Thevuthasan, A. Punnoose, J. Phys.: Condens. Matter. 19 (2007) 266203
[11] Y. Kalyana Lakshmi, K. Srinivas, B. Sreedhar, M. Manivel Raja, M. Vithal, P. Venugopal Reddy, Mater. Chem. Phys., 113 (2009) 749-755
[12] H.-J. Lee, S.-Y. Jeong, C. R. Cho, C. H. Park, Appl. Phys. Lett., 81(2002) 21, 4020-4022
[13] A. Manivannan, M. S. Seehra, S. B. Majumder, R. S. Katiyar, Appl. Phys. Lett., 83 (2003) 1, 111-113

Mater. Res. Soc. Symp. Proc. Vol. 1201 © 2010 Materials Research Society

1201-H05-32

Effects of Growth Parameters on Surface-Morphological, Structural, Electrical and Optical Properties of AZO Films by RF Magnetron Sputtering

Shou-Yi Kuo[1,2], Wei-Ting Lin[1,3], Liann-Be Chang[1,2], Ming-Jer Jeng[1,2]
Yong-Tian Lu[4] and Sung-Cheng Hu[4]

[1] Department of Electronic Engineering, Chang Gung University, Taiwan
[2] Green Technology Research Center, Chang Gung University, Taiwan
259 Wen-Hwa 1st Road, Kweishan, Taoyuan 333, Taiwan
[3] Department of Optics and Photonics, National Central University, Taiwan
[4] Chemical Systems Research Division, Chung-Sung institute of Science & Technology, Taiwan

ABSTRACT

500 nm-thick aluminum-doped zinc oxide (ZnO:Al) thin film is usually used as a front transparent conductive oxide (TCO) contact on photovoltaic devices, and for this application is often deposited by a reactive radio-frequency (r.f.) magnetron sputtering system from a ceramic target. This work reports on the preparation and characterization of AZO thin films on Corning 1737 glass substrates grown by reactive r.f.-magnetron sputtering from a ZnO ceramic target with 2 wt% Al content. It was found that the growth parameters, such as chamber pressure, working power, and deposition temperature, have significant influences on the properties of AZO films. According to the experimental results: (1) Films were polycrystalline showing a strong preferred c-axis orientation. (2) With increasing working power, the resistivity decreased, and mobility and the carrier concentration increased. (3) Lower deposition temperature leads to a decrease in resistivity, with 2.5×10^{-4} Ω-cm representing the lowest resistivity reached.

INTRODUCTION

Transparent conductive oxide (TCO) is an important window contact layer of thin solar cells with low electrical resistivity and high optical transmission [1]. Aluminum-doped zinc oxide (ZnO:Al, AZO) thin films are usually studied on photovoltaic devices because they have a number of advantages, such as non-toxicity, low cost, material abundance, relatively low deposition temperature, good anti-reflection, and high thermal/chemical stability in hydrogen plasma treatment [2-4]. AZO films have been prepared by various deposition techniques, such as sputtering, pulsed laser deposition, chemical vapor deposition, spray pyrolysis, and metal-organic chemical deposition [5-9]. Among these techniques, magnetron sputtering technique has became the potential one because of its advantageous features such as simple apparatus, high deposition rates, and low deposition temperature [10].

In this work, high quality AZO thin films were fabricated and studied as a function of parameters: growth temperature and working power using Al-doped ZnO (2 wt%) ceramic target. The influences of interstitial Zn and Al_2O_3 containments on structural and optoelectronic properties were discussed.

EXPERIMENT

ZnO:Al thin films were deposited on Corning glass (1737, 2×2 cm^2) substrates with a standard thickness of 500 nm by radio frequency magnetron sputtering system using high purity ZnO ceramic target with 2 wt% Al (6-inch diameter) and high purity argon gas with 0.2% oxygen. The target was placed parallel to the substrate with a target-substrate distance of 8 cm, and the substrate was rotated at 12 rpm for thickness uniformity. The substrates were cleaned by acetone, isopropanol and DI water. Before the deposition, the sputter chamber was evacuated by using a cryo-pump to a base pressure of 6×10^{-7} Torr. We studied as a function of the deposition temperature (150-300 °C) and the working (100-400 W), the chamber pressure was controlled at 1 mTorr. Film thickness was measured by α-step surface profiler. The structural property and surface morphology were analyzed by x-ray diffraction (XRD), atomic force microscope (AFM) and scanning electron microscopy (SEM). Electrical properties were measured by four-point probe. Na diffusion was measured by Energy Dispersive Spectrometer (EDS).

DISCUSSION

The X-ray diffraction spectra of AZO films grown at different deposition temperature and r.f.power are shown in the Figure 1(a) and (b), respectively. From figure 1, AZO thin films revealed a polycrystalline hexagonal wurtzite structure and the peaks belonging to (100), (002) and (004) diffraction planes were observed. With decreasing the deposition temperature and increasing r.f. power, the intensity of the all peaks increased. In addition, the decreased full width at half maximum (FWHM) of AZO(002) diffraction peak corresponds to the increase in grain size as deduced from Scherrer formula [11]. The tendency indicates that the crystallization of AZO thin films were improved at lower temperatures and higher r.f. power.

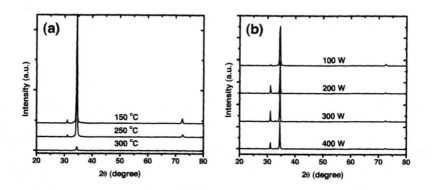

Figure 1. XRD patterns of AZO films with different parameters: (a) deposition temperature (150-300 °C) (b) r.f. power (100-400 W)

Since the surface properties can effect the electrical and optical performance of the optical devices, it is very important to investigate the surface properties of AZO thin films. The surface morphologies of the AZO films change greatly with an increase in r.f. power. Atomic force microscope was used to characterize the surface roughness of AZO films (not shown in here). According to the AFM results, the root-mean-square (RMS) roughness was decreased from 5.96 nm to 2.78 nm with increasing r.f. power. The results obtained from the AFM images were found to be in a good agreement with previous literatures. Besides, increasing r.f. power made the surface roughness smaller. The relatively small roughness of films deposited at higher r.f. power might be ascribed to the improved crystallinity.

Table I indicates the dependence of surface morphology of AZO thin films on various r.f. power and deposition temperature. The root-mean-square (RMS) roughness was decreased with increasing r.f. power and decreasing r.f. power. The relatively small roughness might be ascribed to the improved crystallinity and to the large grain size growth due to the higher r.f. power. The morphology of surface grains shown the sphere shape with different deposition temperature and r.f. power while the chamber pressure at 1 mTorr.

Table I. The dependence of surface roughness on (a) r.f. power ranging from 100 W to 400 W, and (b) substrate temperature ranging from 150 °C to 300 °C .

(a)

r.f. power (W)	100	200	400
RMS (nm)	5.96	4.6	2.78

(b)

Substrate temperature (°C)	150	250	300
RMS (nm)	3.91	4.6	5.47

As high transparency is the most important factor in the application of ZnO:Al films to TCOs, the optical transmittance was determined by a spectrophotometer within the wavelength from UV to IR region. The optical transmittances of AZO thin films deposited at various deposition temperatures and r.f. power are shown in figure 2(a) and 2(b), respectively. And the average transmittance becomes higher while increasing deposition temperature and r.f. power. Regardless of the deposition temperature and r.f. power, AZO films demonstrate transmittance of above 85 % in the range of the visible spectrum, in comparison with the transmittance of ITO film (> 87 %), which is obtained by commercial means. As it could be expected, the increase of the deposition temperature and r.f. power would lead to the enhancement of the structural properties, fact showed previously in the XRD analysis, due to the higher mobility of the energetic particles on the substrate surface. It is worthy to note that the transmittance in the IR region exhibit significant variation. While we decrease the deposition temperature or increase r.f. power, the transmittance decreases progressively. Degrading of infrared transmittance of AZO films might be attributed to the increases of carrier concentration as confirmed by the Hall measurement discussed latter.

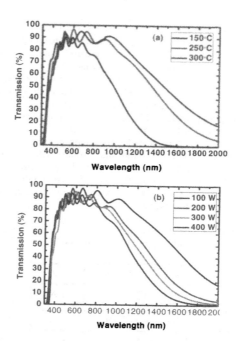

Figure 2. Transmittance patterns of AZO films with different parameters: (a) deposition temperature (150-300 °C) (b) r.f. power (100-400 W).

The electrical properties were investigated by making Hall-effect measurements. Figure 3(a) and 3(b) show the resitivity, carrier concentration and mobility of AZO films deposited at various deposition temperature and r.f. power. While decreasing the deposition temperature from 300 °C to 150 °C, both the carrier concentration and mobility increase with a corresponding decrease in resistivity. Similarly, we can get the same tendency as we decrease r.f. power from 400 W to 100 W as shown in Fig. 3(b). The electrical conductivity of the films is mainly due to intrinsic detects such as interstitial zinc atoms, oxygen vacancies and aluminum doping, and can be decreased up to the semi-metallic regime by increasing the carrier concentration up to 1.5×10^{21} cm^{-3}. The results of carrier concentration is consistent with the observation of optical transmittance in the IR region. In our case, higher temperature might led zinc atoms to get away and enhance the formation probability of Al–O bonds in the AZO films such as Al$_2$O$_3$ phase. To identify the assumption, further experiments are underway.

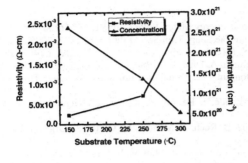

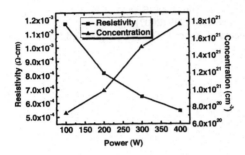

Figure 3. Electrical properties patterns of AZO films with different parameters: (a) deposition temperature (150~300 °C) (b) r.f. power (100-400 W).

CONCLUSIONS

In summary, we found that the deposition temperature and working r.f. power significantly affect the structural, optical and electrical properties of AZO films. From XRD measurement, the FWHM decreased and crystalline size increased with an increase in r.f power and a decrease in deposition temperature. According to our experimental results, energetic Zn atom might depart from the sample surface and enhance the formation probability of Al_2O_3 phase at high deposition temperature. The average transmittance is above 85% and lowest resistivity values can reach 2.5×10^{-4} Ω-cm with 500 nm thickness. Thus, fabricated AZO films are suitable for the application of transparent electrodes as an alternative to ITO.

ACKNOWLEDGMENTS

The authors would like to gratefully acknowledge partial financial support from the Green Technology Research Center of Chang Gung University.

REFERENCES

1. C. G. Granqvist, Solar Energy Mater. Solar Cells 91, 1529 (2007).
2. Jae-Hyeong Lee, Jun-Tae Song, Thin Solid Films 516, 1377 (2008).
3. R. Latz, K. Michael, M. Scherer, Jpn. J. Appl. Phys. 30 (2A), L149 (1991).
4. S. Mandal, R.K. Singha, A. Dhar, S.K. Ray, Mater. Res. Bull. 43, 244 (2007).
5. Jaehyeong Lee, Dongjin Lee, Donggun Lim, Keajoon Yang, Thin Solid Films 515, 6094 (2007).
6. T. Szörényi, L.D. Laude, I. Bertóti, Z. Kántor, Zs. Geretovsky, J. Appl. Phys. 78, 6211 (1995).
7. O. Kluth, G. Schöpe, B. Rech, R. Menner, M. Oertel, K. Orgassa, H.W. Schock, Thin Solid Films 502, 311 (2006).
8. P. Nunes, D. Costa, E. Fortunato, R. Martins, Vacuum 64, 293 (2002).
9. Y. M. Lu, W. S. Hwang, W. Y. Liu, J. S. Yang, Mater. Chem. Phys. 72, 269 (2001).
10. W. F. Liu, G. T. Du, Y. F. Sun, J. M. Bian, Y. Chang, T. P. Yang, Y. C. Chang, Y. B. Xu, Applied Surface Science 253, 2999 (2007).
11. B.D. Cullity, Elements of X-ray Diffraction, second ed., Addison Wesley, Reading, MA, 1978.

Mater. Res. Soc. Symp. Proc. Vol. 1201 © 2010 Materials Research Society

Ab initio investigations of the lattice parameters in ZnMgO alloys

Markus Heinemann, Marcel Giar and Christian Heiliger
I. Physikalisches Institut, Justus Liebig University, 35392 Giessen, Germany

ABSTRACT

We perform density functional theory calculations to determine equilibrium lattice parameters of wurtzite $Zn_{1-x}Mg_xO$ alloys for Mg concentrations x ranging from 0 to 31.25 %. We use the local density approximation (LDA) as well as the generalized gradient approximation (GGA) for the exchange correlation functional. For the lattice constants a and c we find a deviation from Vegard's law and a constant unit cell volume independent of the Mg concentration.

INTRODUCTION

ZnO with a band gap of about 3.4 eV has caught recent attention as a wide band gap semiconductor because it can be used for opto-electronic devices such as light emitting diodes with an operational range from blue to ultraviolet. Therefore, it is advantageous to use band gap modulations which are usually done by substituting bivalent metals (i.e. Mg) for Zinc. Much effort has been made to improve methods for the synthesis of ternary alloys like $Zn_{1-x}Mg_xO$. Recent experimental works have successfully investigated the influence of increasing Mg concentration on $Zn_{1-x}Mg_xO$ thin films grown by pulsed laser deposition (PLD) [1] or plasma assisted molecular beam epitaxy (PAMBE) [2].

The thermodynamic solubility has been found to be within in a range of $x=0$ and $x\approx0.4$ experimentally [2] as well as theoretically [3, 4]. Also the composition dependence of lattice parameters and the band gap has been investigated in detail. Recent theoretical investigations using the LDA [3] and the GGA [5] state a slightly increasing lattice parameter a and a decreasing lattice parameter c with increasing Mg concentration. These results are consistent with the results obtained by PLD [1]. Since LDA underestimates the lattice parameters, we additionally perform first principle calculations using the GGA method and compare both theoretical methods to experiments.

THEORY AND COMPUTATIONAL METHODS

In this work we use a density functional theory (DFT) method to calculate the structural properties of $Zn_{1-x}Mg_xO$ alloys. We use the projector augmented wave method (PAW) [6-8] implemented in the ABINIT programme package [9-11]. For the exchange correlation we choose the Perdew-Wang 92 LDA functional [12] as well as the Perdew-Burke-Ernzerhof GGA functional [13]. The PAW+LDA and PAW+GGA pseudopotentials treat as valence electrons the $3s^2$, $3p^6$, $3d^{10}$ and $4s^2$ electrons for Zn, the $2s^2$, $2p^6$ and $3s^2$ for Mg, and the $2s^2$ and $2p^4$ electrons for O.

The calculations are carried out in a 2×2×2 supercell. For the LDA calculations a 3×3×2 k-point mesh is used in the first Brillouin zone. The energy cut-off for the plane wave basis and for the double grid are set to 820 eV and 1360 eV, respectively. These values ensure convergence of the cell parameters a and c within a range less than a tenth of a percent. For GGA it is necessary to choose a 6×6×4 k-grid and 820 eV and 1900 eV for the plane wave expansion and the double grid to achieve the same accuracy as for the LDA calculations.

We conduct a structural relaxation using the Broyden-Fletcher-Goldfarb-Shanno minimization technique implemented within the ABINIT code. In a first step the atomic positions are relaxed, subsequently a geometrical optimization is carried out in which the cell parameters are changed. We initially perform calculations of the lattice parameters for ZnO in wurtzite and rocksalt structure as well as MgO in rocksalt and wurtzite structure. The wurtzite ZnO lattice parameters attained are taken to construct a 2×2×2 supercell lattice consisting of 16 Zn and 16 O atoms.

In this supercell Zn atoms are successively replaced by Mg, varying the concentration within a range of $0 \leq x(Mg) \leq 31.25$ %. For every single structure a full geometry optimization is done, from which we obtain the corresponding lattice parameters.

RESULTS AND DISCUSSION

Lattice parameters of bulk MgO and ZnO

Most of the group II-VI semiconductors crystallize in either the cubic zinc-blende or the hexagonal wurtzite structure. ZnO can be grown in the wurtzite, zinc-blende and cubic rocksalt structure. The hexagonal wurtzite structure, which is characterized by the lattice parameters a, c and the dimensionless constant u, which gives the relative displacement of the Oxygen sublattice with respect to the Zinc sublattice. Zn is tetrahedrally coordinated by four O atoms and vice versa. MgO however has a rocksalt-type crystal structure with two atoms forming the basis, thus Mg is octahedrally coordinated by O and O is coordinated by six Mg atoms.

Computed equilibrium structural parameters for bulk MgO and ZnO in both wurtzite and rocksalt structure are given in table I along with reported experimental and theoretical data where the theoretical data was obtained by calculations using the PAW-LDA [3] and the Korringa-Kohn-Rostoker method (KKR) [4].

For rocksalt MgO and ZnO as well as for wurtzite ZnO our calculated lattice parameters are in good agreement with the reported experimental and theoretical results. Wurtzite MgO is a hypothetical structure that is not accessible in experiment. However, the comparison with available theoretical calculations shows a good agreement of the lattice parameters.

The relaxed lattice parameters obtained by our LDA calculations are below the experimental values whereas the parameters obtained by using the GGA functional are larger than the experimental results. Therefore, we expect the real parameters to lie within the boundaries given by the LDA and GGA calculations.

Table I. Equilibrium structural parameters of bulk MgO and ZnO in wurtzite (WZ) and rocksalt (RS) structure obtained by LDA and GGA calculations. The lattice parameters a and c are given in Å. Experimental and theoretical data [[a]14, [b]15, [c]16, [d]17, [e]3, [f]4] are given for comparison.

Lattice parameter		LDA	GGA	Experiment	Theory
MgO (WZ)	a	3.2868	3.3821		3.2786[e], 3.2225[f]
	c	4.9508	5.1972		4.8736[e], 5.1593[f]
	u	0.4001	0.3935		0.4046[e], 0.380[f]
MgO (RS)	a	4.1787	4.3210	4.212[a]	4.152[e], 4.1748[f]
ZnO (WZ)	a	3.2187	3.3273	3.258[b], 3.283[c]	3.2032[e], 3.2053[f]
	c	5.1801	5.3534	5.220[b], 5.309[c]	5.1386[e], 5.1317[f]
	u	0.3797	0.3800	0.382[b], 0.3786[c]	0.3814[e], 0.380[f]
ZnO (RS)	a	4.2401	4.3783	4.275[d]	4.2110[e], 4.2269[f]

Structural properties of wurtzite ZnMgO alloys

For the examined $Zn_{1-x}Mg_xO$ alloys the relaxed lattice constants a and c are shown in figure 1 for Mg concentrations ranging from 0 to 31.25 %. This covers the range in which stable ZnMgO has recently been reported grown by PAMBE [2] and PLD [1]. The GGA calculations give about 3 to 5 % higher values than the LDA calculations. From the results of the bulk calculations in table I we expect the experimental results between the LDA and GGA results. This expectation is confirmed by the reported experiments [1,2]. The calculated equilibrium lattice parameters show a slightly deviation from Vegard's law. The lattice parameter a increases with increasing Mg concentration x but stays slightly below the values expected by Vegard's law (see figure 1). For c we find significantly higher values than predicted by Vegard's law especially for concentrations $x>15\%$.

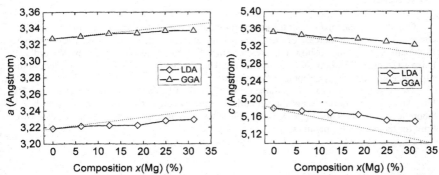

Figure 1. Computed LDA (◊) and GGA (Δ) equilibrium lattice parameters a (left) and c (right) for wurtzite $Zn_{1-x}Mg_xO$ as function of Mg concentration. The LDA and GGA lattice parameters deviate slightly from Vegard's law (dashed lines), particularly for $x > 15$ %.

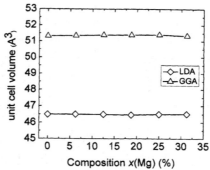

Figure 2. The volume per unit cell for wurtzite $Zn_{1-x}Mg_xO$ alloys as a function of the Mg concentration for LDA and GGA calculations.

In figure 2 we plot the volume per unit cell as a function of the Mg concentration. The volume per unit cell remains fairly constant at a value of 46.5 $Å^3$ for the LDA calculation and 51.4 $Å^3$ for the GGA calculation. This outcome is in good agreement with experimental results from PLD grown thin film ZnMgO alloys [1] where a unit cell volume of about 46.8 $Å^3$ was measured independent of the Mg concentration.

CONCLUSION

Our ab initio calculations of the equilibrium lattice parameters show with increasing Mg concentration a small increase of the lattice parameter a and a decrease of the lattice parameter c. Both parameters, a and c, show a deviation from the linear behavior predicted by Vegard's law in the LDA as well as in the GGA calculations. The volume per unit cell is independent of the Mg concentration in agreement to experiments.

The difference in the lattice parameters between the LDA and GGA results is about 3-5%. Interestingly, the reported experimental lattice parameters are between the results of both theoretical methods. This suggests that the GGA result is an upper and the LDA result a lower limit for the real lattice parameters.

REFERENCES

1. A. Ohtomo and A. Tsukazaki, *Semicond. Sci. Technol.* **20**, S1-S12 (2005)
2. T. A. Wassner, B. Laumer, S. Maier, A. Laufer, B. K. Meyer, M. Stutzmann and M. Eickhoff, *J. Appl. Phys.*, **105**, 023505 (2009)
3. X. F. Fan, H. D. Sun, Z. X. Shen, J.-L. Kuo and Y. M. Lu, *J. Phys. Cond. Mat.* **20**, 235521 (2008)
4. I.V. Maznichenko, A. Ernst, M. Bouhassoune, J. Henk, M. Däne, M. Lüders, P. Bruno, W. Hergert, I. Mertig, Z. Szotek, and W.M. Temmerman, *Phys. Rev. B* **80**, 144101 (2009)
5. T. Ohsawa, Y. Adachi, I. Sakaguchi, K. Matsumoto, H. Haneda, S. Ueda, H. Yoshikawa, K. Kobayashi, N. Ohashi, *Chem. Mater.* **21**, 144-150 (2009)
6. PAW+LDA and PAW+GGA pseudopotentials were generated by AtomPAW (N. Holtzwarth)
7. P. E. Blöchl, *Phys Rev. B* **50**, 17953 (1994)
8. G. Kresse and D. Joubert, *Phys. Rev. B* **59**, 1758 (1999)
9. M. Torrent, F. Jollet, F. Bottin, G. Zerah and X. Gonze, *Comput. Mat. Science* **42**, 337 (2008)
10. X. Gonze, J.-M. Beuken, R. Caracas, F. Detraux, M. Fuchs, G.-M. Rignanese, L. Sindic, M. Verstraete, G. Zerah, F. Jollet, M. Torrent, A. Roy, M. Mikami, Ph. Ghosez, J.-Y. Raty, D.C. Allan. *Comp. Mat. Sc.* **25**, 478-492 (2002)
11. X. Gonze, G.-M. Rignanese, M. Verstraete, J.-M. Beuken, Y. Pouillon, R. Caracas, F. Jollet, M. Torrent, G. Zerah, M. Mikami, Ph. Ghosez, M. Veithen, J.-Y. Raty, V. Olevano, F. Bruneval, L. Reining, R. Godby, G. Onida, D.R. Hamann, and D.C. Allan. *Zeit. Kristallogr.* **220**, 558-562 (2005)
12. J. P. Perdew and Y. Wang, *Phys. Rev. B* **45**, 13244 (1992)
13. J. P. Perdew, K. Burke, and M. Ernzerhof, *Phys. Rev. Lett.*, **77**, 3865 (1996)
14. Landolt-Börnstein, *Numerical Data and Functional Relationships, Crystal Structure Data of Inorganic Compounds*, Springer, Berlin (1975)
15. H. Karzel, W. Potzel, M. Köfferlein, W. Schiessl, M. Steiner, U. Hiller, G. M. Kalvius, D. W. Mitchell, T. P. Das, P. Blaha, K. Schwarz and M. P. Pasternak, *Phys. Rev. B* **53**, 11425 (1996)
16. F. Decremps, F. Datchi, A. M. Saitta, A. Polian, S. Pascarelli, A. Di Cicco, J. P. Itié and F. Baudelet, *Phys. Rev. B* **68**, 104101 (2003)
17. J. M. Recio, M. A. Blanco, R. Luaña, R. Pandey, L. Gerward and J. Staun Olsen, *Phys. Rev. B* **58**, 8949 (1998)

Mater. Res. Soc. Symp. Proc. Vol. 1201 © 2010 Materials Research Society   1201-H05-39

Electrically air-stable ZnO produced by reactive RF magnetron sputtering for thin-film transistor (TFT) applications

D. K. Ngwashi and R. B. M. Cross and S. Paul
Emerging Technology Research Centre, De Montfort University, Leicester LE1 9BH, UK

ABSTRACT

The influence of native point defects on the electrical and optical stability of zinc oxide (ZnO) layers in air produced by reactive RF magnetron sputtering is investigated. ZnO thin films are strongly affected by oxygen (O_2) molecules in ambient atmosphere. For instance, surface defects such as oxygen vacancies act as adsorption sites of O_2 molecules, and the chemisorption of O_2 molecules depletes the surface electronic states and reduces channel conductivity. Thin films of ZnO produced have electrical resistivities between 8.6×10^3 and 8.3×10^8 Ω-cm, and were found to be electrically-stable in air. TFTs fabricated using these films exhibited effective mobilities of ~3 $cm^2V^{-1}s^{-1}$ and the threshold voltage shifts by < 5 V under gate bias stress of 1 MV/cm for up to 10^4 s.

INTRODUCTION

Wide bandgap semiconducting II-VI oxides are an important class of materials in optoelectronics devices. Zinc oxide (ZnO) is cheap, abundant and when produced at room temperature, is compatible with cheap flexible and plastic substrates. The physical properties of sputtered ZnO are strongly dependent on the deposition parameters and on the post-deposition treatment such as post-annealing. Furthermore, a reduction in the concentration of O_2 vacancies in thin films has been found to reduce O_2 chemisorption in an ambient atmosphere, [[1-7]] which could be a major cause of electrical degradation in ZnO-based applications (such as thin-film transistors (TFTs)). TFTs using ZnO as a channel material were first demonstrated by Hoffman et al. [8] and have since then been the focus of a substantial amount of research. However, the stability of these TFTs remains one of the major issues delaying its commercial application [9]. The stability of TFTs under gate bias stress is also of great importance for switching applications where prolonged turn-on is needed such as in static images (e.g. a logo). Threshold voltage instability as a result of bias stressing in TFTs is often attributed to defect creation in the channel material, charge trapping in the gate dielectric material or at the dielectric/channel interface [10,11]. In this work, we investigate the influence of deposition parameters on the stability of sputtered ZnO layers.

EXPERIMENT

ZnO layers deposited by reactive radio frequency (RF) magnetron sputtering at room temperature in an argon (Ar)/O_2 atmosphere have been investigated. The different ZnO thin films obtained are characterized by x-ray diffraction (XRD) using Siemens/Bruker D5000 diffractometer equipped with a 0.1540562 nm (Cu kα) line and UV-Vis spectroscopy using UNICAM UV/Vis spectrometer (UV 2), together with electrical resistivity measurements to investigate the influence of oxygen within the sputtering chamber on the film properties. The film properties were determined for as-deposited films, and electrical and optical measurements were repeated after 18 months to monitor the effects of atmospheric air. The sputtering source used was an 8 inch diameter ZnO target (99.999%, Kurt J. Lesker) at a vertical substrate distance of approximately 6 cm. The base pressure in the chamber prior to deposition was 4.0×10^{-6} mTorr. The O_2 and Ar flow rates into the deposition chamber were controlled by means of mass flow controllers. The flow rates explored were between 0.2 – 3 Sccm and 8 – 20 Sccm for O_2 and Ar respectively and RF power was varied between 25 and 500 W). The films were deposited on soda-lime glass and silicon substrates. Film thicknesses were measured using an AutoEL-III ellipsometer. The current-voltage (IV) characteristics were measured using an HP4140B pico ammeter, and the electrical resistivity extracted using a multi-contact two-terminal transmission line method (TLM) [12] in the dark. All films were post-treated in O_2 (~2.0 mTorr) for one hour at room temperature to further reduce O_2 vacancies at the surface [13]. ZnO TFT test structures were then fabricated using an inverted-staggered structure, with n⁺-silicon as the bottom gate and a 100 nm-thick layer of thermally-grown silicon dioxide (SiO_2) as the gate dielectric. The source/drain electrodes were formed by thermally evaporating 100 nm of aluminium on to the ZnO film using shadow masks (100 μm channel length, and length/width = 10:1). The stability of the devices were investigated under gate bias stress of ±10 V (corresponding to a field of 1 MV/cm) for a maximum stress time of 10^4 s. All measurements were performed at room temperature in the dark.

RESULTS AND DISCUSSION

Optical properties

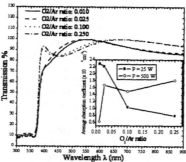

Figure 1 - Optical transmission of films as-deposited at different O_2/Ar ratios at RF power of 500 W and the average absorption in the near band-edge region (2.9 – 3.2 eV) of films produced with different O_2/Ar ratios (insert).

The typical film thickness was 135 nm, and all films produced had an average transparency of over 85% in the visible light (400 – 700 nm) (figure 1). The optical bandgap (E_g) in these films lie between 3.26 – 3.28 eV (similar to CVD and sol-gel ZnO [14-17]). E_g was estimated using Tauc's relation (1) [18] and was unchanged after 18 months.

$$\alpha = \frac{K(h\nu - E_g)^{1/n}}{h\nu}$$

(1)

154

α is the absorption coefficient, υ the frequency of incident photon, E_g the optical bandgap, and, K and h are the proportionality and Planck's constants respectively. The average absorption coefficient near the band edge shows that films deposited at low O_2/Ar ratio exhibit greater absorption at low RF power than high power deposited films (figure 1 inset). This could be attributed to the reduction of atomic oxygen in the plasma via the two processes (2) and (3);

$$O + O_2 + Ar \leftrightarrow O_3 + Ar \tag{2}$$
$$O + O_3 + Ar \leftrightarrow 2O_2 + Ar \tag{3}$$

Where the noble atom, Ar, does not take part in the actual reaction but helps in the conservation of momentum as this reaction involves enormous collision.

At low power, most of the O_2 injected into the deposition chamber is likely to remain undissociated, and so some of the atomic oxygen sputtered from the target is consumed via the reactions (2) and (3). The concentration of atomic oxygen then decreases with increasing O_2 flow at low RF power. At high power deposition, the sputtering process is sufficiently energetic to dissociate most of the O_2 flowing into the chamber, resulting to films with reduced O_2 vacancies.

Structural properties

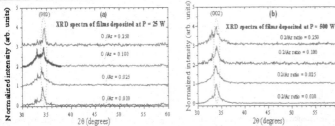

Figure 2 - XRD spectra of films deposited at RF power of (a): 25 and (b): 500 W, at different O_2/Ar ratios (individual plots are up shifted for clarity).

XRD measurements predict crystallite sizes (τ) of all films in the range 5 – 30 nm (figure 2), as estimated by Scherrer's formula, $\tau = 0.9\lambda/\beta\cos(\theta)$; where λ is the wavelength of the incident x-ray and, β and θ are the full-width at half maximum (FWHM) and half of 2θ angle of the (002) peak respectively. Films deposited at high RF power had an average c-axis lattice constant (c) of 0.5282 nm compared to 0.5226 nm for low RF power deposition (Table 1), calculated using Bragg's formula [19].

Table 1 - Variation of full-width τ, c, and FWHM of the (002) ZnO peak with O2/Ar ratio and RF power.

O_2/Ar Ratio	FWHM (degrees) P = 25 W	FWHM (in degrees) P = 500 W	τ (nm) P = 25 W	τ (nm), P = 500 W	Lattice constant c (nm) P = 25 W	P = 500 W
0.010	0.4577	1.0182	20.13	5.35	0.5242	0.5288
0.025	0.2032	1.7653	17.68	5.17	0.5263	0.5298
0.100	0.4115	0.8326	30.00	7.97	0.5204	0.5268
0.250	1.5331	0.7820	27.68	9.01	0.5196	0.5284

Electrical Properties

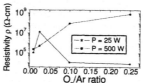

Figure 3 - Variation of resistivities ρ_{25} and ρ_{500} with oxygen for films deposited at 25 and 500 W respectively.

The electrical resistivity of films was seen to increase with oxygen flow (figure 3), for high power deposition. At low RF power, the resistivity of the films is non-monotonic with oxygen flow, likely due to the oxygen reduction processes in (2) and (3). After 18 months exposure to air, films produced at high RF power exhibit greater stability than those produced at low RF power (figure 4). The electrical resistivity increases by a maximum of 1 order in magnitude, whereas lower RF power results in films whose electrical resistivity changes by up to 4 orders in magnitude (figure 4) after 18 months ageing in air.

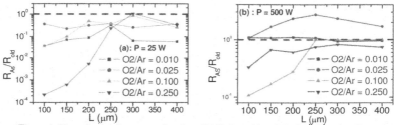

Figure 4 –Normalized resistance as a function of ZnO channel length L of the film after 18 months exposure in air. R_{AS} and R_{old} are the resistances of the as-grown film and of the same film after 18 months respectively, produced at (a): 25 and (b): 500 W RF powers. The dash line indicates the ideal ratio where $R_{AS}/R_{old} = 1$.

TFT applications

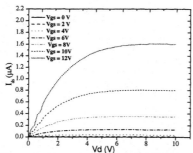

Figure 5 – Output characteristics of as-fabricated ZnO TFT.

ZnO TFTs have been fabricated using the same deposition parameters as those used to produce the most electrically-air stable film ($O_2/Ar = 0.01$). However, in order to fabricate the described TFT structure used in this work, a lower RF power (15 W) was used (this minimizes the damaging effect of energetic sputtering particles on the dielectric layer (SiO_2)). This film had similar resistivity (between $10^5 - 10^6$ Ωcm) to the most stable film deposited at 500 W. Figure 5 shows typical output characteristics for the ZnO TFT device and Figure 6 ((a) and (b)) depicts the transfer characteristics both before and after gate bias stress). This device exhibits hard saturation and operates in enhancement mode, suggesting that the channel is likely only lightly doped n-type). Analysis of the transfer characteristics yield an effective mobility of ~ 3.1 $cm^2/V.s$, an on-off current ratio of over 10^6 and a subthreshold slope of ~ 1.8 V/dec. This demonstrates the optimized film at 500 W in addition to being stable is also suitable for TFT applications.

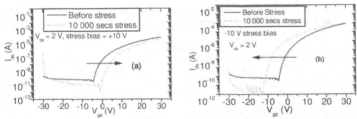

Figure 6 – Transfer characteristics of and ZnO TFT with (a) +10 V and b) -10 V gate bias stresses.

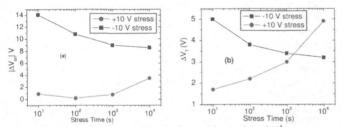

Figure 7 – Change in (a): V_{on} and (b): V_T after a total stress time of 10^4 s.

An application of ±10 V gate bias stress results in both threshold (V_T) and turn-on (V_{on}) (gate voltage corresponding to the minimum drain current on a semilog plot) voltage instabilities (figures 6 and 7). It is observed that positive and negative bias stress shifts the entire characteristic curve to the right and left respectively. In both stress conditions, the curves relax back to their original positions a few hours after stress. Greater shift is observed in the case of negative bias stress. Under both stress conditions, the characteristics remain almost parallel in the subthreshold region; suggesting that the effect of bias stressing on the interface is negligible. This, together with the asymmetry observed in in ΔV_{on} under ±10 V bias stress suggests that charge trapping in the ZnO channel is likely to be the culprit for the bias stress related instability. However, it is hoped that the optimized film at 500 W would produce more stable devices.

CONCLUSIONS

We observed that electrical stability and optical properties of ZnO films are dependent on the process parameters (O_2, Ar flow rate and RF power) in the sputtering process. It is also noted that the conversion of atomic oxygen via ozone to di-oxygen is a possible process to consider for controlling film composition, especially for low power depositions. Furthermore, electrical stability of films in air also exhibits a strong dependence on O_2/Ar ratio and hence on the concentration of O_2 vacancies in the film or at the surface. ZnO TFTs fabricated with the most electrically air-stable film show effective mobilities of ~3.1 cm^2/V.s and On-Off current ratios of over 10^6. ZnO TFT bias stress instability is suspected to be dominated by charge trapping in the ZnO channel material, possibly at the grain boundaries.

ACKNOWLEDGMENTS

The authors gratefully acknowledge Dr. T. L. Makarova for useful discussion on nanostructures and access to XRD equipment. This work was supported by DMU Leicester.

REFERENCES

[1] L. Mingjiao and K. Hong Koo, Applied Physics Letters **84**, 173-5 (2004).
[2] P. Won Il, K. Jin Suk, Y. Gyu-Chul, M. H. Bae, and H. J. Lee, Applied Physics Letters **85**, 5052-4 (2004).
[3] H. Woong-Ki, J. Gunho, K. Soon-Shin, S. Sunghoon, and L. Takhee, IEEE Transactions on Electron Devices **55**, 3020-9 (2008).
[4] V. Khranovskyy, J. Eriksson, A. Lloyd-Spetz, R. Yakimova, and L. Hultman, Thin Solid Films **517**, 2073-8 (2009).
[5] J. Y. Park, Kim, J.-J., Kim, S.S. , Journal of Nanoscience and Nanotechnology **8** (2008).
[6] K. Donghun, L. Hyuck, K. Changjung, S. Ihun, J. Park, P. Youngsoo, and C. JaeGwan, Applied Physics Letters **90**, 192101-1 (2007).
[7] R. Martins, P. Barquinha, A. Pimentel, L. Pereira, and E. Fortunato, Physica Status Solidi A **202**, 95-7 (2005).
[8] R. L. Hoffman, B. J. Norris, and J. F. Wager, Applied Physics Letters **82**, 733-5 (2003).
[9] R. B. M. Cross, M. M. De Souza, S. C. Deane, and N. D. Young, IEEE Transactions on Electron Devices **55**, 1109-15 (2008).
[10] K. Hoshino, D. Hong, H. Q. Chiang, and J. F. Wager, IEEE Transactions on Electron Devices **56**, 1365-70 (2009).
[11] M. J. Powell, IEEE Transactions on Electron Devices **36**, 2753-2763 (1989).
[12] D. K. Schroder, Semiconductor Material and Device Characterization, Second ed. 1998: John Wiley & Sons 143 - 159.
[13] Y. S. Yu, B. I. Kim, S. C. Kim, B. C. Shin, T. S. Kim, and M. Y. Jung, Physica B **376** - **377**, 752-5 (2006).
[14] Y. Natsume and H. Sakata, Materials Chemistry and Physics **78**, 170-6 (2003).
[15] Y. Natsume and H. Sakata, Thin Solid Films **372**, 30-6 (2000).
[16] B. N. Pawar, S. R. Jadkar, and M. G. Takwale, Journal of the Physics and Chemistry of Solids **66**, 1779-82 (2005).
[17] D. Raoufi and T. Raoufi, Applied Surface Science **255**, 5812-17 (2009).
[18] R. G. J. Tauc, A. Vancu, Physica Status Solidi A **15** 627 (1966).
[19] D. N. Neil W. Ashcroft; and Mermin, *Solid State Physics* (Saunders College Publishing 1976).

Mater. Res. Soc. Symp. Proc. Vol. 1201 © 2010 Materials Research Society   1201-H05-41

Current Transport Mechanisms for MSM-Photodetectors on ZnO:N Thin Films

Tingfang Yen, Alan Haungs, Sung Jin Kim, Alexander Cartwright, and Wayne A. Anderson
Electrical Engineering Department, 332 Bonner Hall, University at Buffalo, State University of
New York, Buffalo, NY 14260

ABSTRACT

Metal-semiconductor-metal photodetectors (MSM-PDs) on ZnO:N thin films
deposited by radiofrequency (RF) sputtering and with post N^+ ion implantation
processing were fabricated using a ZnO/Si structure. A 10 times reduction in dark current
was observed compared to the devices on an as-deposited ZnO thin film without ion
implantation. These MSM-PDs gave performances of a photo-to-dark current ratio of 2030 and
responsivity (R) = 2.7 A/W; the pulse response was a 12.3 ns rise time and 15.1 ns fall
time using a femto-second pulse. Temperature-dependent current -voltage (I-V-T)
characteristics of the MSM-PDs were observed and the space charge limited current
(SCLC) theory was applied to determine the current transport mechanisms. In the SCLC
region, $J \sim V^m$ gave m to determine the current transport mechanism and the value of m
changes with temperatures and applied voltages. Current transport is governed by the
ZnO structure rather than the electrodes.

INTRODUCTION

Ultraviolet photodetectors (UV-PDs) operating in the short wavelength are important
devices in various commercial and military applications. Most UV-PDs have been silicon-based.
However, due to the indirect bandgap of 1.2 eV for Si, it requires bulky band filters to block
solar radiation background and it is intolerant of high temperature and caustic environments[1].
ZnO is one of the potential materials to eliminate those disadvantages, and has been of interest in
optical electronics applications for a few years due to its large exciton binding energy, wide
bandgap, and some advantages over GaN, such as low cost of bulk single crystals, lower growth
temperature, tolerance to high energy radiation, and ease of etching in most acids and alkalis.
In this paper, we report MSM-PDs on ZnO thin films. There are many advantages of
using a MSM structure for photodetectors, such as simplicity of fabrication, low capacitance, low
dark currents, good noise suppression, high speed, and high sensitivity. However, the main
drawback of the MSM-PD is high dark current due to the Schottky barrier junction. In this paper,
we have reached lower dark current of MSM-PDs by adopting various post-processing steps to
modify the properties of ZnO thin films [2, 3] and in addition, have investigated the current
transport mechanisms of MSM-PDs.

EXPERIMENT

ZnO thin films were deposited by RF sputtering using an unheated substrate with ratio of
oxygen to argon of 1 to 3 and nitrogen partial pressure of 0.4 mT. A ZnO target with purity of

99.98% was utilized and placed 12 cm from the substrate. Before deposition, a Ti pump was applied to decrease the moisture and contamination in the vacuum system. After sputtering, the samples were implanted by 50 keV N^+ ions with implantation dose of 1×10^{12} cm^{-2} and post annealing in air ambient by rapid thermal anneal (RTA) at 800^0C for 3 min. MSM-PDs were then made with interdigitated patterns having 2 μm spacing and 6 μm width to form a back to back Schottky contact using a Denton Vacuum DV-502A thermal evaporator to evaporate Al.

The I-V curves for conductivity and MSM-PDs were measured by a Kiethley 617 programmable electrometer multimeter with an accuracy of +/- 0.05~0.07% and Kiethley 230 programmable voltage source with an accuracy of +/- 0.15~0.25%. The photo current was tested by P25, P50, P100 which indicates that power density was 25, 50, 100 mW/cm^2 in the full wavelength environment produced by the ELH lamp, a tungsten halogen lamp. A liquid nitrogen cryostat was used for low temperat re measurements. Repeated I-V measurements were reproduced within +/- 2%. The pulse response (PR) measurement used a Verdi-V18 solid state laser with 18 W tuning to 266 nm, emitting 150 fs wide pulses at a repetition rate of 250 kHz.

RESULTS AND DISCUSSION

MSM-PDs with ZnO:N thin film

The MSM-PDs gave a photo-to-dark current ratio of 2030 and R= 2.1 A/W at 20 V as shown in Figure 1 (a). Responsivity versus voltage is shown in Figure 1 (b). Higher built in voltage in the low bias region was observed, which needs higher electric field in the mobility regime to promote carrier flow and collection. The pulse response was tested with a resistor of 50 Ω and gave a result shown in Figure 2 of 12.3 ns rise time, 15.1 ns fall time.

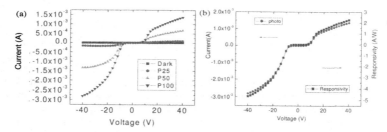

Figure 1. (a) I-V of MSM-PDs fabricated by N^+ ion implanted ZnO film with post-annealing in RTA, air ambient, 800^0C, 3mins; (b) photo current with 100 mW/cm^2 of incident power (P100) and responsivity at each bias voltage.

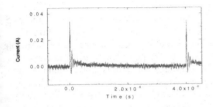

Figure 2. Pulse response of MSM-PDs fabricated by N⁺ ion implanted ZnO film with post-annealing in RTA, air ambient, 800⁰C-3mins.

Compared to MSM-PDs using the as-deposited ZnO thin film, the dark current is suppressed efficiently by the films deposited by RF sputtering with nitrogen ion-implantation and post annealing by RTA in an air-ambient between 750⁰C to 800⁰C for 3 mins. The comparison of MSM-PD dark and photo I-V data is shown in Figures 3 (a) and (b). The dark current of the sample with nitrogen ion-implantation and RTA post-annealing has been improved by one magnitude compared to the as-deposited sample. Therefore, the photon detection sensitivity of the ion-implantation sample is higher than for other samples. The photo I-V comparison showed higher photo I-V collection than for as-deposited samples. This is due to the improvement of lattice structure by post annealing, which activates the N⁺ in the films, and enlarges grain size due to grain agglomeration or growth parallel to the substrate surface. Thus, recombination centers became fewer leading to current leakage suppression.

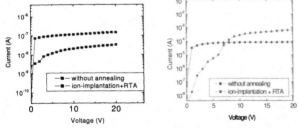

Figure 3. (a) Lower dark I-V for MSM-PDs using nitrogen ion-implantation and RTA post-annealing of ZnO thin films; (b) Photo I-V, with P100, comparison of MSM-PDs for various processes applied to ZnO thin films

I-V-T

In order to decide which current transport mechanisms are determining the I-V characteristic of our devices, we evaluated the temperature-dependence of the slope of the log I-V curve. Figure 4 shows how the forward dark and photo, P100, log I-V characteristics change as a function of temperature for MSM-PDs. The I-V-T measurement was observed with the measuring temperature range from 200 to 400K. The current saturates with the voltage change and is dependent on the temperature change for both dark and photo-current.

The derivative of lnI with respect to voltage is plotted as a function of voltage as Figures 5 (a) and (b). Those figures give clear pictures of the change in the slope of dark and photo I-V curve as a function of both voltage and temperature. As can be seen in Figure 5 (a), we were able to distinguish two voltage regions. In the small bias range, the slope of dark I-V is not constant and the slope for the sample tested above room temperature environment is higher than the one tested below room temperature. In the larger bias region, V> 5V, the slope becomes relatively constant with a value smaller than 1. In Figure 5 (b), slopes of photo I-V decrease rapidly in the low bias voltage range, 0<V<2V. In the middle voltage range, 2V <V<10V, slopes gradually decrease from 2 to 0.5. In the higher voltage range, V>10V, slopes stay consistently closer to 0.

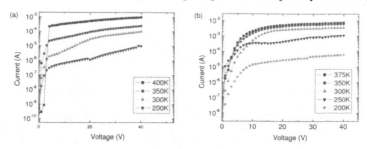

Figure 4. I-V-T curves of MSM-PDs using ZnO thin films deposited by rf sputtering with nitrogen ion implantation post-processing and RTA annealing in a nitrogen ambient at 800°C for 3 mins in (a) dark and (b) photo conditions

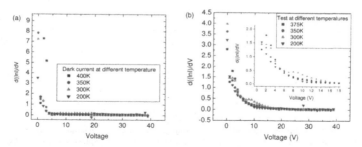

Figure 5. Voltage derivative of I-V characteristic in (a) dark and (b) photo ambient

Most current showed non-exponential dependence of I on V and SCLC theory was applied to determine the current transport mechanisms. In the SCLC region, $J \sim V^m$ gave m to determine the mechanism [4, 5]. Log I-Log V-T was plotted in Figures 6 (a) and (b) for dark and photo current. In Figure 6 (a), the dark current can be separated into three regions, where each region represents a different current transport mechanism. Two slopes were observed at 400K, with m <1, which indicates SCLC dominated by recombination injection [6, 7]. Three slopes were consistently observed at each temperature of 200K, 300K and 350K with m values given in the table below the Figure 6 (a). In the lower bias region 1, m is around 2.5 which indicates SCLC in the mobility regime where the carrier velocity depends on electric field with an exponential distribution of trapping levels. In the region 2, the value of m is smaller than 1 where recombinative injection dominates SCLC. The value of m is close to 1.5 at 300K and 350K in the higher bias region 3 which indicates the SCLC in the ballistic regime when the carrier velocity depends on the injection velocity from the source and is independent of the field and scattering parameters. Also, m >2 is in the region three at 400K.

In the Figure 6 (b), two current transport mechanisms were observed for each photo-current. The m values are given in the table below Figure 6 (b). Current transport mechanisms showed similar photo-curve results in the higher temperature range 375K, 350K, and 300K. Two mechanisms were observed: m > 2 in the lower bias region, indicating SCLC in the mobility regime with exponential distribution of trapping levels and m close to 0.5 in the higher bias above 10V. The lowest temperature 200K showed two mechanisms, m = 3 in region one, and m close to 1 in region two, which indicates SCLC in the velocity-saturation regime where the current is independent of electric field.

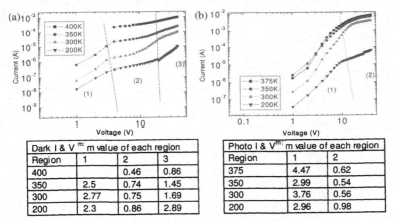

Dark I & V $^{m:}$ m value of each region			
Region	1	2	3
400		0.46	0.86
350	2.5	0.74	1.45
300	2.77	0.75	1.69
200	2.3	0.86	2.89

Photo I & V $^{m:}$ m value of each region		
Region	1	2
375	4.47	0.62
350	2.99	0.54
300	3.76	0.56
200	2.96	0.98

Figure 6. LogI-LogV-T plot of (a) dark and (b) photo current of MSM-PDs and m values of current transport mechanisms in each region of dark and photo current respectively

CONCLUSIONS

MSM-PDs using ZnO:N/Si have been fabricated. Improved photo-sensitivity by reduction in dark current was observed due to the improvement of lattice structure and enlarged grain size to reduce recombination centers. High photo-to-dark current ratio up to 2030, responsivity of 2.7 A/W, and the nano-second response time have been obtained. SCLC determines the current transport mechanisms of MSM-PDs and current transport is governed by the ZnO structure rather than the electrodes.

ACKNOWLEDGEMENT

The research was partially supported by the AFOSR, monitored by Dr. Kitt Reinhardt and a NASA Space Grant, as a subcontract from Cornell University.

REFERENCES

[1]   D. Basak, G. Amin, B. Mallik, G. K. Paul, and S. K. Sen, "Photoconductive UV detectors on sol-gel-synthesized ZnO films," *Journal of Crystal Growth,* vol. 256, pp. 73-77, 2003.
[2]   T. F. Yen, D. Strome, S. J. Kim, A. N. Cartwright, and W. A. Anderson, "Annealing studies on zinc oxide thin films deposited by magnetron sputtering," *Journal of Electronic Materials,* vol. 37, pp. 764-769, May 2008.
[3]   T. Yen, M. DiNezza, A. Haungs, S. J. Kim, W. A. Anderson, and A. N. Cartwright, "Effects of nitrogen doping of ZnO during or after deposition," *Journal of Vacuum Science & Technology B: Microelectronics and Nanometer Structures,* vol. 27, pp. 1943-1948, 2009.
[4]   K. K. Ng, *Complete Guide To Semiconductor Devices*: Wiley Interscience, 2002.
[5]   A.J. Campbell, D. D. C. Bradley, and D. G. Lidzey, "Space-charge limited conduction with traps in poly(phenylene vinylene) light emitting diodes," *Journal of Applied Physics,* vol. 82, pp. 6326-6342, Dec 1997.
[6]   M. Ilegems and H. J. Queisser, "Current transport in relaxation-case GaAs," *Physical Review B,* vol. 12, pp. 1443-1451, 1975.
[7]   G. G. Roberts, "Electron Injection into a p-Type Semiconductor," *Physica Status Solidi (b),* vol. 27, pp. 209-218, 1968.

Mater. Res. Soc. Symp. Proc. Vol. 1201 © 2010 Materials Research Society   1201-H05-46

Comparison of APCVD to LPCVD Processes in the Manufacturing of ZnO TCO for Solar Applications

Wei Zhang[1], Tom Salagaj[1], Jiuan Wei[1], Christopher Jensen[1] and Karlheinz Strobl[1]
[1]CVD Equipment Corporation,
1860 Smithtown Ave., Ronkonkoma, NY 11779

ABSTRACT

A FirstNano EasyTube 3000 CVD system with a 3" diameter horizontal tube furnace was used to investigate the optimization of both APCVD (Atmospheric Pressure Chemical Vapor Deposition) and LPCVD (Low Pressure Chemical Vapor Deposition) processes to grow both boron and fluorine doped ZnO films with a sheet resistance, slice resistance and haze suitable for their potential utilizations as TCO (Transparent Conductive Oxide) layers for photovoltaic applications. Growth rates as high as 100 nm per minute have been obtained in some parameter regions for both processes. In both cases the resulting material property parameters were the same or better than reported in the literature. Although the horizontal hot wall CVD R&D reactor is not optimum for uniform TCO thin film deposition it allowed us to investigate the interrelationship of the most critical parameters with the resulting material properties.

The driving force for this work is the ultimate goal of demonstrating a process parameter solution suggesting that ZnO films (usable for either display system manufacturing and/or photovoltaic applications) can be deposited with optimized material properties that are comparable to LPCVD or sputtering processes, but that the APCVD solution could be more economical for large scale thin film ZnO coating implementation. Ultimately our desire is to transfer such a ZnO deposition process to our proprietary, APCVD CVDgCoat™ platform, which can coat up to 4 meter wide glass sheets and metal foils that move continuously.

INTRODUCTION

TCO thin films and/or materials typically have low electrical resistivity and high transmittance for visible light. Wide band gap semiconducting metal oxides such as Indium Oxide, Tin Oxide and Zinc Oxide can be doped with suitable amounts of Fluorine or Antimony to improve their conductivity while retaining high transparency. Among them Zinc Oxide has prominent advantages over the other TCOs. This is because, firstly, Zinc precursors are generally non-toxic. Secondly, Zinc materials are less expensive than Tin, Indium or Cadmium. Currently, fluorine doped Tin Oxide (FTO) is the most widely used front electrode for thin film based solar module production. However, FTO can be reduced in hydrogen plasma during Silicon deposition process. The resultant elemental Tin at the silicon interface will cause optical loss and unwanted diffusion, which will lower solar cell efficiency. Zinc Oxide, however, is typically more stable than Tin Oxide in a plasma reducing environment. In addition, it has a higher optical transmittance than the other TCOs. All of the above suggests Zinc Oxide a very promising candidate for solar cell applications [1].

Chemical Vapor Deposition (CVD) processes generally have better deposition rate and uniformity than the other deposition processes such as sputtering and physical vapor deposition (PVD). It is more suitable for large-scale commercial thin film deposition applications. Conductive Zinc Oxide films have been deposited by adding impurity dopant such as Fluorine [2], Boron [3, 4], Aluminum [5-8] and Gallium [9-11]. Three different types of CVD processes,

Atmospheric Pressure CVD (APCVD), Low Pressure CVD (LPCVD), and Plasma Enhanced CVD (PECVD), have been investigated for deposition of ZnO layers for solar cell applications and other optoelectronics devices. PECVD is currently in a stage of preliminary investigations [12]. A comprehensive study on APCVD ZnO deposition using different dopant has been conducted by Hu and Gordon [2, 3, 5, 9]. The typical deposition temperature for the APCVD technique is around 400 °C, which is still quite high for thin-film solar module application. Lowering the pressure allows one to reduce the deposition temperature. LPCVD ZnO growth has therefore been investigated to improve the control of the chemical reactions involved in the deposition process to obtain high coating performance at lower temperature [4, 13, 14].

In this study, the EasyTube 3000 CVD system offered by First Nano, a division of CVD Equipment Corporation, was used to investigate the optimization of both APCVD and LPCVD processes to grow both Boron and Fluorine doped ZnO films with a transparency, sheet/slice resistance and haze suitable for their potential utilizations as TCO layers for photovoltaic applications. The resulting material properties will be characterized and compared with the best properties reported in the open literature. The purpose of this investigation is to demonstrate a process parameter solution suggesting that, by using an APCVD process, ZnO films can be deposited with optimized material properties that are comparable to LPCVD or sputtering processes, the APCVD solution will be more economical for large scale thin film ZnO coating implementation. Our goal is to transfer such a ZnO deposition process to the APCVD *CVDgCoat*™ platform which has the ability to coat up to 4-meter wide glass sheets.

EXPERIMENT

The experiment system for this exploration is shown in Figs. 1 (a)-(d). A customized EasyTube™ 3000's system (Fig. 1a), manufactured by First Nano, a division of CVD Equipment corporation, was used for the experiments reported here. The system houses several key process components including a 3" diameter quartz process chamber, a computer controlled, fully automated process control system for maximum safety and process reproducibility, automatic sample loaders, pumps, a ultra high purity (UHP) gas and vapor delivery system and an EasyGas™ EG600 exhaust treatment system. As shown in Fig. 1(b), the three-zone furnace with cascade control provides better temperature stability and uniformity over a large area. Internal thermal couple measures actual temperature over samples. Furnace lid opens automatically during cooling down stage to save operating time and improve productivity. This versatile system has been adjusted to deposit silicon nano wire, silicon dioxide, SiCxOy [15], APCVD Fluorine doped ZnO and LPCVD Boron doped ZnO successfully. Fig. 1 (c) shows the temperature distribution achieved inside the reaction process tube. Computation Fluid Dynamics (CFD) software, ANSYS FLUENT, is used to conduct the 3-D thermal modeling including conduction, convection and radiation. By applying a constant temperature on the inner wall of the three-zone furnace, it is shown in the figure that a very uniform temperature distribution can be achieved on substrate paddle.

Zinc Oxide films were deposited on three 1cm×1cm bare Silicon (100) substrates and another three 1cm×1cm SiC_xO_y coated soda lime glass substrates by APCVD and LPCVD respectively using the ET 3000 system. The detailed locations of the substrates are shown in Fig. 1(d). Nitrogen was used as the carrier gas for diethyl zinc (DEZ), anhydrous ethanol (for APCVD case) and DI water (for LPCVD case). For LPCVD B: ZnO deposition, Diboran (2 % in high purity Argon, Air Products) was used as dopant and mixed with DEZ in nitrogen carrier gas right before flowing into the injector. DI water was injected separately through another injector. For APCVD F: ZnO growth, Hexafluoropropylene (C_3F_6) was used as fluorine dopant and mixed with the DEZ precursor gas before entering the process tube. Because of the high process temperature for APCVD, ethanol, which is weaker oxidant than water was used as oxygen source (to obtain moderate reaction) and injected separately into the process tube. The typical process conditions used can be found in Table I.

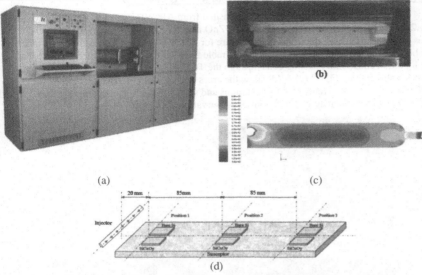

(a)   (c)

(d)

Figure 1: Schematics of (a) ET 3000 system, (b) Three-zone furnace, (c) temperature distribution in the process tube, and (d) over view of sample locations

Table I Typical process conditions for APCVD and LPCVD ZnO deposition

	Reaction Tube		DEZ Bubbler			Ethanol Bubbler			Dopant
APCVD	Temp.	Pressure	Temp.	Pressure	Carrier N_2	Temp.	Pressure	Carrier N_2	C_3F_6
	[C]	[Torr]	[C]	[Torr]	[sccm]	[C]	[Torr]	[sccm]	[sccm]
	375-500	760	25	800 -1000	200 - 400	60	800-1000	200-1000	0-150
	Reaction Tube		DEZ Bubbler			Water Bubbler			Dopant
LPCVD	Temp.	Pressure	Temp.	Pressure	Carrier N_2	Temp.	Pressure	Carrier N_2	B_2H_6
	[C]	[Torr]	[C]	[Torr]	[sccm]	[C]	[Torr]	[sccm]	[sccm]
	140-170	0.3- 0.9	20	300	200-400	25	800-1000	80-200	0-30

Surface topography images have been taken by Scanning Electron Microscopy (SEM) to analyze the surface morphology of the ZnO coatings. Cross-section views were obtained by splitting the Silicon substrate samples to characterize the crystallographic orientation and the microstructure of the ZnO layers. This allowed to measure the coating thickness and to calculate the resulting average deposition rate. The sheet resistance was measured by using a Veeco FPP-100 four point probe. Assuming that the film is homogeneous in the direction perpendicular to the substrate plane, the film slice resistance was deduced from the sheet resistance and coating

167

thickness. The glass transmission and refection were measured by a home-made reflection and transmission measurement setup.

RESULTS & DISCUSSION

Morphology and Microstructure of ZnO from APCVD / LPCVD

Surface topology images for ZnO deposited from APCVD and LPCVD are shown in Fig.2 (a) & (b) respectively. The grain size of ZnO film for LPCVD is smaller than the APCVD ones because of the high deposition temperature for APCVD. Rough surface has been obtained for both cases. Indeed, a rough surface is desirable for solar energy application because it allows the light that enters into the solar cells through the TCO layer to scatter efficiently so as to enhance absorption. Figs.2 (c) & (d) show the cross-section view of ZnO for both APCVD and LPCVD systems. Different crystal orientations and sizes had been obtained for the APCVD and LPCVD process, which agrees well with the observations from other groups [2, 4].

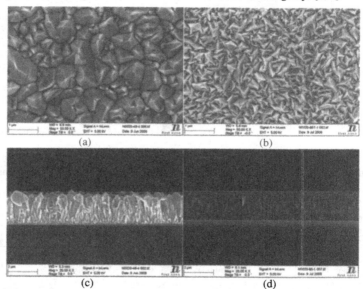

(a)   (b)

(c)   (d)

Figure 2: SEM for (a) Top view of APCVD F:ZnO, (b) Top view of LPCVD B:ZnO, (c) Cross-section view of APCVD F:ZnO and (d) Cross-section view of LPCVD B:ZnO.

Transmittance, Reflectance and Resistance of APCVD / LPCVD ZnO TCO Films

Figure 3(a) shows that a larger than 30% haze was obtained after a B:ZnO dopped coating was applied on top of the SiCxOy barrier layer coated glass. Surface transmission spectra were also taken to study the optical properties of the obtained ZnO films. The total transmittance of ZnO films with optimized process conditions (deposition temperature, pressure, DEZ/H2O ratio and doping) from APCVD and LPCVD process were shown in Fig. 3 (b). In the case of LPCVD B:ZnO, the film thickness was about 2.5 μm deposited in 15 minutes (with a deposition rate of 160 nm/minute) and the sheet resistance was about 5 Ω/– with the deduced film slice resistance of 1.3E-3 Ωcm, which is as low as the lowest value for B:ZnO from LPCVD recorded by other groups [4]. The APCVD F:ZnO has a thickness of 750 nm with a sheet resistivity of 11 Ω/–, corresponding to 8.0E-4 Ωcm slice resistance. As shown in Fig. 3 (b) the drop of total transmittance in the blue region for the dopped ZnO was due to free carrier absorption. For solar cell applications, a low optical absorption is desired to allow more photons to enter the active cell region and generate electron-hole pairs. This confirms the importance of minimizing the amount of dopant needed in the ZnO film and optimizing the mobility to lower the resistivity. In our case, 70 to 80% transmission has been obtained for both APCVD and LPCVD cases, which is comparable with results from other researchers and fulfills the requirement for solar cell applications [1, 2 & 4].

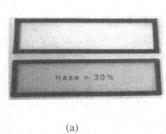
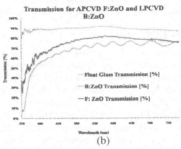

(a)   (b)

Figure 3: (a) Top view of SiC$_x$O$_y$ barrier layer coated soda lime glass in frame (upper one) and top view of B:ZnO coating on top of SiC$_x$O$_y$ barrier layer coated glass (bottom one); (b) Transmission spectra for APCVD F:ZnO and LPCVD B:ZnO coatings on soda lime glass. Deposition conditions for APCVD F:ZnO: growth temperature=400 °C, chamber pressure=760 Torr, Ethanol/DEZ mole ratio=8.5, C$_3$F$_6$/DEZ mole ratio=0.85. For LPCVD B: ZnO: Growth temperature=160°C, chamber pressure =0.85 Torr, H$_2$O/DEZ mole ratio=1.2, B$_2$H$_6$/DEZ mole ratio=0.03.

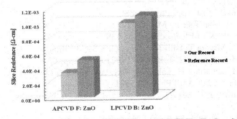

Figure 4: Comparison of resistivity for APCVD and LPCVD ZnO with reference record [1, 2 & 4]. APCVD F:ZnO deposition conditions: growth temperature=400 °C, chamber pressure=760 Torr, Ethanol/DEZ mole ratio=8.5, C$_3$F$_6$/DEZ mole ratio=0.85. LPCVD B:ZnO deposition conditions: growth temperature=160 °C, chamber pressure=0.63 Torr, H$_2$O/DEZ mole ratio=1.2, B$_2$H$_6$/DEZ mole ratio=0.03.

The key factors affecting the coating properties have been theoretically analyzed and experimentally tested for both APCVD and LPCVD ZnO film deposition. These factors includes deposition temperature, chamber pressure, gas velocity, gas Reynolds number, Oxygen source/DEZ mole ratio, Dopant/DEZ mole ratio and deposition time, etc. Based on the theory for design of experiment, various combinations of the factors were further tested to optimize the coating properties. For both APCVD and LPCVD processes, we obtained a respective lower resistance than previously published [1, 2, & 4] despite using the same ET 3000 system. The results and detailed process conditions were shown in Fig. 4. Further investigation for coating properties such as carrier density, hall mobility, dopant concentration, etc. is essential for further optimizing the TCO coating deposition for photovoltaic and solar applications. Integration of the whole investigations will allow us to analyze the interrelationship of the most critical parameters and identify their effects on the resulting material properties.

CONCLUSIONS

In this paper a FirstNano EasyTube 3000 CVD system has been used to deposit F:ZnO and B:ZnO films on Silicon wafer substrates and SiC_xO_y barrier layer coated soda lime glass substrates with both APCVD and LPCVD processes. For both processes, rough surface ZnO coatings had been obtained, which is desirable for solar energy application. High haze factor (larger than 30%) and high deposition rate (higher than 100 nm per minute) were realized for both F:ZnO from APCVD and B:ZnO from LPCVD. A design of experiment study has been conducted for process optimization by using different combinations of key process control parameters in order to achieve a slice resistivity as low as possible for both APCVD and LPCVD processes with good transmittance. For both APCVD and LPCVD processes we report a record-low slice resistivity of 3.0E-04 Ωcm and 1.0E-3 Ωcm, respectively. This investigation provides a promising solution for On-line and Off-line APCVD TCO thin film coating deposition on a very large glass surface area. More follow-up work is necessary to further optimize the process and to transfer such a ZnO deposition process to the APCVD CVDgCoat™ platform which has the ability to coat up to 4-meter wide glass sheets and metal foils moving at high speeds.

REFERENCES

[1].   Liang H, Gordon RG (2007) J Mater Sci 42:6388–6399
[2].   Hu J, Gordon RG (1991) Solar Cells 30:437
[3].   Hu J, Gordon RG (1992) J Appl Phys 71:880
[4].   Fay˙ S et al. (2005) Solar Energy Materials & Solar Cells 86: 385–397 392
[5].   Hu J, Gordon RG (1992) J Electrochem Soc 139:2014
[6].   Sato H, Minami T, Miyata T, Ishii M (1994) Thin Solid Films 246:65
[7].   Minami T, Sato H, Imamoto H, Takata S (1992) Jpn J Appl Phys
[8].   Minami T, Sato H, Sonohara H, Takata S, Miyata T, Fukuda I (1994) Thin Solid Films 253:14
[9].   Hu J, Gordon RG (1992) J Appl Phys 72:5381
[10].   Hirata GA, McKittric J, Cheeks T, Siqueiros JM, Diaz JA, Contreras O, Lopez OA (1996) Thin Solid Films 288:29
[11].   Wang R, King LLH, Sleight A (1996) J Mater Res 11:1659
[12].   Haga K, Kamidaira M, Kashiwaba Y (2000) J. Crystal Growth 214: 77–80.
[13].   Roth A. P., Williams D. F. (1981) J. Appl. Phys. 52/11:6685–6692.
[14].   Takahashi K., Omura A., Konagai (1996) Patent No US5545443.
[15].   Zhang W, Salagaj T, Jensen, C, Strobl K, Davies M, 2009 Materials Research Society (MRS) Fall Meeting, Nov. 30- Dec. 4, 2009 Boston, MA, U.S.A (in press)

Mater. Res. Soc. Symp. Proc. Vol. 1201 © 2010 Materials Research Society   1201-H05-51

Yu-Hao Liao, Yi-Hao Pai, and Gong-Ru Lin

Graduated Institute of Electro-Optical Engineering and Department of Electrical Engineering, National Taiwan University, No.1, Roosevelt Road Sec. 4, Taipei 10617, Taiwan R.O.C.

ABSTRACT

The fabrication of surface nano-roughened zinc oxide (ZnO) is demonstrated by using a zinc/air fuel cell based chemical reactor. The chemical reaction creates the hydroxyl near the Zn and randomizes the pores with accumulated hydroxyls, which causes Zn to be oxidized to form thick and nano-roughened ZnO upon Zn anode or to be transformed into a soluble complex zincate ion in the excess electrolyte. Furthermore, controlling the quantity of $(Zn(OH)_2)$ and $[Zn(OH)_4]^{2-}$ can significantly decrease the defect-related emission and blue-shift PL spectra to the UV range.

INTRODUCTION

Zinc oxide (ZnO) is a wide direct bandgap semiconductor with E_g of 3.36 eV and has a large free exciton binding energy of 60 meV leading to efficient excitonic emission even at room temperature. In particular, it also has good chemical, mechanical, and thermal stability. These properties make ZnO a promising material for fabrication of optical devices [1]. The preparation of surface-roughened ZnO nanostructure is an important issue especially for efficient contact in solar cells. Most of the ZnO nano-structures have been prepared by vapor-based methods. However, these techniques are normally high-cost and need special equipments as well as need to be performed in strict condition, for example high-vacuum. To solve these problems, some techniques have been developed such as wet chemical method [2], aqueous solution route and rapid thermal processing [3], and solution growth method [4]. Nevertheless, weak crystalline and defect properties are normal questions for the material prepared by these techniques. In this work, we demonstrate a novel technique for preparation of nano-roughened ZnO by zinc/air fuel cell based chemical reactor at room temperature. The crystalline characterization, defect and surface morphology analyses of ZnO and growth mechanism will be discussed.

EXPERIMENT

The zinc/air fuel cell based chemical reactor is shown in Figure 1. The zinc plate is used as the anode in a zinc/air cell employing 7M KOH as electrolytes. The ZnO samples with area of 10 cm^2 are prepared at room temperature with a constant reactive time of 100 min by zinc/air fuel cell based chemical reactor discharging at resistance of 100, 25, and 1.6 Ω for sample A, B, and C, respectively. The room temperature photoluminescence (PL) measurements with excitation at 325 nm of He–Cd laser are performed in order to investigate defects of the nano-roughened ZnO structure. Furthermore, the XRD measurements are applied to analyze formation of ZnO crystals under different preparation condition.

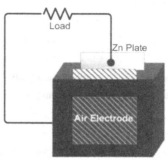

Figure 1 the zinc/air fuel cell based chemical reactor.

RESULTS AND DISCUSSION

The morphology of ZnO prepared in different reactive conditions are shown in Figure 2. It shows that the ZnO nano-roughened structure synthesized with resistance of 100 Ω is tip-like and the tip diameter is ~100 nm with a height of 600 nm. When the resistance is decreased to 1.6 Ω, the tip diameter increases to 900 nm and its height simultaneously augments to 2 μm. Mohamad [6] have demonstrated that the oxygen and H_2O molecules may create hydroxyl near the Zn anode and randomize the pores with accumulated hydroxyls.

Therefore, the Zn is oxidized to form a nano-roughened ZnO upon Zn anode or is transformed into a soluble complex zincate ion ($[Zn(OH)_4]^{2-}$) in the excess electrolyte. In addition, lower resistance should promote the higher discharged current leading to enough Zn(OH)$_2$ transforms into [Zn(OH)$_4$]$^{2-}$ and induce larger tip-like structure.

Figure 2 The SEM micrograph of the ZnO prepared under the fuel cell discharging at (A) resistance of 100 Ω for sample A, (b) resistance=25 Ω for sample B, and (c) resistance=1.6 Ω for sample C.

PL spectra are shown in the Figure 3. The band peaking at 580-595 nm for all the samples are most-likely related to oxygen interstitial defects in nano-roughened ZnO. In particular, it obtains obvious blue- shifted and weak PL in wavelength of 580-595 nm when decreasing the load of resistance up to 1.6 Ω due to the reaction speed increase, which implies that the origin of the visible emission related defects might be decay.

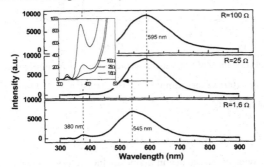

Figure 3 PL spectra of the ZnO on zinc electrode synthesized under the fuel cell discharging at (A) resistance of 100 Ω for sample A, (b) resistance=25 Ω for sample B, and (c) resistance=1.6 Ω for sample C.

Such a blue-shifted wavelength and decreased PL intensity are corresponding to the oxygen interstitial defect disappear, which is mainly attributed to the compensation effects of complex zincate ion aggregation and re-oxidation occurred at the outer surface of Zn anode. In addition,

It also finds that the sample grown at resistance of 1.6 Ω shows the obvious band-edge UV PL peak of ~380 nm (figure 3 inset).

Figure 4 shows the XRD results. Except the most intense XRD signal corresponding to (101 orientation) at 2θ of 31.6°, it can also observe the others at (002) and (100) corresponding to 34.3° and 36.2°, respectively. The XRD peak intensity increases by decreasing the resistance from 100 to 1.6 Ω, which implies that the ZnO crystallizes well and expands grain size. In more detail, the nano-roughened ZnO with containing oxygen interstitial defect prefer forming the weak crystal due to the incomplete lattice structure. Therefore, the result was also confirmed that blue-shifted wavelength and decreased intensity of PL as the oxygen interstitial defect decays.

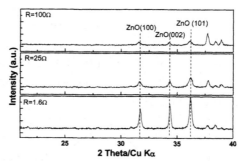

Figure 4 XRD spectra of the ZnO on zinc electrode synthesized under the fuel cell discharging at (A) resistance of 100 Ω for sample A, (b) resistance=25 Ω for sample B, and (c) resistance=1.6 Ω for sample C.

CONCLUSION

Electrochemical behavior of Zn anodes in a zinc/air fuel cell based chemical reactor for nano-roughened ZnO structure synthesis has been successfully demonstrated. The SEM results indicated that the nano-roughened ZnO structure, including morphology and structure size is dependent on the resistance. The sample grown controlled with resistance of 1.6 Ω have largest tip structure due to the strong or quick reaction promoting a larger quantity of $Zn(OH)_2$. Both XRD and room temperature PL spectra indicate that the ZnO crystallizes well and expands grain size could be created at lower resistance.

REFERENCES

1. Ü. Özgür, Ya. I. Alivov, C. Liu, A. Teke, M. A. Reshchikov, S. Doğan, V. Avrutin, S.-J. Cho, and H. Morkoç, *J. Appl. Phys.* vol. **98**, 041301 (2005) .

2. Jin-Ho Choy, Eue-Soon Jane, Jung-Hee Won, Jae-Hun Chung, Du-Jeon Jang, and Young-Woon Kim, *Adv. Mater.*, vol. **15**, pp. 1911 (2003).

3. Oleg Lupan, Lee Chow, Guangyu Chai, Beatriz Roldan, Ahmed Naitabdi, Alfons Schulte, and Helge Heinrich, *Mater. Sci. Eng.*, vol. **145**, pp. 57 (2007).

4. Chenglin Yan, and Dongfeng Xue, *J. Cryst. Growth*, vol. **310**, pp. 1836 (2008).

5. Anderson Janotti, and Chris G. Van de Walle, *J. Cryst. Growth*, vol. **287**, pp. 58 (2006).

6. A. A. Mohamad, *J. Power Sources*, vol. **159(1)**, pp.752 (2006).

Growth

Mater. Res. Soc. Symp. Proc. Vol. 1201 © 2010 Materials Research Society   1201-H06-02

Al₂O₃ as a Transition Layer for GaN and InGaN Growth on ZnO by MOCVD

Nola Li[1], Shen-Jie Wang[1], Will Fenwick[1], Andrew Melton[1], Chung-Lung Huang[2], Zhe Chuan Feng[2], Christopher Summers[3], Muhammad Jamil[1], and Ian Ferguson[4]

[1]Electrical and Computer Engineering, Georgia Institute of Technology, Atlanta, GA 30332

[2]Graduate Institute of Photonics and Optoelectronics, and Department of Electrical Engineering, National Taiwan University, Taipei, Taiwan 106-17, ROC

[3]Materials Science and Engineering, Georgia Institute of Technology, Atlanta, GA 30332

[4]Electrical and Computer Engineering, University of North Carolina Charlotte, Charlotte, North Carolina, 28223-0001

ABSTRACT

GaN and InGaN layers were grown on annealed 20 and 50nm Al_2O_3/ZnO substrates by metalorganic chemical vapor deposition (MOCVD). GaN was only observed by high resolution x-ray diffraction (HRXRD) on 20 nm Al_2O_3/ZnO substrates. Room temperature photoluminescence (RT-PL) showed the red shift of the GaN near band-edge emission, which might be from oxygen incorporation forming a shallow donor-related level in GaN. HRXRD measurements revealed that (0002) InGaN layers were also successfully grown on 20nm Al_2O_3/ZnO substrates. In addition, thick InGaN layers (~200-300nm) were successfully grown on Al_2O_3/ZnO and bare ZnO substrates. These results are significant as previous studies showed decomposition of the layer at InGaN thicknesses of 100nm or less.

INTRODUCTION

ZnO is an ideal substrate for epitaxial growth of GaN and InGaN. ZnO has the same wurtzite structure as GaN with a small c-plane lattice mismatch of only 1.8% [1,2]. It offers structural similarities to GaN over the non-native substrate materials, such as sapphire and SiC, which have large mismatches in lattice constant. InGaN, with a composition of 18% indium, has a perfect lattice-match with ZnO in the a-axis direction, which allows for the possible growth of InGaN layers without misfit dislocations [1-3]. Furthermore, ZnO and GaN have a similar thermal expansion coefficient which allows for almost zero thermal strain [4]. ZnO is a conductive material allowing it to be utilized in vertical structures to form electrodes on both sides. It is also easily etched chemically allowing for a thin GaN structure [5].

Metalorganic chemical vapor deposition (MOCVD) is the most widely used epitaxial growth method in industry; and therefore, studies in this report will be done by MOCVD. MOCVD growth is performed at high temperatures where ZnO substrates decompose causing diffusion of Zn and O into the epilayers. NH_3, as the nitrogen precursor, decomposes at a high temperature contributing H_2 to the atmosphere, which back etches the ZnO surface forming etched pits and surface peel off [6]. Previously, dimethylhydrazine (DMHy) was used as a nitrogen source for the GaN growth on sapphire due to higher cracking efficiency at low temperature (420°C) compared with NH_3 (15% decomposes at 950°C) [7, 8]. More recently, a ZnO film was investigated as a buffer layer for GaN growth on sapphire by MOCVD at a relatively low temperature using DMHy [5]. However, low temperature decomposition of DMHy was found to cause high levels of carbon contamination making high quality GaN growth difficult [7]. Hence, high temperature growth of GaN on ZnO substrates with NH_3 as a nitrogen source is an issue that needs to be addressed. Zn diffusion is another problem, which has been demonstrated by secondary ion mass spectrometry (SIMS) depth profile of the GaN/ZnO interface [9, 10]. The diffusion can cause poor epitaxial growth and degrade the film quality. It has been reported that the Zn doping in the GaN layer forms a deep level, red-shifting the bandedge by 0.5eV [11]. Moreover, the same observation has also been reported for Zn diffusion into InGaN layers from the ZnO substrate [3].

In this study, atomic layer deposition (ALD) is used to provide an Al_2O_3 transition layer on ZnO substrates before GaN or InGaN growth in order to prevent Zn and O diffusion, protect the ZnO surface from H_2 back etching due to pyrolysis of NH_3, and promote nitride growth. Al_2O_3 was chosen as the transition layer due to its excellent thermal stability and because it is a readily available material that can be deposited uniformly by ALD. ALD was chosen as the growth method due to its ability to obtain a smooth surface and more accurate thickness control on ZnO. This method allows for smooth layer deposition with low pinhole density and good uniformity over large area substrates. Furthermore, the thickness is accurately controlled just by the number of growth cycles rather than temperature, etc [12]. Numerous studies have reported that amorphous Al_2O_3 films grown on Si substrates by ALD can be transformed to crystalline Al_2O_3 films [12, 13]. So far, to our knowledge, there are no papers that refer to Al_2O_3 films grown on ZnO substrates as a preparation layer for epitaxial GaN. This paper will address GaN growth on 20 and 50nm Al_2O_3/ZnO substrates as well as InGaN growth on 20nm Al_2O_3/ZnO substrates. Thick InGaN layers of ~200-300nm were also grown for the first time on Al_2O_3/ZnO substrates.

EXPERIMENTAL DETAILS

Al_2O_3 films were grown on the Zn face of ZnO (0001) substrates by ALD at a temperature of 100°C with TMAl and H_2O as the precursors. The film growth took place in a cyclic manner at a base pressure of 0.5Torr. The thickness of the Al_2O_3 film used in this study was 20 and 50nm. Post-annealing of the samples was performed in a furnace at 1100°C at different crystallization times in a N_2 ambient.

The GaN layers (buffer plus main layer) were grown with identical growth conditions on annealed 20 and 50nm Al_2O_3/ZnO substrates by MOCVD in a modified commercial rotating disk reactor with dual injector blocks. The GaN buffer layer for the GaN epitaxial growth was grown at 770°C with a thickness of 30nm using trimethylgallium (TMGa) and NH_3 as the gallium and nitrogen sources, respectively. Two investigated GaN epi-layers were grown at a thickness of 200nm at different growth temperatures of 800 and 850°C at a growth rate of 0.25 and 0.4µm/h, respectively.

InGaN layers were grown by MOCVD on 20nm Al_2O_3/ZnO substrates with various annealing times. The GaN buffer layer for InGaN epitaxy was grown at 530°C with a thickness of 30nm using TMGa and NH₃ as the gallium and nitrogen sources, respectively. InGaN layers were grown at a thickness of 100nm at 700°C with a growth rate of 0.18μm/h by introducing trimethylindium (TMIn) and triethylgallium (TEGa) into the reactor. N₂ carrier gas was used during the whole growth process. The structures were characterized by high resolution x-ray diffraction (HRXRD). Optical properties were measured by room temperature photoluminescence (RT-PL) with a HeCd laser of a 325nm monochromatic beam.

DISCUSSION

Epitaxial growth of GaN layers on Al₂O₃/ZnO substrates

The 2θ/w HRXRD scan for the epitaxial GaN layers grown at 800°C on 20 and 50nm Al₂O₃/ZnO substrates with 10 and 20 minutes annealing are shown in Figure 1(a). The peaks with the higher intensity are the diffraction from the ZnO (0002) plane of ZnO bulk materials. Moreover, a second peak can only be seen in the 20nm sample. This result shows that the (0002) plane of the wurtzite GaN was successfully grown on the 20nm Al₂O₃/ZnO substrate. However, the growth of a GaN layer is not realized on the 50nm Al₂O₃/ZnO substrate. It is possible that the thicker Al₂O₃ film has a weaker structural relationship with the underlying ZnO substrate after crystallization. Therefore, the polycrystalline 20nm α- Al₂O₃ may possess highly textured orientation with hexagonal ZnO compared to 50nm samples to assist the nucleation of the GaN layer at the initial stage.

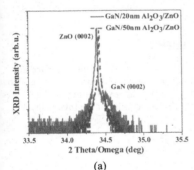 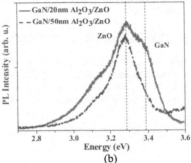

(a)   (b)

Figure 1. (a) HRXRD 2θ/ω scan and (b) RT-PL spectra of GaN layers grown at 800°C on 20 and 50nm annealed Al₂O₃/ZnO substrates by MOCVD.

RT-PL was performed for GaN layers grown at 800°C on 20 and 50nm Al₂O₃/ZnO substrates, Figure 1(b). The emission of ZnO substrates used for GaN growth is at 3.28eV, which is the same value as the emission of bare ZnO substrate in our experiment. Only a shoulder appeared at 3.39eV in the 20nm sample which is attributed to the near-edge emission of GaN. This emission is a 30meV red-shift from the typical band-edge emission of GaN at 3.42eV. Moreover, a significant near-edge emission of GaN for 850°C growth on the 20nm sample is found to be around 3.36eV; whereas, the 50nm sample does not show any GaN peak,

Figure 2. This 20nm sample still shows a red-shift of 60meV. It was reported that the zinc doping into the GaN or zinc diffusion into the InGaN from the ZnO substrate forms a deep level inducing a red-shift emission of ~0.4-0.5eV [11, 14, 15]. Furthermore, oxygen substituting nitrogen on lattice sites has been believed to act as a shallow donor in unintentionally doped GaN [16]. It is possible that the oxygen in the GaN layers is from the ZnO substrate as ZnO is highly unstable at high temperatures. Therefore, these slight red-shifts may be caused by oxygen contamination in the GaN layers instead of zinc. As a result, the oxygen diffusion into GaN could form an oxygen-related shallow donor level below the conduction band that induces the red-shift.

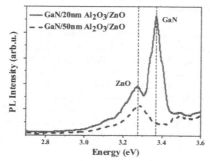

Figure 2. RT-PL spectra of GaN layers grown at 850°C on 20 and 50nm annealed Al_2O_3/ZnO substrates by MOCVD.

Epitaxial growth of $In_xGa_{1-x}N$ layers on Al_2O_3/ZnO substrates

InGaN (0002) diffraction peaks were obtained on 20nm Al_2O_3/ZnO substrates annealed at 1100°C for various annealing times of 0, 10, 20 and 40 minutes, as seen in Figure 3. It was also seen that the intensity of the InGaN peaks decreased with increasing annealing time. The annealing time of the Al_2O_3/ZnO substrates was found to affect the quality of the InGaN layer.

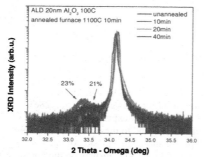

Figure 3. HRXRD $2\theta/\omega$ scan of InGaN layers grown on 20nm Al_2O_3/ZnO substrates at 1100°C with various annealing times for 0, 10, 20, and 40 minutes.

Moreover, no InGaN (0002) peak was seen on the un-annealed sample. The highest InGaN peak was observed for the annealing time of 10 minutes, which possessed the highest quantity of Al_2O_3 crystallization peaks as seen by HRXRD. Shifts seen in the InGaN peaks denote different indium incorporation. This could be due to the different surfaces being grown on, which will change the nucleation density as well as the growth mode of the subsequent GaN and InGaN epilayer. Therefore, the indium composition can be changed due to the surface differences. The detail mechanism still needs to be studied.

MOCVD growth of thick $In_xGa_{1-x}N$ on Al_2O_3/ZnO substrates

In addition, thick InGaN layers (up to 300nm) were successfully grown on Al_2O_3/ZnO and bare ZnO substrates by MOCVD. A LT-GaN buffer (580°C) was grown prior to the thick InGaN growth. These layers are a significant improvement over previous work, which were limited to very thin epilayers (<100nm). The results for thick InGaN grown on Al_2O_3/ZnO are shown below. HRXRD for a 300nm InGaN layer on Al_2O_3/ZnO is shown in Figure 4(a). An InGaN peak at ~33.5 degrees is similar to what was seen for the 300nm InGaN grown on a bare ZnO substrate. However, the InN peak at ~31.5 degrees is reduced by half a magnitude compared to growth on a bare ZnO substrate. RT-PL for a typical InGaN epilayer on Al_2O_3/ZnO is shown in Figure 4(b). The peak at ~2.3eV shows a more pronounced InGaN peak than previous growths on bare ZnO substrates had shown.

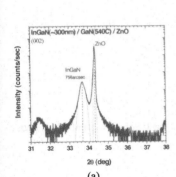 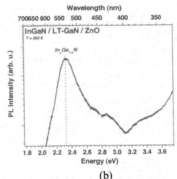

(a)   (b)

Figure 4. (a) HRXRD and (b) PL of a thick InGaN layer on ALD-Al_2O_3/ZnO.

CONCLUSIONS

This study presented the MOCVD growth of GaN layers on annealed Al_2O_3/ZnO substrates by using NH_3 as a nitrogen source at high growth temperature. HRXRD identified that GaN was only seen on the 20nm Al_2O_3/ZnO substrate and not on the 50nm Al_2O_3/ZnO substrate. RT-PL showed the red-shift of the GaN near band-edge emission, which may be caused by oxygen diffusion into the GaN layer forming an oxygen-related level of a shallow donor. InGaN layers of 100nm were also successfully grown on 20nm Al_2O_3/ZnO substrates after

10min annealing at 1100°C in a high temperature furnace. In addition, thick InGaN layers of ~200-300nm were also grown on ZnO substrates. Results showed distinct peaks from the thick InGaN layers. The increased thickness of the layers showed a significant improvement over previous InGaN growths, which could only grow InGaN layers of 100nm or less. The ALD-grown Al_2O_3 interlayer is a promising layer for GaN growths on ZnO substrates.

ACKNOWLEDGMENTS

This work was supported by the Department of Energy (DE-FC26-06NT42856). The work at National Taiwan University was supported partially by funds of NSC 95-2221-E-002-118- and NSC96-2221-E002-166. ZnO substrates were provided by CERMET Inc.

REFERENCES

1. A. Kobayashi, J. Ohta, H. Fujioka, J. Appl. Phys. **99**, 123513 (2006).
2. G. Namkoong, S. Burnham, K. Lee, E. Trybus, W. A. Doolittle, M. Losurdo, P. Capezzuto, G. Bruno, B. Nemeth, J. Nause, Appl. Phys. Lett. **87**, 184104 (2005).
3. N. Li, S. J. Wang, E. H. Park, S. C. Lien, Z. C. Feng, A. Valencia, J. Nause, and I. Ferguson, J. Appl. Phys. **102**, 106105 (2007).
4. F. Hamdani, M.Y., D. Smith, H. Tang, W. Kim, A. Salvador, A.E. Botchkarev, J.M. Gibson, A.Y. Polyakov, M. Skowronski, and H. Morkoc, J. Appl. Phys. **83**, 2 (1998).
5. D.J. Rogers, F.H. Teherani, A. Ougazzaden, S. Gautier, L. Divay, A. Lusson, O. Durand, F. Wyczisk, G. Garry, T. Monteiro, M.R. Correira, M. Peres, A. Neves, D. McGrouther, J.N. Chapman, M. Razeghi, Appl. Phys. Lett. **91**, 071120 (2007).
6. N. Li, E. H. Park, Y. Huang, S. J. Wang, A. Valencia, B. Nemeth, J. Nause, and I. Ferguson Proc. SPIE **6337**, 63370Z (2006).
7. E. H. Park, J. S. Park, and T. K. Yoo, J. Cryst. Growth **272**, 426 (2004).
8. R.T. Lee and G.B. Stringfellow, J. Elect. Mater. **28**, 963 (1999).
9. W.K. Popovici, A. Botchkarev, H. Tang, H. Morkoc, and J. Solomon, Appl. Phys. Lett. **71**, 23 (1997).
10. C.H. Suzuki, H. Goto, T. Minegishi, A. Setiawan, H. J. Ko, M. W. Cho, and T. Yao, Curr. Appl. Phys. **4**, 643 (2004).
11. S. Nakamura, J. Cryst. Growth **145**, 911 (1994).
12. S. Jakschik, Thin Solid Films **425**, 216-220 (2003).
13. L Zhang, H C Jiang, C Liu, J W Dong, and P Chow, Journal of Physics D: Applied Physics **40**, 3707-3713 (2007).
14. N. Nepal, M.L. Nakarmi, H.U. Jang, J.Y. Lin, H.X. Jiang, Appl. Phys. Lett. **89**, 192111 (2006).
15. S.J. Wang, N. Li, E.H. Park, S.C. Lien, Z.C. Feng, A. Valencia, J. Nause, I. Ferguson, J. Appl. Phys. **102**, 106105 (2007).
16. J. Neugebauer and C.G. Van de Walle, Phys. Rev. B **50**, 8067 (1994).

Mater. Res. Soc. Symp. Proc. Vol. 1201 © 2010 Materials Research Society   1201-H06-07

Crystal polarity control of ZnO films and nonlinear optical response

Jinsub Park[1,2], Y. Yamazaki [2], Y. Takahashi [2], T. Minegishi[1], S. H. Park[1], S. K. Hong[3], J. H. Chang[4], T. Fujiwara[2] and T. Yao[1]

1 Center for Interdisciplinary Research, Tohoku University, Aramaki Aoba 6-3, Aoba-ku, Sendai 980-8578, Japan

2 Department of Applied Physics, Tohoku University 6-6-05, Aoba, Sendai 980-8579, Japan

3 School of Nanoscience and Technology, Chungnam National University, Daejeon 305-764, South Korea

4 Major of Nano Semiconductor, Korea Martine University, Pusan 606-791, South Korea

ABSTRACT

We report that the nonlinear optical response of polarity controlled ZnO films grown by selective growth technique of Zn-polar and O-polar ZnO layers on sapphire substrate using Cr-compound buffer layers. ZnO layers grown on CrN/sapphire show Zn polar, while those grown on Cr_2O_3/sapphire result in O-polar ZnO films. In order to verify the origin of nonlinear optical response of ZnO, the polarity-controlled ZnO thin films grown on different buffer layers were investigated as nonlinear optical materials for second harmonic generation (SHG). The effective nonlinear optical coefficient (d_{eff}) of ZnO grown on Cr-compound buffer layers showed a higher value than that of ZnO grown on MgO buffer layers. Finally, by combining suggested in-situ polarity control technique with photolithography technique, we have fabricated 1D and 2D periodically-polarity-inverted (PPI) heterostructures with periodicity ranging from 60 µm to 4 µm. The lateral polarity inversion is confirmed by piezo response microscopy. Such PPI ZnO heterostructures show the enhancement of SHG intensity comparing with the ZnO films.

INTRODUCTION

In terms of crystal structures, ZnO is expected to have nonzero second-order susceptibility due to the lack of inversion symmetry in wurtzite crystal structures. Recently, ZnO has attracted interest as a nonlinear optical material with high nonlinear optical response [1,2]. In fact, there has been a continuous requirement of nonlinear optical (NLO) materials for potential applications in integrated optics [3]. Among the various approaches, the periodical poling method involving $LiNbO_3$, $LiTaO_3$, and $KTiOPO_4$ is paving the way toward the development of second-order nonlinear optical devices [4,5]. However, such bulk materials have drawbacks such as low flexibility and high costs in real applications. Therefore, research and development of thin films with high nonlinear optical response are important for understanding the characteristics of response of materials to nonlinear optical properties and its applications to nonlinear optical devices. Measurements of second harmonic generation (SHG) were conducted in ZnO crystals [6], thin films,[1,2,7] and nanostructures [8,9]. Although many variations of SHG in ZnO materials have been reported in literatures [5–9], such as the deposition conditions[7], films thickness[2], grain shape [10], and orientation of the crystallites[11], detailed analyses of nonlinear optical response properties and the origin of SHG have been controversial and have not yet yielded satisfactory results.   One of the reasons for the controversy would result from the samples with not well controlled polarities. It is obvious that ZnO films with well-defined crystal polarity are prerequisite to understand the origin of the nonlinear optical response in thin films.

We have previously reported a method for controlling the polarity of ZnO films on c-Al_2O_3 substrates by using Cr-compound buffer layers [12]. The ZnO films grown on rocksalt structure (111) CrN showed Zn polarity, while those grown on rhombohedral (0001) Cr_2O_3 showed O polarity. On the basis of the polarity control methods, the fabrication of quasi-1D periodically polarity-inverted (PPI) ZnO structures with various combinations of Zn- and O-polar regions was demonstrated [13,14,15]. Thus far, the suggested PPI ZnO structures are (1) Zn-ZnO/CrN and O-ZnO/Al_2O_3 [13,14] and (2) Zn-ZnO/MgO and O-ZnO/Al_2O_3 [15]. In this paper, we report the fabrication of periodically and laterally polarity-inverted ZnO structures by using Zn-ZnO/CrN and O-ZnO/Cr_2O_3 buffer layers. The fabrication of ZnO-based 1D and 2D PPI structures is expected to result in their various applications to optical devices such as nonlinear optical devices via quasi-phase-matched waveguides or photonic crystals [16,17].

EXPERIMENT

Polarity-controlled ZnO films with different buffer layers were grown on commercial, two-side polished (0001) Al_2O_3 substrates by means of plasma assisted molecular beam epitaxy (PA MBE). In order to compare the nonlinear optical response of ZnO films with different polarity and buffer layers, we prepared four types of ZnO samples: (i) Zn-polar ZnO films grown on (S1) 7 nm-thick MgO and (S2) CrN buffer layer, and (ii) O-polar ZnO films grown on (S3) 1.5 nm-thick MgO and (S4) Cr_2O_3 buffer layers.

Selective growths of Zn-polar and O-polar ZnO were achieved on (0001) Al_2O_3 by introducing MgO and Cr-compound buffer layers[12,18]. The polarity of ZnO films grown on rocksalt structure CrN was determined to be Zn polar, and O-polar ZnO films are grown on rhombohedral Cr_2O_3 layers [12]. In addition, Zn-polar ZnO growth occurred on MgO layers with thickness greater than 3 nm, whereas O-polarity growth occurred on MgO layers with thickness less than 2 nm [18].

We used Cr, Mg, nitrogen, and oxygen or nitrogen plasma as sources for CrN and MgO buffer layers on c–plane sapphire. Cr_2O_3 layers were formed through oxidation of CrN/ c–plane sapphire by exposing oxygen plasma at 650°C for 10 min. ZnO films were grown at 750°C by using Zn cell and oxygen plasma.

Based on the suggested polarity control method, in order to fabricate the PPI ZnO structure on c-sapphire substrate, growth and patterning sequence with lithography and etching steps as illustrated in Fig. 1 are conducted. For the preparation of template, a CrN was grown on (0001) Al_2O_3. At first, 15 nm-thick low temperature (LT) ZnO layers are grown on a Zn exposed CrN/Al_2O_3 templates by employing the standard growth procedures reported elsewhere [12]. This step shall result in the growth of Zn-polar ZnO on the CrN intermediate layer. The second step for the fabrication of a PPI ZnO structure is a periodical patterning on the LT ZnO. Standard photolithography is conducted in order to make stripes along the ZnO [11-20] direction. The directions of stripes are selected by consideration of flat surface and etching rate. Periods of the stripe patterns were varied from 4 µm to 60 µm. Next, wet etching was conducted, which completely etched the LT ZnO till the CrN layers. This process made opening of CrN layers through the stripes as shown in Fig. 1(b). In order to form the Cr_2O_3, O_2 plasma irradiation was conducted which is essential for the O-polar ZnO growth. Finally, a high temperature (HT) ZnO layer was grown on LT ZnO with suppression of rotational domain at 750°C, which shall grow the Zn-polar ZnO on CrN/Al_2O_3 and the O-polar ZnO on $Cr_2O_3/CrN/sapphire$. The surface morphology and period of 1D and 2D-patterned ZnO films were evaluated by contact mode atomic force microscopy (AFM) with an Au-coated Si tip. Periodical polarity inversion was confirmed not only by measuring the surface voltage-piezo response (V-Z) curves but also by observing piezo response microscopy (PRM) images. The second-order nonlinear optical coefficient was determined by employing the Maker fringes technique of measuring the incident-angular-dependant SH signal in the transmission mode [19]. In our study, a 1.06-μm output of a Q-switched Nd:YAG laser (repetition rate = 10 Hz; pulse width = 8 ns) was used as the fundamental beam, and the two half-wave plates were used to control the polarization of the fundamental and second harmonic beams.

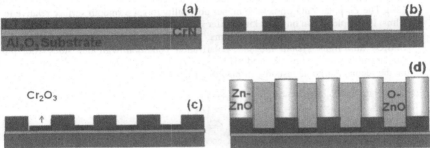

Figure 1. Fabrication procedure for the PPI ZnO structure. (a) LT ZnO growth on the Zn pre-exposed CrN buffer. (b) Stripe patterned ZnO by photolithography and etching. (c) O_2 plasma treatment for the formation of Cr_2O_3 on CrN. (d) ZnO re-growth with the Zn-polar on the CrN and O-polar ZnO on the Cr_2O_3, resulting in the fabrication of the PPI ZnO structure.

DISCUSSION

Nonlinear optical response for polarity controlled ZnO films

Figure 2(a) and (b) show the intensity of second harmonic signal which are obtained in the transmission mode from the ZnO with different buffer layers as a function of the incident angle of the fundamental beam. The Zn-polar films show the 2.5 times higher intensity than those of the O-polar films in both buffer layer i.e. MgO and Cr-compound buffer layers. In terms of buffer layer effect without the consideration of polarity, the absolute intensity of SH signal of ZnO films grown on Cr-compound buffer layers is higher than that of ZnO grown on MgO buffer layers. In current stage, the reason of the difference of SH signal intensity by the kind of buffer layers is not clear but it can be expect to be used in NLO materials of ZnO films grown on Cr-compound buffer layers.

To get into inside of the polarity dependence of ZnO films on the nonlinear optical response, the measurements of the SHG signal from a ZnO single crystal grown hydrothermally (Tokyo

Denpa) with two-side polished Zn-face and O-face are conducted. The SH intensity of both face ZnO crystal did not show the distinguished difference (not show here). Therefore, we think that the nonlinear optical response of thin films has a different mechanism from the single crystal. The speculation of stacking sequence fault generation model by flipping of stacking sequence during the polar surface growth was suggested by Wang *et al.* They reported that the second-order susceptibility for ZnO films depend on thickness of films [2]. The reduction in the second-order susceptibilities of thick films can be explained by the model in which the polar axis, which is initially aligned with respect to the substrate, flips as the increase of films thickness. Unfortunately, this simple explanation can not support our results because ZnO films have different polarity and buffer layers. For the possible explanation, let us to discuss the surface morphology dependence on the polarity of ZnO films. Yu *et al.* suggested that the easy island nucleation contributes to the smooth morphology in Zn-polar films while the reduced nucleation kinetics on the O-polar surface causes the three-dimensional island growth by phase-field model [20]. The Zn-polar ZnO films have higher possibility to get a smooth surface relative to O-polar films. In sample grown on same kind buffer layer, the Zn-polar surface shows the smooth surface comparing with that of the O-polar films. The rough surface decreases the SHG intensity in Si/SiO_2 thin films system in same polarization condition with our study (i.e, p-polarized fundamental beam and p-polarized SH beam)[21]. Therefore, the surface morphology can effect on the second harmonic response in ZnO thin films.

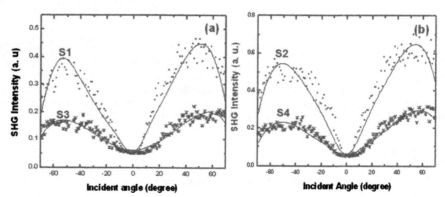

Figure 2. Second harmonic signal intensity from (a) S1(closed circles)and S3(crosses) and (b) S2(closed circles) and S4 (crosses) as a function of the incident angle of the fundamental beam, when the fundamental and SH beam are p-polarized. The solid curves correspond to theoretical fitting.

In order to find out the correlation between crystallinity and effective nonlinear optical coefficient (d_{eff}), the comparison of the full width at half maximums (FWHMs) of (0002) and (10-11) XRD ω-rocking curves (XRCs) are accomplished as shown in the Fig. 3. of XRD ω-rocking curves from (0002) plane (a) and (10-11) plane of ZnO (b). The name of sample is denoted at each point. The calculated typical value of deff for ZnO samples are 3.98±0.2 and 4.76±0.2 for the Zn-polar ZnO and 3.31±0.2 and 4.74±0.2 for the O-polar ZnO with MgO and Cr-compound buffer layer, respectively. Threading dislocations in (0001) ZnO typically form as edge, screw, or mixed component dislocations with respective Burgers vectors b = 1/3 <11-20>, <0001>, and 1/3 <11-23> [22]. Here, (0002) XRCs related the screw component dislocation along with burgers vector b=±[000c] and (10-11) XRCs related to the edge component dislocation along the b= a/3<11-20>.

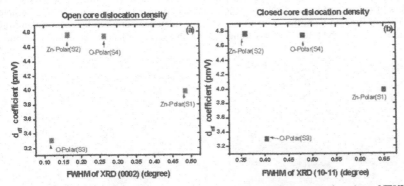

Figure 3. Dependence of effective second order nonlinear coefficient as a function of FWHM (a) (0002) and (b) (10-11) plane of ZnO

In terms of interactions between X-ray and crystal, the FWHM off (0002) XRD of ZnO film is affected by the tilt component dominated by the screw dislocations. As following this concept, the interactions between fundamental wave and plane of ZnO films will be increased resulting in increase the possibility to produce the second order harmonic.

Moreover, Elsner *et al.* reported that the screw dislocations exist as open-core dislocations, whereas the threading edge dislocations have filled cores in wurtzite structured GaN, which is structurally equivalent to ZnO, with theoretical calculation by density functional calculation [23]. In terms of surface or interface area, the large open core threading dislocations density makes more surface area in the grains than threading edge dislocation. Moreover, the c- axis aligned open core dislocation line will increase the interaction with the incident fundamental beam. Therefore, the increase of the interface or surface added up by open core defect line strong effects on the second harmonic signal from thin films. This consideration well explains the more strong correlation of FWHM of (0002) XRCs with the SH signal than that of (10-11) XRC of ZnO films. In the case of ZnO on MgO buffers, the dependence of the SH coefficient on the crystal quality well agreed with the previously reported results that the crystal quality have inverse relationship with the nonlinear optical response of films [1]. In contrary to this, ZnO on Cr-compound buffers, there is no correlation crystallinity with nonlinear optical response. Without consideration of different buffers, the XRCs of (0002) ZnO shows the slight dependence on the effective SH coefficient but the XRCs of (10-11) ZnO don't show the distinguish dependency. These results can be explained by the generation of surface or interface area by generation of threading dislocations.

In another possible explanation, the scattering possibility of light with the dislocation will be increased at some maximum point. Therefore, the intensity of SH signal decreased by scattering

189

of the light with the dislocations over the maximum point which can explain the decrease of the S1 sample.

Figure 4 shows a reciprocal space mapping of (0002) diffraction spots of a ZnO films grown on (0001) Al$_2$O$_3$ substrate with MgO (Fig. 4(a)) and CrN buffer layer (Fig. 4(b)), respectively. The thickness of buffer layers (i.e., CrN and MgO) and ZnO films are almost same in both samples. In the case of ZnO on CrN, the growth axis is a little titled comparing to that of ZnO on MgO. As mentions before, the tilting of films along the growth axis can effect on the SH generation.

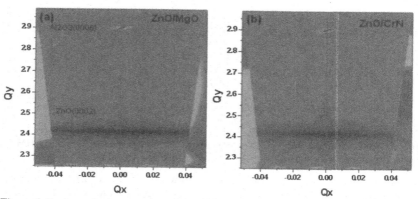

Figure 4. Reciprocal space mapping of the (0002) diffraction spots of ZnO films grown on MgO (a) and CrN buffer (b), respectively.

In previous, we discussed the surface morphology dependence on the polarity of ZnO films which effects on the nonlinear optical response of ZnO films. As suggested the high possibility of the smoother morphology in Zn-polar films relative to O-polar films. In term of surface morphology of ZnO films without the consideration of polarity, we compare the dependence of d_{eff} as a function of surface root mean square (RMS) values as shown in the figure 5(a). For SH d_{eff} coefficient curve, the strong correlation between surface roughness addressed by RMS value and d_{eff} are shown in Fig. 5(a). The second order susceptibility of ZnO with flat surface is larger than that of ZnO with rough surface. In here, we want to discuss the effect of the surface scattering and defects as like stacking fault of ZnO films. The relatively large scattering loss of SHG in AlGaAs is reported that unequal domain dimensions, and grain boundary in surface effects are all accountable for reduction in the SHG conversion efficiency [24]. Therefore, the flat surface and interface roughness is important in SHG. Moreover, the SH intensity of ZnO films are affected by the fundamental wave intensity as like the equation 1.

$$I_{2\omega}(\theta) = \frac{128\pi^3}{cA} \frac{(t_{af}^{1y})^4 (t_{fs}^{2p})^2 (t_{sa}^{2p})^2}{(n_{2\omega}\cos\theta_{2\omega})^2} \times I_\omega^2 (\frac{2\pi L}{\lambda})^2 (\chi_{eff}^{(2)})^2 \frac{SIN^2\Phi}{\Phi^2}$$

(1)

$$\Phi = \frac{2\pi L}{\lambda}(n_w \times \cos\theta_\omega - n_{2w}\cos\theta_{2\omega})$$

here, and $t_{af}^{1y}, t_{fs}^{2p} \cdot t_{sa}^{2p}$ are the transmission coefficient of the fundamental beam from air to film, film to substrate, and substrate to air, respectively.

From the above equation 1, the intensity of SH signal can be reduced by reduction of fundamental wave intensity resulting from surface scattering.

In second, the possibility of generation of stacking defaults and planar defect in films become lower in film with flat surface relative to rough surface. Therefore, the strong response of ZnO films with lower surface RMS value related to the stacking faults generation. This result well agreed with the crystal quality dependence.

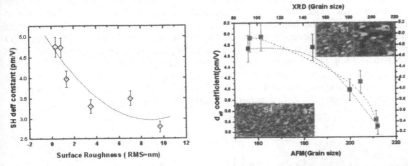

Figure 5. Dependence of effective second order nonlinear coefficient as a function of (a) surface roughness addressed by RMS values and (b) grain size measured by AFM (low x-axis, blue dot and line) and XRD scan of 2θ-ω (upper x-axis, red dot and line). The inset of figure shows the surface image of each ZnO films.

Figure 5 (b) shows the measured d_{eff} as a function of grain size of different ZnO films. For the determination of grain size, two kinds of measurement were conducted. One is the surface grain size from the ZnO surface by investigation with the atomic force microscopy (AFM), the other is calculation from XRD 2θ-θ measurement by using the Scherrer's equation ($t = 0.9\lambda / B\cos\theta$: here, t=domain size, B= FWHM, θ=brag angle, and λ=1.54056 Å). The average grain size for all the films are in the range 100~190 nm. Surface dimensions determined by AFM are in the 150~200 nm range. This value is consistent with the result of above XRD. The AFM surface images of each ZnO films are shown in the inset of figure 5(b). For both curves, the correlation between grain size and deff well agreed with the same tendency. The second order susceptibility of ZnO with smaller grain size is larger than that of ZnO with larger grain size. Xiang *et al.* showed that smaller nanowires to have larger piezoelectric response constants due to their large surface to volume ratio [25]. Our experimental result strongly implies that the relatively large amount of interface between grains significantly effect on the second order response of ZnO films with assumption of almost same responsibility in bulk characteristics.

In terms of different growth rate depending on polarity, Zn-face ZnO films shows 1.5 times higher growth rate than O-face one [12]. Therefore, it can be expected that the grain size of Zn-face ZnO film is smaller than that of O-face films, since higher growth rate means shorter migration length of adatom under the same supplying ratio of species.

Lateral polarity inversion in PPI ZnO structures

In order to assess the periodical polarity inversion, PRM measurements were performed using a Seiko Instruments SPA400 attached with atomic force microscope (AFM). PRM technique has been successfully used to determine the lateral polarity change in PPI GaN structures with patterned Ga- and N-face polarities [26]. The ZnO films will be expanded when a positive electric field is applied along the [0001] direction, while contracted when reversing the direction of the electric field, which results in contrast change of PRM images. The image of piezoelectric response for the fabricated structure clearly showed the different brightness as shown in Fig. 6(a). A line scan along the yellow line of the PRM images of the PPI ZnO structures in Fig. 6(a) revealed the periodical change of response voltage (phase) by applied AC voltage as shown in Fig. 6(b). Here, the brighter region and darker region correspond to the O-polar and the Zn-polar regions, respectively. The contrast of PRM image and the piezo response voltage profile are clear evidences for the successful formation of array and grating by using the periodically selective growth of ZnO structure with a flipping of polarity.

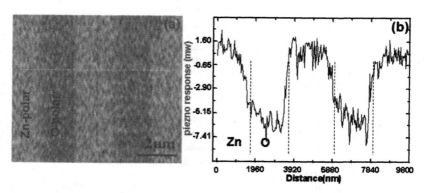

Figure 6. (a) Piezo-response microscopy images of Zn- and O-polar regions in the 1D PPI ZnO structures and (b) corresponding cross-sectional piezo response voltage profile.

Enhancement of second harmonic generation in PPI ZnO

In order to compare the SH signal intensity of ZnO films and PPI ZnO structures, the same experimental measurements are conducted with transmission mode. Figure 7 shows the SH intensity variation as a function of incident angle from the PPI ZnO structures and Zn- and O-polar films grown on Cr-compound buffer layers. The enhancement of SH intensity from 1D PPI ZnO structures without the consideration of periodicity which changed from 4 to 60 μm.

192

From these results, we can clearly know that the SH intensity from PPI ZnO has been enhanced relative to the ZnO films. Moreover, the dominant enhancement comes from the Zn-polar ZnO films grown on CrN buffer layers.

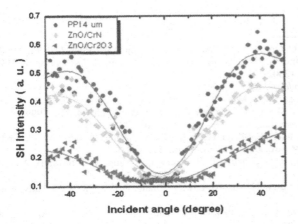

Figure 7. The dependence of SH Intensity from PPI ZnO and polarity controlled ZnO films as a function of incident angle. PPI ZnO with 4 μm periodicity.

CONCLUSIONS

The nonlinear optical response of polarity controlled ZnO thin films grown by suggested selective polarity control methods were investigated. Effective second order nonlinear optical coefficients of ZnO grown on Cr-compound buffer layer are higher than those of one grown on MgO buffer layers. The crystallinity of films shows the slight correlation with nonlinear susceptibility, especially, the FWHM of XRD from symmetry plane has more strong relationship relative to the asymmetry plane. Correlation was found between the grain size and the values of d_{eff} as the d_{eff} of ZnO film increased with the decreased the grain size. Finally, we demonstrated the PPI ZnO structures by using the in-situ polarity control methods. The SH intensity from PPI ZnO has been enhanced relative to the ZnO films which means the PPI ZnO can be used in applications to nonlinear optical devices including SHG and photonic crystals.

ACKNOWLEDGMENTS

This study was partially supported by the Research Fellowships for Young Scientists Program of the Japan Society for the Promotion of Science (JSPS).

REFERENCES

1. H. Cao, J. Y. Wu, H. C. Ong, J. Y. Dai, and R. P. H. Chang, *Appl. Phys. Lett.* **73**, 572 (1998).
2. G. Wang, G. T. Kiehne, G. K. L. Wong, and J. B. Ketterson, X. Liu, and R. P. H. Chang, *Appl. Phys. Lett.* **80**, 401 (2002).
3. M. M. Fejer, *Phys. Today* **47** (**5**), 25 (1994).
4. S. N. Zhu, H. Y. Zhu, and N. B. Ming. *Science* **278**, 843 (1997).
5. K. Fradkin, A. Arie, A. Skliar, and G. Rosenman, Appl. Phys. Lett. **74**, 914 (1999).
6. G. Wang, G. K. L. Wong, J. B. Ketterson, *Appl. Opt.* **40**, 5436 (2001).
7. M. C. Larciprete, D. Passeri, F. Michelotti, S. Paolini, C. Sibilia, M. Bertolotti, A. Eraldini, F. Sarto, F. Somma, S. Lo Mastro, *J. Appl. Phys.* **97**, 023501 (2005).
8. J. C. Johnson, H. Yan, R. D. Schaller, P. B. Peterson, P. Yang, R., J. Saykally, *Nano Lett.* **2**, 279 (2002).
9. J. C. Johnson, K. P. Knutsen, H. Yan, M. Law, Y. Zhang, P. Yang, R., J. Saykally, *Nano Lett.* **4**, 197 (2004).
10. U. Neumann, G. Gruwald, U. Griebner, G. Steoinmeyer, W. Seeber. *Appl. Phys. Lett.* **84**, 170 (2004).
11. U. Neumann, R. Grunwald, U. Griebner, G. Steinmeyer, M. Schmidbauer, W. Seeber, *Appl. Phys. Lett.* **87**, 171108 (2005).
12. J. S. Park, S. K. Hong, T. Minegishi, S. H. Park, I. H. Im, T. Hanada, M. W. Cho, and T. Yao, J. W. Lee, and J. Y. Lee, *Appl. Phys. Lett.* **90**, 201907 (2007).
13. J.S. Park, J.H. Chang, T. Minegishi, H.J. Lee, S.H. Park, I.H. Im, T. Hanada, S.K. Hong, M.W. Cho, and T. Yao, *J. Electron Mater.* **37**, 736 (2008).
14. T. Minegishi, A. Lshizawa, J. Kim, S. Ahn, S.Park, J. Park, and I. Im, D.C.Oh, H. nakano, K. Fujii, H. Jeon, and T. Yao, *J. Vac. Sci. Technol. B* **26(3)** 1120 (2008).
15. J. S. Park, T. Minegishi, S. H. Lee, I. H. Im, S. H. Park, T. Hanada, T. Goto, M. W. Cho,T. Yao, S.K. Hong, and J.H. Chang, *J. Vac. Sci. Technol. A* **26(1)**, 90 (2008).
16. A. Chowdhury, H. M. Ng, M. Bhardwaj, and N. G. Weimann, *Appl Phys. Lett.* **83**, 1077 (2003).
17. R. Katayam, Y. Kuge, T. Kondo, K. Onabe, *J. Cryst. Growth* **301-302**, 447 (2007).
18. H. Kato, K. Miyamoto, M. Sano, and T. Yao, *Appl. Phys. Lett.* **84**, 4562 (2004).
19. P. D. Maker, R. W. Terhune, M. Nisenholf, and C. M. Savage, Phys. Rev. Lett. **8**, 21 (1962).
20. Y. M. Yu and B. G. Liu, *Phys. Rev. B* **77**, 195327 (2008).
21. J. I. Daap, B. Doris, Q. Deng, M. C. Downer, J. K. Lowell, A. C. Deibold, *Appl. Phys. Lett.* **64**, 2139 (1994).
22. M. W. Cho, A. Setiawan, H. J. Ko, S. K. Hong, and T. Yao, *Semicond. Sci. Technol.* **20**, S13 (2005).
23. J. Elsner, R. Jones, P. K. Sitch, V. D. Porezag, M. Elstner, Th. Frauenheim, M.I. Heggie, S. Oberg, and P. R. Briddon. *Phys. Rev. Lett.* **79**, 3672 (1997).
24. S. J. B. Yoo, R. Bhat, C. Canequ, and M. A. Koza, *Appl. Phys. Lett.* **66**, 3410 (1995).

25. H. J. Xiang, J. Yang, J. G. Hou, and Q. Zhu, *Appl. Phys. Lett.* **89**, 223111 (2006).
26. B. J. Rodriguez, A. Gruverman, A. I. Kingon, and R. J. Nemanich, and O. Ambacher, *Appl. Phys. Lett.* **80,** 4166 (2002).

Mater. Res. Soc. Symp. Proc. Vol. 1201 © 2010 Materials Research Society   1201-H06-08

MBE-Grown Ni$_y$Mg$_{1-y}$O and Zn$_x$Mg$_{1-x}$O Thin Films for Deep Ultraviolet Optoelectronic Applications

Jeremy W. Mares[1], Ryan C. Boutwell[1], Matthew T. Falanga[1], Amber Scheurer[1] and Winston V. Schoenfeld[1]

[1]CREOL, The College of Optics and Photonics, University of Central Florida
4000 Central Florida Blvd, Bldg. 53
Orlando, FL, 32816

ABSTRACT

We report on the heteroepitaxial growth of high-quality single crystal cubic Zn$_x$Mg$_{1-x}$O and Ni$_y$Mg$_{1-y}$O thin films by radio frequency oxygen plasma-assisted molecular beam epitaxy (RF-MBE). Film compositions over the ranges x = 0 to x = 0.65 and y = 0 to y = 1 have been grown on lattice-matched MgO (100) and characterized optically, morphologically, compositionally, and electrically. Both of these ternary materials are shown to have bandgaps which vary directly as a function of transition metal (Ni or Zn) concentration. Optical transmission measurements of Ni$_y$Mg$_{1-y}$O show the bandgap to shift continuously over the approximate range 3.5 eV (for NiO) to 4.81 eV (for y=0.075). Similarly, the bandgap of cubic Zn$_x$Mg$_{1-x}$O is shifted from about 4.9 eV (for x = 0.65) to 6.25 eV (for x=0.12). Films exhibit good morphological quality and typical roughness of Ni$_y$Mg$_{1-y}$O films is 5 Å while that of Zn$_x$Mg$_{1-x}$O is less than 15 Å, as measured by atomic force microscopy (AFM). X-ray diffraction (XRD) is employed to confirm crystal orientation and to determine the films' lattice constants. Film compositions are interrogated by Rutherford Backscattering (RBS) and electrical characterization is made by room-temperature Hall measurements.

INTRODUCTION

Here we report on a cubic oxide material family that exhibits some compelling material properties including a variable, direct deep ultraviolet bandgap, visible transparency and the potential to realize p-n structures. The compounds under investigation are the cubic ternary oxides Zn$_x$Mg$_{1-x}$O and Ni$_y$Mg$_{1-y}$O which bear some similarities to one another and comprise a structurally and electronically compatible materials system. Both of these materials are direct, wide bandgap semiconductors whose functional spectral ranges span from the near ultraviolet to the potential maximum bandgap of pure MgO, 7.8 eV [1-3]. They both also exhibit cubic rocksalt structure (conditionally, for ZnMgO) and, each has a lattice mismatch less than 1.5% to MgO, which is a widely available and affordable substrate material. Because the compounds are cubic and therefore symmetric under inversion, they have no strain-induce piezoelectric or material polarization fields such as those present in wurtzite AlGaN or ZnO.

The NiO and MgO binaries are mutually miscible and therefore can be grown with an arbitrary Ni:Mg ratio [3]. Thus the attainable bandgaps of cubic NiMgO range continuously from that of NiO, ~3.5 eV, to 7.8 eV. Zn$_x$Mg$_{1-x}$O, on the other hand, can be grown in the cubic phase only when the Zn concentration (x) is less than approximately 50% - a value which depends somewhat upon the deposition technique employed and growth conditions. With Zn concentrations greater than approximately one half, the wurtzite phase is generally favored by the compound (though it is important to note that mixed-phase growth often occurs for values $0.35 < x < 0.63$). The bandgap of the cubic form ranges from less than 5 eV (for $x \approx 0.5$) to 7.8 eV while that of wurtzite ZnMgO spans from the value for ZnO, 3.37 eV, to greater than about 4.3 eV (also at about x=0.5) [4-8].

The reason for investigating both compounds in tandem relates to the dichotomy in the carrier type propensities of the two materials. ZnMgO is known to be readily n-type doped; a quality which is presumed to be inherently related to the tendency of the parent binary, ZnO, to be intrinsically n-type [9-12]. In fact, it is well supported by literature that ZnO and ZnMgO are very challenging to stably and consistently p-type dope. NiO, on the other hand, is intrinsically p-type and can also be extrinsically p-doped [13-19]. The p-type nature of NiO is attributed largely to excess oxygen (much like the n-type character of ZnO is associated with oxygen deficiency) which is ubiquitously observed in NiO [17, 20]. It is therefore expected that with increasing Ni concentration (y) in $Ni_yMg_{1-y}O$, the p-carrier concentration may also increase.

The two ternary compounds may therefore be utilized in juxtaposition in order to engineer DUV p-n or p-i-n heterostructures which exploit NiMgO as the p-region and ZnMgO as the n-region. Even used independently, however, these compounds may serve well in metal-semiconductor-metal (MSM) detectors. Because of the very wide bandgaps of these materials, they could serve to detect wavelengths shorter than even pure Aluminum Nitride.

In this report we discuss initial research conducted towards assessing the viability of these materials for such applications. The work carried out has been primarily concerned with characterizing and improving film quality to ascertain whether sufficiently pristine films can be grown, which is an implicitly critical prerequisite to proceeding with photonic device fabrication attempts.

EXPERIMENT

The oxide films were grown by radio frequency oxygen plasma-assisted molecular beam epitaxy (RF-MBE). The growth system maintained a minimum base pressure below 10^{-10} Torr between growths. Electronic-grade oxygen (>5N purity) was used as well as high purity Mg, Zn and Ni (4N, 5.5N and 5.5N purities, respectively). The cation materials were evaporated from standard hot-lipped and high-temperature (for Ni only) Knudsen cells. Monatomic oxygen was generated from a typical inductively coupled plasma injector (f_0 = 13.56 MHz) with RF powers ranging from 350 to 500 W. Oxygen flux was varied between 1 and 3 standard cubic centimeters per minute. Oxygen flow rates set the maximum growth pressures (which were reduced during growth by Zn and Mg gettering) to values ranging between approximately 5 and 15 μTorr. The cell temperatures of the metals were systematically varied during many successive growth refinements to achieve best integration of the metals. In brief, the temperatures of the Zn, Mg and Ni cells were varied over the ranges 320-375 °C, 330-360 °C and 1300 - 1450 °C, respectively. The films were grown on 1 cm^2 (100)-oriented MgO substrates (MTI Crystal). Prior to growth, all crystals were cleaned in ultrasonic baths of acetone, isopropyl alcohol and deionized water and then dried under Nitrogen flow. Substrates were degassed at 600 °C for more than 3 hours in a separate ultra-high vacuum preparation chamber. Substrate temperature for the ZnMgO series was 300 °C and that for the NiMgO series was varied from 300 to 700 °C.

The material characterization techniques presented here include Rutherford Backscattering (RBS), XRD, AFM and optical transmission. The RBS was carried out with He^{2+} ions at an energy of 2.25 MeV and the backscattered energy spectra were fitted with standard RBS analysis software to determine film compositions and thicknesses. X-ray diffraction measurements were made with a Rigaku D/MAX x-ray diffractometer (Cu-Kα radiation, λ = 1.54056 Å) and diffraction peaks fitted to asymmetric pseudo-Voigt peaks for best results. For AFM, a Veeco Nanoscope D3100 atomic force microscope was used in tapping mode. Optical transmission measurements were taken using a Cary 500 UV-VIS spectrophotometer. Prior to transmission measurements, the previously rough undersides of all substrates were polished to optical smoothness with diamond lapping films used in decreasing grain size (30 to 0.3 μm) to ensure accurate determination of optical properties.

RESULTS AND DISCUSSION

The results of materials characterization of 13 films are presented herein. Five of these are cubic ZnMgO films, six are NiMgO, one film is binary NiO and one is homoepitaxial MgO. The RBS spectra obtained (currently part of publications under review) revealed concentrations of the ZnMgO films to be x = {0.12, 0.33, 0.44, 0.58, 0.65} and those of the NiMgO films to be y = {0.075, 0.12, 0.16, 0.25, 0.32, 0.56}. Figure 1a shows the XRD data for the entire series over the angular range 42° ≤ 2θ ≤ 44°, which contains the greatest intensity diffracted peak, the cubic (002) peak. All films exhibited strictly cubic, (001) oriented films growth with the exception of the highest Zinc content film. It was observed for this film (x = 0.65) that, in addition to the predominately cubic rocksalt diffraction pattern, there was also indication of low intensity peaks that are attributed to wurtzite components of the film. In particular, the wurtzite structure ZnMgO (002) and (101) diffraction peaks were detected at 34.65° and 36.32°, respectively, and were found to be lower in intensity than the cubic-(002) peak by a factor of approximately 500. Thus, for the highest Zn concentration film, phase segregation into wurtzite structure occurred.

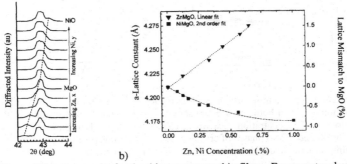

a)   b)

Figure 1. (a) X-ray diffraction results for the thirteen ternary thin films. Four spectra show the distinct (002) peaks of the ternary films in addition to the MgO (002) one; locations of the other ternary peaks are numerically fit. Dashed lines connect the ternary (002) peaks to illustrate the trend. (b) Lattice constants as determined from XRD measurements ZnMgO and NiMgO films.

In the spectra shown, the MgO substrate (002) peak (lattice constant a = 4.2115 Å) is present for all films and has the greatest intensity. Four of the films exhibited very distinct cubic-(002) diffraction peaks which are attributed to the ZnMgO and NiMgO films (x = {0.44, 0.58, 0.65} and y = 1). These peaks are easily fit to determine the film lattice constants. Fitting for the remaining films was subject to parameter initialization based on adjacent sample concentrations. The lattice constants of the films as a function of Zn or Ni concentration are shown in Figure 1b. The data is in good overall agreement with literature. Notably, the lattice constant of ZnMgO is approximately a linear function of Zn concentration (x) while that of NiMgO exhibits greater bowing and is better fit by a 2nd order fit. On the right axis, Figure 1b shows the lattice mismatch to the substrate, MgO. The greatest mismatch exhibited by a strictly cubic film is that of the x = 0.58 film, which has mismatch of ~1.3%. Pure NiO has a mismatch of only 0.85 % with MgO, per our measurements.

The results of transmission measurements for the films are shown in Figure 2a. The top pane of the figure shows the transmission curves for the ZnMgO films while those for the NiMgO films are shown on the bottom. The absorption edges for the films in both categories are shown to shift to longer wavelengths with increased integration of the transition metal (Ni or Zn). For NiMgO, the absorption edges span the entire region from approximately 180 nm to 320 nm. The values for ZnMgO range from 180 to about 240 nm for the strictly cubic films. Interestingly, the phase segregated sample exhibited two distinct absorption edges. It is likely that the higher energy absorption edge (which is stronger) is due to cubic regions of the film while the lower energy absorption feature ($\lambda \approx 345$ nm) could be due to hexagonal regions of the film. All films exhibit good transparency (T > 70%, including interference modulation) in the visible region.

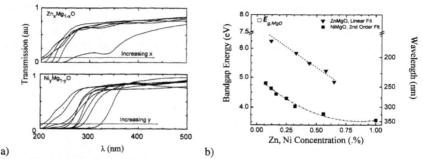

Figure 2. (a) Optical transmission measurements for ZnMgO (top) and NiMgO (bottom). Absorption edges for both ternaries are shown to shift to longer wavelengths with increased transition metal concentration. The highest Zn concentration film shows two distinct absorption edges, attributed to two phases of crystal growth. (b) Bandgap values for NiMgO and ZnMgO thin films, as extracted from optical transmission measurements.

The bandgaps of the films were determined by deriving absorption, α, from the transmission spectra and plotting $(\alpha h v)^2$ as a function of hv (plots also under review for publication). Figure 2b shows the bandgap values derived for the series as a function of transition metal concentration. As evidenced by the transmission spectra, the bandgap energies of the films clearly decrease as a function of Ni or Zn concentration. Including binary MgO, the values for NiMgO range from 3.56 to 7.8 eV and those of ZnMgO range from 4.85 to 7.8 eV. However, it is important to note that the instrument used cannot measure wavelengths shorter than 175 nm (energies greater than 7.08 eV) and therefore cannot measure directly the band-edge absorption of pure MgO. In addition, when transmitting through the thick substrates (~ 500 μm), the absorption of known defects (such as F and F+ centers) [21, 22] is strong and can be misconstrued for the true bandgap absorption of the binary. It should be pointed out that the bandgaps of both compounds seem to be discontinuous between the minimum transition metal concentration and that of pure MgO. This may be attributed in part to experimental error, however, ZnMgO has exhibited this discontinuity in other works as well [1, 5]. The source of the seemingly abrupt drop in bandgap with Zn (or Ni) integration is not known but the authors hypothesize that the measured bandgap values may be slightly skewed due to defect contributions to absorption that are not yet quantified.

AFM scans reveal all strictly cubic films to have root-mean square roughness (R_q) less than 30 Å, taken over 1 μm² regions. Typical R_q values for NiMgO with y > 0.25 were approximately 5-7 Å, while values for ZnMgO were 15-20 Å. The roughness of the phase-segregated film was approximately 44 nm. Most films exhibited clearly cubic structure, with pyramidal, (100)-oriented surface structures becoming more evident at lower concentrations of Ni or Zn.

Hall measurements revealed all films to be relatively high in resistivity, as is expected from an undoped, wide bandgap semiconductor at room temperature. Importantly, the majority carrier type of NiMgO was found to be holes while that of ZnMgO was electrons (though for the highest resistance films, the sign of the carrier type was subject to some vacillation between measurements). The resistance of the ZnMgO is found to be greater than that of the NiMgO films by one to two orders of magnitude for 0 < {x,y} < 1. Pure NiO has the lowest resistance of the films at a value of $\rho = 1.1 \times 10^3$ Ω·cm while the greatest resistance is exhibited by pure MgO, 3.8×10^8 Ω·cm. Figure 3 summarizes the film resistivities measured.

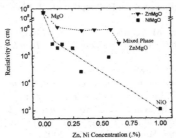

Figure 3. Resistivity measurements of the two ternary compounds as a function of transition metal concentration. Films exhibit high resistivity which decreases, in general, with increased transition metal concentration.

CONCLUSION

In summary, the growth and characterization of ZnMgO and NiMgO single-crystal thin films on MgO substrates was presented. General growth methods were described and initial optical, structural, morphological and electrical characterizations of the two ternary compounds were presented, and related to one another. Films were interrogated by RBS to determine transition metal concentrations, which ranged over the entire compositional range of the two materials, as restricted by phase separation in the case of ZnMgO. XRD measurements were presented which confirmed the single crystal, (100) character of the films and allowed for extraction of the films' lattice constants. Optical transmission measurements showed the bandgaps of both materials to shift to lower values with increased integration of Zn or Ni. Bandgaps of NiMgO ranged from approximately 3.5 to 7.8 eV while those of cubic ZnMgO ranged from 4.85 to 6.25 eV. Results of AFM investigations were described and electrical characterization of the films, as assessed by Hall measurements, was presented. Films were found to be of excellent morphological quality and high resistivity, corroborating overall quality of the films.

ACKNOWLEDGMENTS

The authors would like to express appreciation of the support provided by the U.S. Army Research Office (Dr. Michael D. Gerhold) under the auspices of the U.S. Army Research Office Scientific Services Program administered by Battelle (Delivery Order 0188, Contract No. W911NF-07-D-0001).

REFERENCES

1. N. B. Chen and C. H. Sui, *Mater. Sci. Eng., B*, **126**, 16 (2006).
2. J. W. Kim, H. S. Kang, J. H. Kim, et al., *J. Appl. Phys.*, **100**, 5 (2006).
3. A. Kuzmin and N. Mironova, *J. Phys. Condens. Matter*, **10**, 7937 (1998).
4. C. Bundesmann, A. Rahm, M. Lorenz, et al., *J. Appl. Phys.*, **99**, 113504 (2006).
5. S. Choopun, R. D. Vispute, W. Yang, et al., *Appl. Phys. Lett.*, **80**, 1529 (2002).
6. S. S. Hullavarad, N. V. Hullavarad, D. E. Pugel, et al., *J. Phys. D: Appl. Phys.*, **40**, 4887 (2007).
7. R. Schmidt, B. Rheinlander, M. Schubert, et al., *Appl. Phys. Lett.*, **82**, 2260 (2003).
8. R. Schmidt-Grund, M. Schubert, B. Rheinländer, et al., *Thin Solid Films*, **455-456**, 500 (2004).
9. D. C. Look, *Mater. Sci. Eng., B*, **80**, 383 (2001).
10. D. C. Look, D. C. Reynolds, J. R. Sizelove, et al., *Solid State Commun.*, **105**, 399 (1998).
11. U. Ozgur, Y. I. Alivov, C. Liu, et al., *J. Appl. Phys.*, **98**, 1 (2005).
12. S. Pearton, D. Norton, K. Ip, et al., *J. Vac. Sci. Technol., B*, **22**, 932 (2004).
13. X. Chen, K. Ruan, G. Wu, et al., *Appl. Phys. Lett.*, **93**, 112112 (2008).
14. J.-M. Choi and S. Im, *Appl. Surf. Sci.*, **244**, 435 (2005).
15. U. S. Joshi, Y. Matsumoto, K. Itaka, et al., *Appl. Surf. Sci.*, **252**, 2524 (2006).
16. B. Lalevic, B. Leung, and Fuschill.N, *Bulletin of the American Physical Society*, **17**, 246 (1972).
17. S. Lany, J. Osorio-Guillen, and A. Zunger, *Phys. Rev. B: Condens. Matter*, **75**, 241203 (2007).
18. A. M. Salem, M. Mokhtar, and G. A. El-Shobaky, *Solid State Ionics*, **170**, 33 (2004).
19. Y. Vygranenko, K. Wang, and A. Nathan, *Appl. Phys. Lett.*, **89**, 172105 (2006).
20. K. Oka, T. Yanagida, K. Nagashima, et al., *J. Appl. Phys.*, **104**, 013711 (2008).
21. Y. Chen, J. L. Kolopus, and W. A. Sibley, *Physical Review*, **186**, 865 (1969).
22. G. P. Summers, T. M. Wilson, B. T. Jeffries, et al., *Phys. Rev. B: Condens. Matter*, **27**, 1283 (1983).

Mater. Res. Soc. Symp. Proc. Vol. 1201 © 2010 Materials Research Society   1201-H06-10

Melt Grown ZnO Bulk Crystals

Detlev Schulz, Steffen Ganschow, Detlef Klimm
Leibniz Institute for Crystal Growth, D – 12489 Berlin, Max-Born-Str. 2, GERMANY

ABSTRACT

Bulk crystals of zinc oxide can be grown from the melt by a Bridgman technique under pressure. This new technology using an iridium crucible shows the potential to yield large single crystals of good crystalline perfection. Crystals with diameters up to 33 mm and a length of up to 50 mm have been demonstrated. The impurity content can be strongly reduced by using the crucibles repeatedly.

INTRODUCTION

Commercial zinc oxide wafers are fabricated almost exclusively from hydrothermally grown ZnO bulk crystals. The hydrothermal technology rests on decades of experience mainly with the growth of α-quartz crystals for piezoelectric applications. From alkaline hydrothermal solutions of zinc oxide, ZnO crystals sized up to 3 inch can be grown within several weeks. It is typical for crystal growth processes from solutions, however, that traces of the solvent (here mainly H, Li, K) are incorporated to the grown bulk. This is uncritical for a piezoelectric material, but can have substantial influence on a semiconductor.

Crystal growth technologies that rely on pure melts without solvent (e.g. Czochralski, Bridgman) are desirable alternatives, and are used almost exclusively for the mass production of other electronic materials (Si, Ge, GaAs). For a long time it was assumed that the high melting point of ZnO (T_f = 1975 °C), together with the high oxygen partial pressure that is needed to stabilize ZnO at T_f, would not allow to find any crucible material that can withstand molten ZnO. Recently we could show that this claim is wrong, instead a reactive atmosphere containing carbon dioxide can deliver by the equilibrium reaction $CO_2 \leftrightarrow CO + \frac{1}{2} O_2$ a "self adjusting" oxygen partial pressure which stabilizes ZnO melt in an iridium crucible [1]. By this way zinc oxide bulk crystals can be grown from iridium crucibles e.g. by the Bridgman method.

Now boules with 33 mm diameter and ca. 50 mm length can be grown that are suitable for the production of wafers that were successfully used for the deposition of (Zn,Mg)O and (Zn,Cd)O epilayers by MBE [2]. As expected, the concentration of most impurities is lower, compared with hydrothermal samples [3].

Despite these encouraging results, two major problems arise:
a) Thermal stresses during crystallization and cooling of the ZnO boule might be so large that grain boundaries or even cracks will form.
b) Some trace impurities that are still found in the melt grown crystals (e.g. Fe), are presumably introduced from the crucible: The chemical purity of the iridium metal is typically only 99.9%.

This contribution compares the status of ZnO melt growth with other technologies, preferably hydrothermal growth, and discusses prospects of future improvements.

EXPERIMENT

A Bridgman-like setup is used for the growth of zinc oxide from the melt [4]. The assembly comprises a metallic crucible which is surrounded by insulating media, e.g. zirconia, alumina etc. The iridium crucible is directly heated by an rf-coil, this means, that no external heater is present. The whole process is conducted in a high pressure chamber to apply an oxygen-containing atmosphere at elevated pressure. This is necessary in order to decrease the evaporation rate of zinc oxide. The dissociation is described by :

$$ZnO \leftrightarrow Zn + \tfrac{1}{2} O_2 \tag{1}$$

Thermodynamical calculations have shown, that the total pressure of zinc oxide is $p_{tot} = 1.06$ bar at $T_f = 1975$ °C [5]. In addition, the evaporation of oxygen leads to remaining zinc that could easily react with iridium. Consequently, the overpressure is supplied by carbon dioxide, that decomposes at these high temperatures and results in a variable partial pressure of oxygen. In contrast to a pure oxygen atmosphere, when only one specific partial pressure of oxygen is present, the dissociation of carbon dioxide is also dependent on temperature and may lead to an "self-adjusting" oxygen partial pressure by choosing an appropriate composition of the atmosphere, e.g. by mixing carbon dioxide with argon [1]. Using a carbon dioxide atmosphere the iridium crucible remains stable even in an oxidizing atmosphere and the decomposition rate of zinc oxide is simultaneously reduced. This is one key parameter for melt crystal growth of zinc oxide in an iridium crucible.

Since the starting material offers a rather low powder density it has to be compressed for crystal growth in a crucible. This is also required because pre-melting is no option in presence of a seed crystal. The zinc oxide powder is first subjected to cold isostatic pressing in a mould at a pressure up to 2000 bar. To increase the density of the green compact it is subsequently heated at ca. 1100 °C for one day. Finally, a density of the compact of about 95 % of the theoretical value of zinc oxide can be obtained. Both polar directions have been already used for the seeds and other direction will be tested in future. The growth process can be divided into three basic steps:
1) melting and homogenization of the starting material,
2) adjusting the axial temperature gradient and partial melting of the seed,
3) crystallization.
During the first step heating proceeds up to a temperature above the melting point. Since temperature measurements are difficult at around 2000 °C, the event of melting is indirectly detected by a sudden change in temperature. The second step is established by changing the position of the crucible with respect to the rf-coil. This is accompanied by a constant temperature at the control point leading to a partial melting of the seed. After a second homogenization period the crystallization is initiated by a closed loop control of the temperature.

Currently single crystals of zinc oxide of 33 mm in diameter and up to 50 mm in length can be grown by the Bridgman technique. The limitation in diameter is only due to the use of a "standard" size. The crystalline properties were measured by X-ray diffraction and are represented by the rocking curve.

One critical issue in growth from the melt with the help of a crucible is contamination of the growing material by the crucible itself. Since purification of iridium is either difficult or expensive, we tried multiple usage of the crucible instead of single-use. This is complicated by the fact that the cylindrical part of the crucible has to separated from the cone in order to reveal the single crystal. After removal of the crystal the crucible is welded again and can be used for the second growth run. The impurity content has been investigated by secondary ion mass spectrometry (SIMS).

DISCUSSION

Crystal growth from the melt exhibits a number of features, which makes it superior to other growth techniques as far as single crystals are concerned. First, the ratio in density between the nutrient (e.g. vapor phase, solution) and the solid is closer to one when crystallizing from a melt compared to e.g. hydrothermal growth. This leads to a comparingly high crystallization rate. As a result the crystal volume produced per time makes melt growth the most economic process.

Large single crystals of zinc oxide are known to originate from growth methods from the vapor, hydrothermal growth as well as melt growth. Whereas the first two methods have been extensively studied for more than four decades [6], the first successful report on melt growth was released in 1999 [7]. To date actually only two groups worldwide are dealing with melt growth of zinc oxide.

Hydrothermal method

Comparing the transport velocity of the above mentioned methods the lowest values are reported for hydrothermally grown crystals [8]. In alkaline aqueous solutions the crystal is free to grow in any direction and therefore, the growth rate of different faces has to be considered. Dependent on the choice of solvents as well as their concentration ratios the growth rate is normally below 0.2 mm/day for the [0001] direction [6, 9]. This low growth rate might be compensated by simultaneous growth of many crystals. Faster growth is observed in the directions perpendicular to the c-axis leading to typically platelet-like crystals. This strongly anisotropic growth may lead to growth sectors of different impurity content or defect density. Large single crystals up to 3 inch in diameter with excellent structural quality have been already demonstrated.

Growth from the vapor

Physical vapor transport has been recognized as providing very low transport rates already in the 1970s [10]. Meanwhile numerous transport agents (e.g. H_2, C, Cl_2) have been tested and shown to not only increase the transport rate but also to result in high quality crystals [11, 12, 13]. Usually crystals grow in ampoules made of silica under wall contact. Crystals grown without wall contact show a needle-like habit with fastest growth along the c-axis. The growth velocity may reach up to a few hundred microns per hour. Single crystals up to two inch in diameter and 1 cm in thickness were reported [14].

Melt growth

Melting of zinc oxide is still a sophisticated task due to several obstructive properties of the molten material. The relatively high total pressure (see Eq. 1) leads to a strong evaporation and requires an overpressure. Furthermore, there are only a few materials that can withstand more than 2000 °C in an oxidizing environment. To avoid the use of a crucible the "pressurized melt growth" has been described [15]. A skull melting equipment is used under elevated pressure up to 100 bar. This technique is based on induction heating, where eddy currents directly flow inside the material. The starting material is placed inside water-cooled copper fingers, preventing the rim from being melted. Finally, molten zinc oxide is contained in solid zinc oxide, that is also denoted as the "skull". Owing to the absent contact to a foreign material, except for the atmosphere, the purity should be as high as that of the starting material. Critical issues of this methods are temperature control and temperature gradients. Electromagnetic forces inside the melt influence strongly the flow pattern of the liquid. It is difficult to establish directional solidification, either by lowering the crucible or by using a pulling arrangement. Crystals as large as 2 inch in diameter have been prepared.

Several companies offer zinc oxide substrates based on the hydrothermal method and skull melting. By contrast, the Bridgman technique developed by the authors is still on a level of academic research. Single crystals up to 33 mm in diameter have been demonstrated with good crystalline perfection [4]. The full width at half-maximum (FWHM) of the X-ray rocking curve can be lower than 50 arcsec. However, a more realistic upper limit of the FWHM value is 300 arcsec, which is mainly due to low angle grain boundaries (Fig. 1). There is no indication of strain due to contact to the iridium crucible.

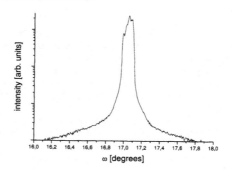

Figure 1: X-ray rocking curve for {0002} reflection, spot size 10 x 2 mm^2, open detector, log-scaling

During the growth process the crucible is nearly closed and there is no direct observation possible. So far the solid-liquid interface could not be detected in the grown crystals by different diagnostic techniques. The growth rate is estimated to be in the order of a few millimeters per hour, being one hundred times larger as for hydrothermal growth.

The purity of the grown crystal depends on the purity of the starting material and on the purity of the crucible. In order to identify the main sources for impurities three different crystals have been investigated by SIMS. In Table 1 a considerable improvement for most elements can be found when changing the starting material from 4N to 5N. Although to a lower extent this can also be seen when using a crucible for a second growth run. For this crystal two data are shown, from the seed end and the tail end respectively.

Table 1. Impurity analysis of ZnO crystals by SIMS (4N and 5N: purity of starting material, crucible re-use: 5N starting material and crucible used second time (upper row: seed end, lower row: tail end), HT: hydrothermal sample)

	Li	Na	Mg	Al	K	Ca	Cu	Ga	Si
ZnO 4N	1E18	2E19	3E18	2E19	3E18	3E18	1E19	1E18	5E19
ZnO 5N	8E15	3E17	1E17	1E18	2E17	4E17	7E17	5E17	2E18
ZnO crucible re-use	4E15	4E17	1E17	5.1E17	2.3E17	1E17	<5E17	<1E16	1.7E18
	8.3E15	1.2E17	3E16	3.2E17	4.2E16	2.6E16	<5E17	<1E16	3.4E17
HT	3E17	3E17	2E17	5E18	2E17	1E17		6E18	5E18

Most of the elements, e.g. Li, Na, Mg, K, Si, do not change in concentration by using the crucible for the second time. This means, that the crucible is probably not contaminated by these impurities. On the other hand, there is a difference for Al and Ga, the latter being below the detection limit. Since these elements act as donors in ZnO, their concentration should be well defined in terms of doping. For almost all impurities segregation is observed, leading to a lower concentration at the tail end compared to the seed end. For comparison a commercially purchased sample grown hydrothermally is added. Except for Li, Al and Ga one can find similar levels of contamination. Since for the hydrothermal method e.g. LiOH is used as a mineralizer, the concentration of Li is much higher in the crystals than in the melt grown ones. The high amount of group-III impurities is in good agreement with electrical data [16]. Even though the electron concentration at room temperature for hydrothermal crystals is usually lower than that for Bridgman grown crystals, there is indication for a high degree of compensation in hydrothermal crystals.

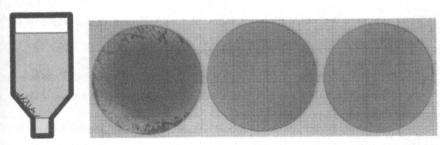

Figure 2. left: Schematic of the iridium crucible including seed and crystal, with iridium precipitation in the cone; right: three as-grown wafers from the bottom (left), middle (center) and the tail (right) part of the crystals

The analysis of impurities did not show the occurrence of iridium in the Bridgman grown crystal. Though iridium is even visible to the naked eye in some regions of the grown boule, it was not detected by SIMS in the investigated samples. Iridium can be found in the lower part, this means first grown part, of the crystal (Fig. 2). Probably the zinc oxide melt slowly dissolves the crucible, but the precipitation only occurs in the lower part of the setup. Therefore, it can be assumed, that the rate of which iridium is dissolved in the melt, decreases with time. Otherwise, it had to be found along the whole boule. Since the lower part, mainly the cone, is normally not used for wafer production, the iridium precipitation has only a low impact on the yield.

CONCLUSIONS

Single crystal growth of zinc oxide is still under investigation, since commercially available substrates need further improvement and may play an essential role towards the development of p-type material. The Bridgman method is compared with the hydrothermal method, skull melting and growth from the vapor phase. The developed process offers several advantages e.g. up-scaling or control of dopants, which could either be introduced to the starting material or supplied via the vapor phase. The growth of 2 inch crystals is currently under investigation and improvements of the crystal quality can be expected by adjustment of the thermal conditions during growth.

ACKNOWLEDGMENTS

The authors gratefully acknowledge the contribution by S. Lautenschläger from Justus Liebig University Giessen for the SIMS measurements and would like to thank A. Kwasniewski for doing XRD.

REFERENCES

1 D. Klimm, S. Ganschow, D. Schulz, R. Bertram, R. Uecker, P. Reiche and R. Fornari, J. Crystal Growth 311, 534 (2009).
2 S. Sadofev, P. Schäfer, Y.-H. Fan, S. Blumstengel, F. Henneberger, D. Schulz and D. Klimm, Appl. Phys. Lett. 91, 201923 (2007).
3 G. Brauer, W. Anwand, D. Grambole, J. Grenzer, W. Skorupa, J. Cizek, J. Kuriplach, I. Prochazka, C.C. Ling, C.K. So, D. Schulz and D. Klimm, Phys. Rev. B79, 115212 (2009).
4 D. Schulz, S. Ganschow, D. Klimm, M. Neubert, M. Roßberg, M. Schmidbauer and R. Fornari, J. Crystal Growth 296, 27 (2006).
5 D. Klimm, S. Ganschow, D. Schulz, and R. Fornari, J. Crystal Growth 310, 3009 (2008).
6 L. Dem'yanets and V. Lyutin, J. Crystal Growth 310, 993 (2008).
7 J.E. Nause, III-Vs Review 12, 28 (1999).
8 D. Ehrentraut, H. Sato, Y. Kagamitani, H. Sato, A. Yoshikawa and T. Fukuda, Progress Crystal Growth Charac. Mater. 52, 280 (2006).
9 W. Lin, D. Chen, J. Zhang, Z. Lin, J. Huang, W. Li, Y. Wang and F. Huang, Crystal Growth & Design 9, 4378 (2009).
10 R. Helbig, J. Crystal Growth 15, 25 (1972).
11 J. Ntep, S. Said Hassani, A. Lusson, A. Tromson-Carli, D. Ballutaud, G. Didier and R. Triboulet, J. Crystal Growth 207, 30 (1999).
12 M. Mikami, S.-H. Hong, T. Sato, S. Abe, J. Wang, K. Masumoto, Y. Masa and M. Isshiki, J. Crystal Growth 304, 37 (2007).
13 K. Grasza, P. Skupinski, A. Mycielski, E. Lusakowska and V. Domukhovski, J. Crystal Growth, 310, 1823 (2008).
14 D.C. Look, D.C. Reynolds, J.R. Sizelove, R.L. Jones, C.W. Litton, G. Cantwell and W.C. Harsch, Solid State Comm. 105, 399 (1998).
15 J. Nause and B. Nemeth, Semicond. Sci. Technol. 20, S45 (2005).
16 D. Schulz, S. Ganschow, D. Klimm and K. Struve, J. Crystal Growth 310, 1832 (2008).

Nano Structures

Mater. Res. Soc. Symp. Proc. Vol. 1201 © 2010 Materials Research Society   1201-H07-05

Yao Cheng[1], Yao Liang[2], Ming Lei[1], Sui Kong Hark[2] and Ning Wang[1]
[1]Physics Department, The William Mong Institution of Nano Science and Technology, The Hong Kong University of Science and Technology, Hong Kong, China.
[2]Physics Department, The Chinese University of Hong Kong, Hong Kong, China.

ABSTRACT

Based on the focused ion beam (FIB) technology, we have prepared ZnO containing periodic nano-sized structures by an ultra thin Ga ion beam. ZnO nanowires can keep a good crystal quality after Ga ion bombardment. The cathodoluminescence (CL) spectroscopy study of the Ga-doped ZnO nanowires at low temperatures shows that the Ga doping effect can largely suppress the green emission that may mainly originate from the defects on the surfaces of ZnO nanowires.

INTRODUCTION

Zinc oxide has attracted great attentions because of its interesting properties of wide band-gap (3.37eV) and large exciton binding energy (60meV). In recent years, novel morphologies of ZnO structures, such as nanocantilever arrays [1], nanohelixes [2, 3], nanorings [4] and nanosheets [5], have been synthesized. By controlling the sizes, shapes, crystal structures, and surface structures, these ZnO nanostructures are exploiting a wide range of technological application in chemical, optical, electronic and mechanical fields. Because most of these novel nanostructures were directly grown from solution or vapor process, their morphologies largely relied on growth conditions. Therefore, development of nanostructural modification methods for desiring properties is timely and of extremely high importance in current nanomaterials research.

In addition to the development of methods for fabricating various ZnO nanostructures, much effort has been devoted to investigating the optical properties of these nanostructures. Besides the UV excitonic emission peak, ZnO nanostructures often exhibit different peaks in the visible spectra region, which have been attributed to defect emissions. For example, the green emission [6-14] has been commonly observed in ZnO nanostructures. Several mechanisms have been proposed in order to explain the origin of the green emission such as the vacancies and oxygen (or zinc) interstitials [6-11], antisite oxygen [12], surface defects [13] and Cu impurities [14]. So far, the origin of the green emission still remains an open and controversial question and requires further studies. In the meantime, however, some methods have been developed to influence the green emission. For example, coating a surfactant layer on ZnO nanostructures may suppress the green emission [13]. In this work, we report a method for creating periodic nanostructures based on nanowires templates by the FIB technique. We observed the damaging effect induced by the high-energy Ga beam and the optical property changes resulted from the selective Ga doping effect in ZnO nanostructures.

EXPERIMENT DETAILS

ZnO nanowires were grown by direct oxidation of pure zinc powder placed in a quartz tube and heated at about 1000 °C without using any catalysts [15]. The nanowires were first scratched from the substrates and ultrasonicated in alcohol solution. Then, the nanowires were dispersed on Si substrates and carbon holey films for Ga ion beam treatment and structure modification and characterization. The nanostructure fabrication was carried out in a SII SMI FIB system using Ga ion energy of 30keV with beam current 1pA. The Ga ion doping was carried out on a FEI 611 FIB system under a voltage of 5keV and a beam current of 2pA. Their crystal structures were investigated by an analytical transmission electron microscope (TEM) (Philips CM 120) equipped with an energy-dispersive X-ray spectroscopy (EDS) and a high-resolution TEM (JEOL 2010F). The cathodoluminescence (CL) spectra were measured in a scanning electron microscope (SEM) using an Oxford Instrument MonoCL system with an accelerating voltage of 5kV and a beam current of 460pA.

DISCUSSION

Traditionally, high-energy ions have been widely used to etch or implant materials on specific areas of wafer surfaces in semiconductor industry. The regions which are not intended to be implanted are coated with photoresist patterns or masks [16]. The FIB technique offers advantages for controlling ion milling and implantation at specific areas without using masks or photoresist. This technique also provides new capabilities to control gradients in doping and to control the depth of doping or implantation.

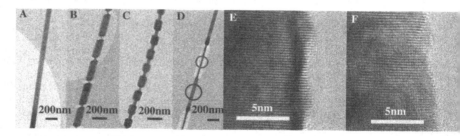

Figure 1. (A) TEM image of a typical as-grown ZnO nanowire; (B), (C) and (D) are the distinct periodic ZnO nanostructures fabricated by FIB milling. (E) and (F) are HRTEM images of FIB non-milling and milling parts, corresponding to the big and small circle in (D).

Figure 1A is a bright-field TEM image of an as-synthesized ZnO nanowire, which is the "template" used for fabricating periodic nanostructures by FIB milling process. The diameters of the nanowires used for this work were about 100nm and the lengths were from several to hundreds micrometers. These ZnO nanowires are single crystalline and exhibit smooth surfaces. Figure 1B, 1C and 1D show ZnO nanostructures with different periodic lengths prepared by FIB. The thinner parts along the nanowire were etched directly with the FIB milling process. Compared to conventional grown periodic nanowires from a solution or vapor process, the FIB method provides a precise way to design the length of each period (Figure 1B, 2C and 1D) and structural modification at a specific area on the nanowire. Although the high-energy Ga beam shows excellent milling of the nanostructures, however, structural damaging often occurs. Figure

1E and 1F are HRTEM images showing the atomic structures of ZnO nanowire with and without Ga ion milling along the same nanowire respectively, which correspond to the big and small circle positions in Figure 1D. In Figure 1E, the as-synthesized ZnO nanowire (grown along the most preferable [0001] direction) shows excellent crystalline quality. As shown in Figure 1F, no obvious damage occurs in the ZnO crystal lattice after Ga ion milling. In the nanowires shown in Figure 1F, Ga has been detected by the X-ray energy dispersive spectroscopy (EDS), indicating that Ga has been doped into the nanowire.

In the ZnO periodic nanostructures fabricated by FIB, we have observed large changes of their optical properties. Using a low energy Ga-beam, the green emission in ZnO nanowires can be modified without changing the shape or structure of the nanowire. This is a doping process which is done under a high tension of 5keV with an exposure time of several seconds. Figure 2 shows the CL spectra recorded at liquid nitrogen temperature (77K) with an electron beam accelerated at 5 kV. The inset images of Figure 2A show the SEM image and the corresponding CL image of the periodically Ga doped nanowire. The doping effects can be clearly observed from the CL image. The bright green emission comes from the undoped part of the ZnO nanowire and the light green emission represents the doped part along the same nanowire. In the CL spectra of Figure 2A, two emission bands from the doped and undoped ZnO nanowires can be observed at about 377nm and 524nm respectively. Very obviously, the defect-related green emission was significantly suppressed by the Ga doping effect, while the UV emission has a little decrease on the emission intensity. At the liquid nitrogen temperature, three emissions peaks have been recorded with positions at 3.33eV, 3.28eV and 3.20 eV respectively, which have been reported in previous work [17]. Ga doping did not affect these bound-exciton emissions of ZnO nanowires.

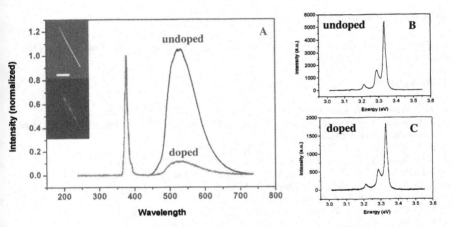

Figure 2. (A) The normalized CL spectra recorded from doped and undoped parts along a single ZnO nanowire at 77K. Inset images are the SEM image of the doped ZnO nanowire and its corresponding CL image. The scale bar is 5µm. (B) and (C) are the corresponding UV peaks recorded at 77K.

213

According to the theoretical analyses [18], for Ga^+ implantation, the mean implantation depth in ZnO should be less than 15nm for 5keV [19]. This indicated that the doping effect in the ZnO nanowires prepared in our experiment should occur only on their surface range. This may also prove that the defects responsible for the green emission are mainly located at the surfaces of ZnO [13, 20], because the surface recombination layer responsible for defect emission in ZnO nanowires can be estimated to be about 30nm thick [21]. Hence, we further deduce that the defect-related emission can be influenced greatly by gallium surface implantation. The detailed mechanism for Ga doping effect is unclear. One of the possible models is that the gallium ions may occupy the octahedral interstitial positions in the ZnO nanostructure and this can quench the green emission in ZnO [22]. Another possible model is the screening of defects by electrons. This is because that the Ga atoms may substitute the Zn sites. In this case, a shallow donor level can form and the free electron density will increase. Therefore, the free electron can compensate positive oxygen vacancy defects, which causes the oxygen vacancy formed in the ZnO lattice to be suppressed [23-25]. We believe that the present work provides useful information for better understanding the defect-related emission in ZnO nanomaterials.

CONCLUSIONS

We have developed a new method to modify ZnO nanostructures and their optical properties using the focused ion beam (FIB) technology. Because of the high accuracy of Ga beam, periodic ZnO and Si nanowire structures have been fabricated. The cathodoluminescence (CL) spectroscopy study at low temperatures shows that Ga doping can largely suppress the green emission.

ACKNOWLEDGMENTS

This work is supported by the Research Grants Council of the HKSAR, China (Project No. CityU5/CRF/08 and 603006).

REFERENCES

1. Z. L. Wang, X. Y. Kong and J. M. Zuo, *Phys. Rev. Lett.* **91**, 185502 (2003).
2. Z. L. Wang, X. Y. Kong, Y. Ding, P. X. Gao, W. L. Hughes, R. S. Yang and Y. Zhang, *Adv. Funct. Mater.* **14**, 943 (2004).
3. X. Y. Kong and Z. L. Wang, *Nano Lett.* **3**, 1625 (2003).
4. X. Y. Kong, Y. Ding, R. Yang and Z. L. Wang, *Science* **303**, 1348 (2004).
5. J. H. Park, H. J. Choi, Y. J. Choi, S. H. Sohn and J. G. Park, *J. Mater. Chem* **14** 35 (2004).
6. X. Q. Meng, D. Z. Shen, J. Y. Zhang, D. X. Zhao, Y. M. Lu, L. Dong, Z. Z. Zhang, Y. C. Liu and X. W. Fan, *Solid State Commun.* **135**, 179 (2005).
7. Y. Q. Chen, J. Jiang, Z. Y. He, Y. Su, D. Cai and L. Chen, *Mater. Lett.* **59**, 3280 (2005).
8. R. C. Wang, C. P. Liu, J. L. Huang and S. J. Chen, *Appl. Phys. Lett.* **87** , 053103 (2005).
9. X. Liu, X. H. Wu, H. Cao and R. P. H. Chang, *J. Appl. Phys.* **95**, 3141 (2004).
10 Y. W. Heo, D. P. Norton and S. J. Pearton, *J. Appl. Phys.* **98**, 073502 (2005).
11. Q. X. Zhao, P. Klason, M. Willander, H. M. Zhong, W. Lu and J. H. Yang, *Appl. Phys. Lett.* **87**, 211912 (2005).

12. Q. Yang, K. Tang, J. Zuo and Y. Qian, *Appl. Phys. A: Materials Science & Processing* **79**, 1847 (2004).

13. A. B. Djurisic, W. C. H. Choy, V. A. L. Roy, Y. H. Leung, C. Y. Kwong, K. W. Cheah, T. K. G. Rao, W. K. Chan, H. T. Lui and C. Surya, *Adv. Funct. Mater.* **14**, 856 (2004).

14. D. C. Look, C. Coskun, B. Claflin and G. C. Farlow, *Physica B: Condensed Matter* **340-342**, 32 (2003).

15. Y. W. Wang, L. D. Zhang, G. Z. Wang, X. S. Peng, Z. Q. Chu and C. H. Liang, *J. Cryst. Growth* **234**, 171 (2002).

16. J. Olrloff, M. Utlaut and L. Swanson, *High Resolution Focused Ion Beams: FIB and Its Applications*, Kluwer Academic/Plenum Publishers, New York, NY (2003).

17. J. Yang, S. Li, Z. W. Li, K. McBean and M. R. Phillips, *J. Phys. Chem. C* **112**, 10095 (2008).

18. L. G. H. J. P. Biersack, *Nucl. Instrum. Methods* **174**, 257 (1980).

19. D. Weissenberger, M. Duerrschnabel, D. Gerthsen, F. Perez-Willard, A. Reiser, G. M. Prinz, M. Feneberg, K. Thonke and R. Sauer, *Appl. Phys. Lett.* **91**, 132110 (2007).

20. A. Djuriscaroni, cacute and Y. H. Leung, *Small* **2**, 944 (2006).

21. I. Shalish, H. Temkin and V. Narayanamurti, *Phys. Rev. B* **69**, 245401 (2004).

22. A. R. Kaul, O. Y. Gorbenko, A. N. Botev and L. I. Burova, *Superlattices and Microstructures* **38**, 272 (2005).

23. S. Shirakata, T. Sakemi, K. Awai and T. Yamamoto, *Thin Solid Films* **451-452**, 212 (2004).

24. H. Matsui, H. Saeki, H. Tabata and Kawai. T, *Jpn. J. Appl. Phys.* **42**, 5494 (2003).

25. M. J. Zhou, H. J. Zhu, Y. Jiao, Y. Y. Rao, S. K. Hark, Y. Liu, L. M. Peng and Q. Li, *J. Phys. Chem. C* **113**, 8945 (2009)

Optical Properties and Devices

Mater. Res. Soc. Symp. Proc. Vol. 1201 © 2010 Materials Research Society   1201-H08-11

Room temperature ferromagnetism in Al-doped/Al$_2$O$_3$-doped ZnO film

Y. W. Ma[1], J. Ding[1], M. Ran[1], X. L. Huang[1] and C. M. Ng[2]
[1]Department of Materials Science and Engineering, National University of Singapore, Singapore
[2]Chartered Semiconductor Manufacturing Ltd, 60 Woodlands Industrial Park D, Singapore

ABSTRACT

In this manuscript, we study the magnetic property of Al-doped/Al$_2$O$_3$-doped ZnO films. We found that metallic Al-doped ZnO film shows room temperature ferromagnetism (RTFM). RTFM is correlated with the interaction of Al metallic clusters and ZnO matrix. The charge transfer has been observed between metallic Al and ZnO matrix. Therefore, RTFM in metallic Al doped ZnO may be highly probable due to charge transfer between metallic Al clusters and ZnO matrix. For Al$_2$O$_3$-doped ZnO film (denoted as $(Zn_{1-x}, Al_x)O$), RTFM was found in $(Zn_{1-x}, Al_x)O$ film with a certain Al concentration range (16 mol%<x<50 mol%). The saturation magnetization is maximized in $(Zn_{0.70}, Al_{0.30})O$ film. The mechanism of RTFM can be explained as the interaction of ZnO nanocrystals (NCs) embedded in the amorphous phase and defects surrounding them.

INTRODUCTION

Recently, ZnO receives extensive attentions due to its potential applications in ultraviolet light emitting devices (LED), piezoelectric transducers and even spintronics applications [1-4]. Diettl et al theoretically predicted that transition metals (TMs) doped ZnO possess room temperature ferromagnetism (RTFM) [5]. Moreover, such RTFM has been found experimentally in TMs doped ZnO, especially Co-doped ZnO [6-7]. However, some research groups even reported that pure ZnO films/nanoparticles show room temperature ferromagnetism [8-10]. Garcia et al found that ferromagnetism in ZnO nanocrystals (NCs) when they are coated with organic molecules [9]. The charge transfer between ZnO NCs and organic molecules was observed in this system. Sundaresan et al observed the ferromagnetism in many inorganic NCs including ZnO [10]. The presence of surface defects on ZnO NCs is the origin of RTFM. Xu et al reported that both ferromagnetism and anomalous Hall Effect (AHE) were found in ZnO film, which was fabricated in the N$_2$ ambient atmosphere by pulse laser deposition [11]. Previously, we reported that non-magnetic elements doped ZnO films showed room temperature ferromagnetism after high vacuum annealing. The ferromagnetism disappeared when the magnetic film was further annealed in air atmosphere. The mechanism to cause ferromagnetism can be explained as the interaction of non-magnetic metallic clusters and ZnO matrix [12]. In this manuscript, we reported that both metallic Al doped and Al^{3+} doped ZnO films show RTFM. The mechanisms to cause ferromagnetism in these systems are different. Through the detailed analysis (X-ray photoelectron spectrum (XPS), transmission electron microscope (TEM) and SQUID system), we found that RTFM in Al^0 (Al^0 denoted metallic Al) doped ZnO film is mainly contributed by the interaction of Al metallic clusters and ZnO matrix surrounding them. On the other hand, RTFM in Al^{3+} doped ZnO film is caused by the defects surrounding ZnO nanocrystals (NCs) which are embedded into alumina-based amorphous phase.

EXPERIMENT

To fabricate the Al^0 doped ZnO films, ZnO film (120 nm) was first deposited on highly-textured quartz (x-cut) at 800°C under an oxygen partial pressure of 10^{-4} torr. The metallic Al film of 10 nm (Good Fellow, 99.99%) was then deposited on the ZnO film under a high vacuum of 10^{-8} torr. The deposited film was annealed under a high vacuum of 10^{-8} torr (continuous pumping with a turbo-pumping system in our PLD system) at a temperature of 700 °C for 30 min. Through the high temperature vacuum annealing, Al was diffused into ZnO underlayer. On the other hand, Al^{3+} doped ZnO films were fabricated by single $(Zn_{1-x}, Al_x)O$ PLD target. The $(Zn_{1-x}, Al_x)O$ targets were prepared by mixing ZnO and Al_2O_3 powders (Sigma, 99.99% in purity). $(Zn_{1-x}, Al_x)O$ films were deposited onto quartz (x-cut) substrate by pulse laser deposition (PLD) at 400°C under an oxygen partial pressure of 10^{-4} torr. X-ray diffraction (XRD) (Bruker, Advance D8), x-ray photoelectron spectroscopy (XPS) (Kratos AXIS Ultra DLD) and transmission electron microscopy (TEM) (JEOL, Jeol 3010) were used for element identification, composition analysis and microstructure investigation. The magnetic properties were studied by a superconducting quantum interference device system (SQUIDs) (Quantum Design, MPMS, X5).

DISCUSSION

(1) RTFM of Al^0 doped ZnO film

The target of ZnO, the deposited ZnO film, Al target and deposited Al films were all nonmagnetic, which were examined by SQUIDs. Figure 1(a) shows the hysteresis loops of the Al/ZnO film. Al/ZnO film in the as-deposited state is not ferromagnetic. However, if the as-deposited film was annealed under vacuum at 700°C, room temperature ferromagnetism is induced. The coercivity of the loop approximates to 100 Oe. Figure 1(b) illustrates that the relation of saturation magnetization (M_s) and annealing temperature of Al/ZnO film. It is clear that the magnitude of M_s is highly dependent on annealing temperatures and M_s is maximized (M_s=1.8 emu/cm³) at T=700°C. With increase in annealing temperature, M_s drops again. Each magnetic result was an average from at least three measurements and all the experiments were repeated. For Al/ZnO film upon 700°C vacuum annealing, FC and ZFC curves were measured to find Curie temperature (T_c) of the film. The result shows that T_c is above room temperature (300K), as shown in the inset of Fig. 1(b). The film was analyzed by XRD measurement and there was no other impurity phase such as (Zn, Al) oxide detected. It has been further confirmed by the glazing angle XRD scan and the high resolution XRD (HRXRD). Hence, it excludes that the origin of ferromagnetism is induced by the impurity phase in the Al/ZnO film. If the magnetic Al/ZnO film was further annealed at 700 °C in air, RTFM was disappeared and no hysteresis loop was observed.

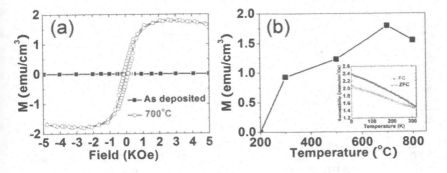

Figure 1. (a) The in-plane hysteresis loops of Al/ZnO films in the as deposited state and upon vacuum annealing (the substrate signal is deducted). (b) M_s in the dependence on vacuum annealing temperature of Al/ZnO films. The inset of (b) is temperature dependence of magnetization at field-cooled (FC) and zero-field-cooled (ZFC) conditions at an applied field of 1000 Oe for the film upon 700°C vacuum annealing.

In order to understand the mechanism to cause room temperature ferromagnetism in Al/ZnO film, XPS was used for the investigation of charge transfer between Al and ZnO. XPS analysis was performed on the Al(10nm)/ZnO(120nm) film upon 700°C vacuum annealing to investigate the mechanism of ferromagnetism. Figure 2(a) and 2(b) shows the XPS spectra of Al $2p$ peaks and Zn $2p$ peaks of the film, respectively. The metallic Al peaks at the binding energy (BE) of 72.9 eV indicates that the presence of Al metallic phase from the surface to the depth of 40 nm [13]. Zn peaks are not symmetry from the surface to the depth of 30 nm, with the peak shift to some extents. At the depth of 40 nm of the film, the Zn 2p peak becomes symmetry, with BE of 1021.8 eV [13]. The inset of Fig. 2(b) shows the peak fitting for the Zn peak at the depth of 10 nm. The experimental peak can be fitted into two peaks: $Zn^{(2-x)+}$ and Zn^0. It is noted that in the pure ZnO film, only Zn^{2+} peak is presented. Due to Al diffusion into ZnO matrix after the vacuum annealing, some Al atoms are oxidized into Al^{3+}. Concurrently, some Zn^{2+} ions are reduced into Zn^0 or Zn^{1+}, which leads to the Zn peak shift and asymmetry. Hence, the electron charge transfer takes place in between Al and ZnO matrix. This may lead to ferromagnetism, as shown in Fig. 1(a). After the subsequent air annealing, the metallic Al peaks disappeared and all metallic Al phase changed to Al oxide phase (not shown here). No charge transfer was occurred in between Al and ZnO. Furthermore, ferromagnetism in the film disappears after air annealing. The RTFM may be induced when there is charge transfer between the metal cluster and the ZnO matrix. The charge transfer creates oxygen non-stoichiometry and/or defects in the film [9]. The charge transfer may alter the electronic structure of metallic Al^0 doped ZnO.

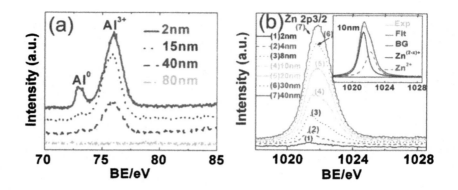

Figure 2. XPS spectra of (a) Al peaks and (b) Zn peaks of Al/ZnO upon 700°C vacuum annealing respectively.

(2) RTFM of Al^{3+} doped ZnO film

The magnetic property of $(Zn_{1-x}, Al_x)O$ film was analyzed by SQUID system. The $(Zn_{1-x}, Al_x)O$ targets and quartz substrate were non-magnetic. As the Al doping concentration is low (x=0.02), there is no ferromagnetic behavior in the film. The room temperature ferromagnetism appears for the sample with the Al molar concentration of 30 mol%. The saturation magnetization (M_s) of the film is 3.6 emu/cm^3 with a coercivity of 100 Oe at room temperature. The variation of M_s with Al molar concentration has been investigated. It can be seen that there was a threshold in Al molar concentration (x=0.16) to induce ferromagnetism in the $(Zn_{1-x}, Al_x)O$ films. $(Zn_{0.70}, Al_{0.30})O$ film has highest M_s. When Al molar concentration increases further (x=0.70), the ferromagnetism disappears.

The origin of ferromagnetism was analyzed by TEM, XPS and photoluminescence (PL) system. The high resolution TEM (HRTEM) and XRD were used to investigate the microstructure. TEM microscopy of the pure ZnO film deposited on the quartz substrate showed that ZnO has a uniform lattice. The calculated d-spacing (0.26 nm) indicates that the film lattice is in the ZnO-(002) direction, which is consistent to the XRD results with the diffraction peak of $2\theta=34.4°$. Therefore, the pure ZnO film is highly textured in the (002) direction. With increase in Al contents in ZnO film, amorphous phase of $(Zn_{1-x}, Al_x)O$ is formed along the grain boundary of ZnO. When x=0.30, microstructure of $(Zn_{0.70}, Al_{0.30})O$ film is shown in Fig. 3(a). It is clearly that ZnO NCs are quite uniformed embedded into alumina based amorphous phase. The average ZnO NCs is around 5 nm. The inset of figure 3(a) shows selected area diffraction pattern (SAED) of $(Zn_{0.70}, Al_{0.30})O$ film, ZnO NCs have polycrystalline structure. Figure 3(b) shows XPS spectrum of $(Zn_{0.70}, Al_{0.30})O$ film. Al exists as Al^{3+} with binding energy (BE) of 75.6 eV and Zn also presents as Zn^{2+} with BE=1021.8 eV in the $(Zn_{1-x}, Al_x)O$ film [13]. Through detailed analysis of the XPS spectrum, there is no metal cluster (either Zn or Al) found in the film. XPS depth profile analysis showed that only Al, O, and Zn elements present in the film. Therefore, with the detection limit, we did not find any impurities inside the film. Therefore, ZnO NCs are playing a crucial role to cause ferromagnetism. Hall measurement was to determine carrier concentration of Al^{3+} doped ZnO films. The carrier concentration in pure ZnO is $1.2 \times 10^{21} cm^{-3}$ and it increases with Al doping concentration. The carrier concentration is maximized at 3×10^{23} cm^{-3} with Al doping concentration of 4 mol%. The increment of carrier concentrations is highly due to defects formation in the ZnO film by doping with Al^{3+}. With further increment of Al concentration, the carrier concentration drops. The carrier concentration cannot be measured when Al concentration >20 mol% as the resistivity is too high and it is above our Hall measurement system limit. PL system was used to investigate the amount of defects in the film. It was found that defects peak (around 550 nm) was enhanced with Al concentration. The relative intensity of defects peak was quite high in $(Zn_{0.70}, Al_{0.30})O$ film and it decreases with further increment of Al contents [14]. When $(Zn_{0.70}, Al_{0.30})O$ film was further annealed in vacuum for 1 hour, the ferromagnetism was enhanced. Sundaresan *et al* suggested that the origin of ferromagnetism is due to the unpaired electron spins which have origin in the oxygen vacancies on the surfaces of ZnO NCs [10]. These electrons located at the oxygen vacancies are polarized and can induce ferromagnetism. In this framework, we propose that ferromagnetism in $(Zn_{1-x}, Al_x)O$ film is due to interaction of ZnO NCs and defects surrounding it.

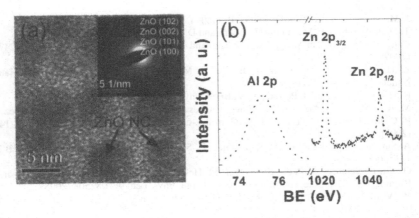

Figure 3. (a) HRTEM images of $(Zn_{0.70}, Al_{0.30})O$ film. The black arrows indicate the ZnO NCs in the amorphous matrix. (b) XPS spectrum of $(Zn_{0.70}, Al_{0.30})O$ film.

CONCLUSIONS

In summary, for Al doped ZnO film, RTFM is attributed to the charge transfer between the Al metal clusters and ZnO matrix. For Al^{3+} doped ZnO film, RTFM is only observed in certain Al^{3+} concentrations in ZnO film (from 16 mol% to 50 mol%). RTFM is probably contributed by ZnO NCs interacting with defects surrounding them.

ACKNOWLEDGMENTS

J. Ding would like to thank the support of Singapore National Research Foundation (NRF-G-CRP 2007-05). Y. W. Ma thanks the financial support from President Graduate Fellowship (PGF, Singapore).

REFERENCES

1. X. Jiang, F. L. Wong, M. K. Fung, and S. T. Lee, Appl. Phys. Lett. **83**, 1875 (2003).
2. H. Ohno, H. Munekata, T. Penney, S. von Molnár, and L. L. Chang, Phys. Rev. Lett. **68**, 2664 (1992).
3. M. Venkatesan, C. B. Fitzgerald, J.G. Lunney, and J.M. D. Coey, Phys. Rev. Lett, **93**, 177206 (2004).
4. J. M. D. Coey, M. Venkatesan, and C. B. Fitzgerald, Nat. Mater. **4**, 173 (2005).
5. T. Dietl, H. Ohno, F. Matsukura, J. Cibert, and D. Ferrand, Science, **287**, 1019 (2000).
6. Congkang Xu, Kaikun Yang, Liwei Huang, and Howard Wanga, J. Appl. Phy. **130**, 124711 (2009).
7. Z. H. Zhang, Xuefeng Wang, J. B. Xu, S. Muller, C. Ronning, and Quan Li, Nature Nanotech. **4**, 523-527 (2009).
8. N. H. Hong, J. Sakai, N. Poirot, and V. Brizé, Phys. Rev. B, **73**, 132404 (2006).
9. M. A. Garcia, J. M. Merino, E. F. Pinel, A. Quesada, J. de la Venta, M. L. Ruíz González, G. R. Castro, P. Crespo, J. Llopis, J. M. González-Calbet and A. Hernando, Nano Lett. **7**, 1489 (2007).
10. A. Sundaresan, R. Bhargavi, N. Rangarajan, U. Siddesh, and C. N. R. Rao, Phys. Rev. B **74**, 161306(R) (2006).
11. Q. Xu, H. Schmidt, S. Zhou, K. Potzger, M. Helm, H. Hochmuth, M. Lorenz, A. Setzer, P. Esquinazi, C. Meinecke, and M. Grundmann, Appl. Phys. Lett. **92**, 082508 (2008).
12. Y. W. Ma, J. B. Yi, J. Ding, L. H. Van, H. T. Zhang, and C. M. Ng, Appl. Phys. Lett. **93**, 042514 (2008).
13. C. D. Wagner, W. M. Riggs, L. E. Davis, and J. F. Moulder, Handbook of X-Ray Photoelectron Spectroscopy (Perkin-Elmer Corporation, Minesota, 1979).
14. Y. W. Ma, J. Ding, D. C. Qi, J. B. Yi, H. M. Fan, H. Gong, A. T. S. Wee, and A. Rusydi, Appl. Phys. Lett, **95**, 072501 (2009).

TCO and TFT

Mater. Res. Soc. Symp. Proc. Vol. 1201 © 2010 Materials Research Society   1201-H09-07

ZnO Thin Film Transistors for RF Applications

Burhan Bayraktaroglu, Kevin Leedy and Robert Neidhard
Air Force Research Laboratory, Sensors Directorate, AFRL/RYD
WPAFB, OH 45433, U.S.A.

ABSTRACT

Nanocrystalline ZnO thin films grown by the pulsed laser deposition technique were used to fabricate high performance thin film transistors suitable for RF applications. It was shown that drain current on/off ratios of higher than $1x10^{12}$, sub-threshold voltage swing values lower than 100 mV/decade and hysteresis-free operation could be maintained with films grown across a wide temperature range (25°C to 400°C). Films grown at 200°C have the lowest surface roughness and result in devices with the highest current density operation. Devices with 1.2 μm gate lengths and Au-based gate metals had record current gain and power gain cut off frequencies of $f_T = 2.9$ GHz and $f_{max} = 10$ GHz, respectively.

INTRODUCTION

Although thin film transistors (TFT) have been in use as long as the conventional transistors, they are still regarded as poorly performing devices that are only suitable for low speed applications. Very low electron mobilities (0.1-1 cm^2/V.s) typically associated with conventional TFTs based on amorphous Si and organic semiconductors coupled with threshold instabilities due to poor grain boundary and interface charge control issues have prevented this technology from advancing to more demanding applications [1]. Recently, alternative TFT technologies based on heavy metal oxide compounds such as ZnO [2-3], GIZO [4], $InGaO_3$ [5] and zinc tin oxide [6] have been investigated to overcome these limitations.

As a binary compound, ZnO represents the simplest form of the metal oxide semiconductor. It can be fabricated as a high performance thin film easily by various deposition techniques including RF sputtering, pulsed laser deposition (PLD), atomic layer deposition, and solution based approaches. Most of these approaches produce very similar films that result in excellent thin film transistor characteristics. For example, using PLD, devices with 2μm gate lengths produced record figure-of-merit numbers such as on/off ratio of $1x10^{12}$, current density of >400 mA/mm and field effect mobility of 110 cm^2/V.s [7].

ZnO is one of the most versatile semiconductors available today and offers solutions in more classes of applications than any other semiconductor technology. The advantages of this semiconductor stem from the combination of two features that are unique to ZnO. First, it is a wide bandgap semiconductor ($E_g = 3.37$ eV) which makes it transparent to IR and visible light [8]. Wide bandgap enables low leakage current and high temperature operations. Also, it can support high electrical fields for high voltage device applications. Second, it maintains most of its intrinsic single crystal optical and electronic properties in a nanocrystalline thin film (i.e. not single crystal) form. This high level of tolerance to defects makes it suitable for a wide range of thin film applications including conformal and flexible electronics. A summary of ZnO TFT features and their implications in various applications are shown in Table I.

Table I. Summary of various applications resulting from unique properties of ZnO.

ZnO Properties	Applications
Wide Bandgap	High voltage, high temperature electronics
Transparent Contacts	Transparent electronics
Nanocrystalline	Flexible and printable, large area, 3D integration
High Speed	Microwave amplifiers, fast logic circuits
High Current Density	Control electronics
Low Leakage Currents	Ultralow power electronics

Overall, ZnO brings together the superior electronic and optical benefits of wide bandgap semiconductors and the application diversity of thin film electronics without sacrificing performance. Because of this successful merging of previously incompatible technology approaches, nanocrystalline ZnO TFTs are suitable for a wider range of applications than almost any other semiconductor technology. Looking at its advantages from a thin film technology point of view, nanocrystalline ZnO films can be used for the same applications that are currently addressed by organic, polycrystalline Si, or amorphous Si. These applications include transparent electronics, printed electronics, large area, flexible and conformal electronics. Because of its higher speed, high current density, high on/off ratio, and low leakage current properties, ZnO TFTs can also be used in high speed logic circuits, microwave power amplifiers, and ultra-low power electronics applications.

High speed operation of PLD-grown nanocrystalline ZnO TFTs was demonstrated on GaAs [7] and Si [9] substrates. Small-signal current and power gain cut-off frequency values of 2.45 GHz and 7.45 GHz, respectively were demonstrated using 1.2 μm gate length devices. Building on these results, we have refined materials preparation, device design and fabrication techniques to extend the f_T and f_{max} values to 2.9 GHz and 10 GHz, respectively. Extensive thin film growth studies were undertaken to better understand the impact of growth conditions on device performance. It was shown that nanocrystalline ZnO films suitable for RF applications can be prepared at substrate temperatures ranging from 25°C to 400°C.

EXPERIMENT

Thin Film Preparation and Characterization

ZnO films were deposited in a Neocera Pioneer 180 pulsed laser deposition system with a KrF excimer laser (Lambda Physik COMPex Pro 110, λ=248 nm, 10 ns pulse duration). The base pressure of the chamber was 4×10^{-8} Torr. The following processing conditions were used to deposit ZnO films: laser energy density of 2.6 J/cm^2, laser repetition rate of 30 Hz, deposition temperature of 25°C to 400°C, oxygen partial pressure of 10 mTorr during the deposition, and substrate-to-target distance of 9.5 cm. The wafer was rotated at a constant speed of 20°/sec and the target was rotated at a constant speed of 40 °/sec. A deposition rate of 0.3 nm/sec was achieved. The target was a 50 mm diameter by 6 mm thick sintered ZnO ceramic disk (99.999%).

The ZnO crystal structure was determined by using a PANalytical X'Pert Pro MRD x-ray diffractometer. The film morphology was analyzed with an FEI DB235 scanning electron microscope (SEM) and a JEOL 4000EX transmission electron microscope (TEM) operating at 400 kV. Surface roughness was measured with a Veeco Dimension 3000 atomic force microscope (AFM) and analyzed with SPIP image processing software. Optical measurements were made on films deposited on quartz substrates using a Varian Cary 5000 spectrophotometer.

Thin Film Transistor Fabrication and Characterization

The devices were fabricated on 20 nm-thick SiO_2-covered high-resistivity (> 2000 Ω. cm) Si wafers. As shown in Figure 1, a bottom-gate configuration was used for device fabrication with a Ni/Au (5 nm/120 nm) gate below 20-30 nm thick PECVD-deposited SiO_2 gate insulator. The device fabrication sequence was typical of bottom-gate transistors and involved the fabrication of gate metal by evaporation and lift off techniques. The gate insulator and ZnO thin film were deposited over the gate metal in sequence. The ZnO layer thickness was 50 nm to 100 nm. As mentioned above, the ZnO deposition temperature ranged from 25°C to 400°C. Some wafers were annealed in clean room air ceramic ovens at 400°C for 10 min. The device area was defined by mesa etching of the ZnO film in dilute HCl solution. Via holes were opened in the SiO2 layer over the gate contact pads using reactive ion etching before the fabrication of Ti/Pt/Au (20/30/350 nm) source/drain contacts by evaporation and lift of techniques. No passivation layers were used over the top surface of the devices in this study.

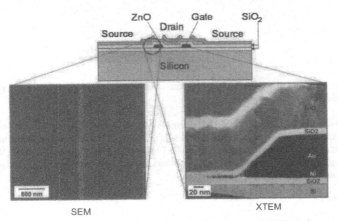

Figure 1. Cross sectional drawing of ZnO TFT together with SEM and TEM pictures of the gate contact area.

As shown in Figure 1, the gate insulator and ZnO films cover the gate metal conformably. Nanocolumns of the ZnO film can be seen in this picture to follow the contours of the gate metal with no physical gaps between the columns. The ZnO film thickness also appears to be unchanged over the same contours.

High frequency (RF) devices were fabricated using the same fabrication sequence described above. In the design of the RF devices, however, parasitic elements were minimized as much as possible. The important parasitic, whose minimization had the greatest impact on the device performance were the substrate resistivity, gate resistance and the overlap capacitance between gate and source/drain electrodes. In a 2-gate finger configuration device design, as shown in Figure 2, the minimum gate length was 1.2 µm (defined as the gap between source and drain contacts). The source and drain contacts had various amounts of overlap to the gate electrode in three designs. The overlap amounts were approximately 1.5, 1, and 0.5µm, as measured by SEM, in designs designated as Design A, Design B, and Design C, respectively. The gate series resistance was minimized by using a gold-based gate metal, whose thickness was limited to 150 nm in this study.

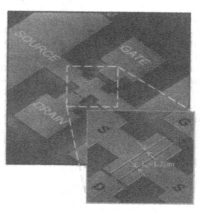

Figure 2. SEM image of RF ZnO TFT. The gate length and gate width were 1.2 µm and 2x50 µm, respectively.

The devices were dc characterized using an Agilent 4156C Precision Semiconductor Parameter Analyzer. Microwave performance was measured at room temperature using on-wafer coplanar microwave probes in an Agilent 8364B Precision Network Analyzer. From the measured s-parameters, current gain, $|h_{21}|^2$, and maximum available gain (MAG) values were determined as a function of frequency.

DISCUSSION

The SEM images in Figure 3 show polycrystalline ZnO surfaces with uniform, densely packed grains approximately 25 nm in size. Films deposited at 400°C exhibited grains with the most contrast. Resolution of grains decreased with decreasing deposition temperature, as shown for a 100°C deposition in Figure 3a. This surface morphology was consistently observed over the entire deposited area.

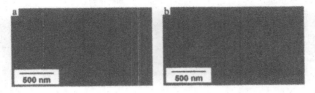

Figure 3. Scanning electron microscope images of the surface morphology of ZnO thin films deposited at a) 100°C and b) 400°C.

X-ray diffraction scans of the ZnO films in Figure 2 exhibited a highly oriented c-axis orientation with only the ZnO (002) peak present. The relative intensities of the (002) peak in films deposited between 25°C and 100°C were low compared to films deposited at 200°C to 400°C with the highest intensity observed at a 300°C deposition. The (002) peak positions ranged from 33.87° for a 25°C deposition increasing to 34.37° for a 400°C deposition. All 2Θ positions were less than the 34.421° 2Θ (002) peak from the JCPDS #36-1451 powder diffraction file indicating strained lattice structures. The highly textured c-axis orientation is consistent with other studies of PLD ZnO films [10-13]. The full-width at half-maximum value for the (002) peak decreased from 1.14° at 25°C deposition to 0.87° at 400°C deposition indicating improved film crystallinity with increasing temperature.

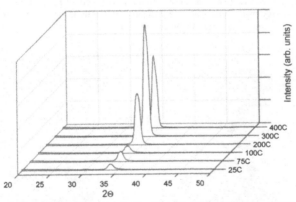

Figure 4. X-ray diffraction patterns of ZnO thin films deposited at 25°C to 400°C.

Cross sectional TEM images reveal densely compacted, highly faulted columnar-shaped grains as shown in Figure 5. Minimal differences in grain structure were observed as a function of deposition temperature from 100°C to 400°C. Similar faulted polycrystalline ZnO deposited by PLD at 500°C was observed on Si substrates [12]. A smooth interface exists between the ZnO and the underlying SiO_2 layer while the tops of the ZnO grains are dome-shaped. The nominal 25 nm ZnO grain size and approximately 100 nm thick TEM sample complicated grain boundary resolution due to overlapping grains.

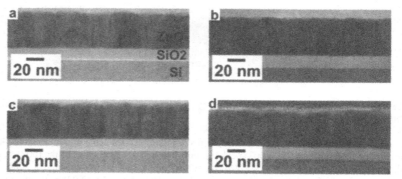

Figure 5. Cross sectional TEM images of ZnO thin film deposited on 20 nm SiO$_2$/Si at a) 100°C, b) 200°C, c) 300°C and d) 400°C.

AFM images with 500 nm x 500 nm collection areas in Figure 6 show that all films were predominantly smooth. RMS roughness values ranged from 0.65 nm to 1.65 nm. Although the lowest roughness occurred in a film deposited at 200°C, no trend in roughness as a function of deposition temperature was observed. The RMS roughness is similar to other reported values of ZnO deposited by PLD [12,13]. Grain size calculations based on grain boundary intercepts in a scanning probe image processing software indicated 25 nm to 35 nm ZnO grains, again with no trend observed as a function of deposition temperature.

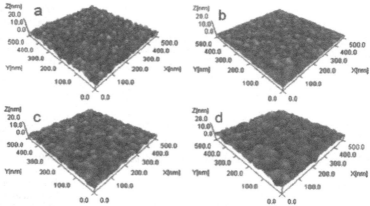

Figure 6. AFM images of ZnO thin film deposited at a) 100°C, b) 200°C, c) 300°C and d) 400°C.

The influence of postgrowth annealing in ambient air on film transparency is shown in Figure 7. The average transmittance was 90% from about 390 to 2500 nm wavelengths for as-grown films and showed no change in transmittance in the same wavelength range after a 600°C 1 hr. air anneal. A slight shift in absorption edge was observed after annealing corresponding to a bandgap increase from 3.2 eV in as-grown films to 3.22 eV in annealed films.

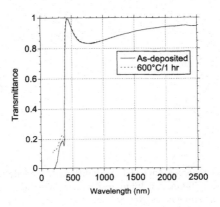

Figure 7. Optical transmittance of 100nm thick ZnO films deposited on quartz: as-deposited and after a 600°C/1 hr air anneal.

Common source I-V characteristics of devices with various gate lengths are shown in Figure 8. The gate width and gate insulator thickness values were kept constant at 400 μm and 30 nm, respectively. The drain voltage was swept in both directions at each gate bias value. As expected, devices with longer gate lengths exhibit more saturated current characteristics but lower current values at the same gate voltage compared to shorter gate length devices. At the same gate and drain bias values, the drain current was inversely proportional to the gate length. This is better illustrated by the summary plot on the right side of the figure. I_D*L_G vs. V_D characteristics of three different size devices are almost identical in the linear region, indicating normal device scaling. In the saturated region, shorter gate length devices show somewhat higher current values due to short gate effects (note: gate insulator thickness is constant for all devices). It is also worth noting the absence of hysteresis in device characteristics.

233

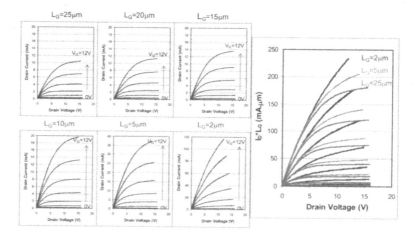

Figure 8. I-V characteristics of ZnO TFTs with different gate lengths. Gate width and gate insulator thickness were 400μm and 30nm, respectively. The figure on the right shows a composite of 3 devices with gate lengths of 2, 5, and 25 μm. Note that the y-axis of this figure is I_D*L_G.

The absence of hysteresis can also be observed in device transfer characteristics shown in Figure 9 where the gate voltage was swept in both directions. Figure 9a shows the transfer characteristics of devices with constant gate width (400 μm) and varying gate lengths, whereas Figure 9b shows the transfer characteristics of 5 mm gate length devices with different gate widths. The drain bias was 1 V for all measurements. Some of the notable features observable from these figures are the high on/off ratios (>$1x10^{12}$), low sub-threshold voltage swings (60-100 mV/decade), and the almost identical device characteristics in the exponential region with all size devices. It is obvious from these figures is that devices scale well with gate length and width within the range investigated. We attribute this well-behaved device scaling to the uniformity of the ZnO film and ZnO-insulator interface properties.

Impact of Film Growth Temperature on Device Performance

The basic characteristics of the device were unaffected by the film growth temperature. As shown in Figure 10, the transfer characteristics of devices using films grown at temperatures ranging from 25°C to 400°C are very similar. All devices had excellent on/off ratios, the same sub-threshold voltage swings, and no hysteresis characteristics. The only discernable growth temperature dependent parameter was the current density, which will be discussed in more detail below. The threshold voltage, as measured from the linear portion of $I_D^{1/2}$ vs. V_G curves, was almost constant at 1 V, as shown in Figure 11. These results indicate that the ZnO film and the ZnO-insulator interface qualities are not strongly impacted by the film growth temperature.

234

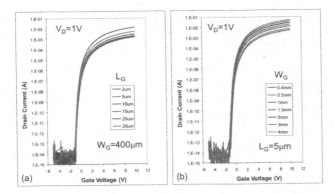

Figure 9. ZnO TFT transfer characteristics as a function of device size indicating excellent device size scaling.

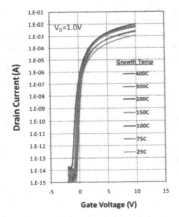

Figure 10. Transfer characteristics of nanocrystalline ZnO TFTs grown at different temperatures.

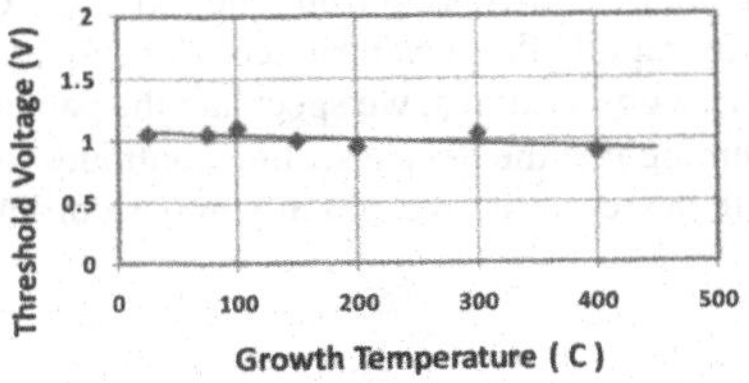

Figure 11. Threshold voltage dependence ZnO film growth temperature

The only device parameter that seems to be impacted by the film growth temperature is the current density, which is related to the film conductivity and the mobility of electrons in the channel. Since the current density is also dependent on the gate length, we have examined growth temperature impact on current density on various gate length devices. The results shown in Figure 12 indicate that the same current density behavior holds true for each device size. The current density peaks for films grown at 200°C, and then shows a slow decline at higher temperature growth conditions. The maximization of the current density at 200°C may be related to improved film surface roughness obtained at this temperature, as discussed above. It is also

235

possible the nanocolumns in the film are more closely packed at that temperature and cause a reduction in charges at grain boundaries. The increase current density may be a result of reduced scattering at the grain boundaries.

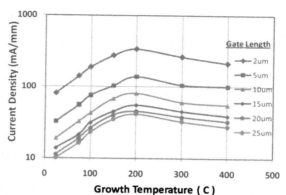

Figure 12. Current density dependence on film growth temperature and device gate length.

Figure 13 shows that current density dependence on gate length is not impacted by the film growth temperature. In other words, the same device scaling rules apply for films grown at different temperatures as indicated by a parallel shift of characteristics. This shift is significant from room temperature to 200°C (approximately 4X increase in current density) as seen in Figure 13a. On the other hand, a 25% drop in current density is observed when the film growth temperature is increased from 200°C to 400°C (see Figure 13b). Beyond 400°C, the current density rapidly drops to near zero as nanocolumns begin to be separated from each other. Based on these observations, we speculate the packing of nanocolumns is maximized at 200°C. Assuming that the device scaling continues to hold, the expected current density of 1 μm gate length devices is above 500 mA/mm according to Figure 13a.

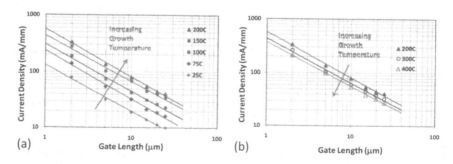

Figure 13. Current density dependence on device dimensions as a function of film growth temperature.

236

The microwave performance of devices with $L_G = 1.2$ μm and $W_G = 2\times50$ μm were determined at various drain and gate bias values. The calculated maximum available gain (MAG) and current gain ($|h_{21}|^2$) values from the measured s-parameters are shown in Figure 14. The bias conditions for these devices were $V_G=6V$ and $V_D=11V$, which resulted in a drain current of $I_D=32$ mA (i.e. 320 mA/mm current density). All three device designs (with varying amounts of gate electrode overlap with source/drain contacts) showed an identical current gain cut off frequency of 2.9 GHz.

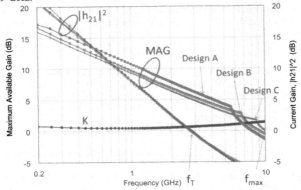

Figure 14. Small signal microwave characteristics of nanocrystalline ZnO TFTs.

The power gain characteristics of the same devices were considerably different. At lower frequencies where the stability factor, K, was less than unity, the MAG declined with frequency at -3 dB/octave rate. For K>1, the decline rate was -6 dB/octave. Design A, with gate-source/drain overlap distance of approximately 1.5 mm, showed f_{max}=7.5GHz. This value is comparable to previously reported value of 7.45 GHz for the same size device.[9] Designs B and C, on the other hand showed f_{max} values of 8.5 GHz and 10 GHz, respectively. These results compare favorably with the RF performance obtained (f_T= 180 MHz, f_{max}= 155 MHz) with other metal oxide (InZnO) thin film transistors on glass substrates [14]. To the best knowledge of the authors, these are the highest cutoff frequencies obtained with thin film transistors, and indicate the excellent potential of nanocrystalline ZnO TFTs for RF circuit applications.

CONCLUSIONS

It was shown that nanocrystalline ZnO thin films prepared by the pulsed laser deposition technique across a wide temperature range (25°C to 400°C) are suitable for the fabrication of high performance transistors. Most critical device parameters including drain current on/off ratios, sub-threshold voltage swings, threshold voltages and hysteresis-free operation were shown to be independent of film growth temperature. The only parameter that had dependence on growth temperature was the current density, which exhibited a peak for films grown at 200°C. ZnO nanocolumn packing density variation with growth temperature is the likely cause of current density variations. Devices fabricated on Si substrates with 1.2 μm gate lengths and Au-based gate metals had record current gain and power gain cut off frequencies of f_T = 2.9 GHz and f_{max} = 10 GHz, respectively. To our knowledge, these are the highest frequency operation of any thin film transistors.

ACKNOWLEDGMENTS

This work was supported in part by Air Force Office of Scientific Research under LRIR 07SN03COR (Dr. K. Reinhart) and Defense Advanced Research Projects Agency (Dr. J. Albrecht). The authors thank D. Tomich and J. Brown for ZnO film characterization.

REFERENCES

1. C. R. Kagan and P. Andry, "Thin Film Transistors", Marcel Dekker Publishing, New York, 2003.
2. P.F. Carcia, R. S. McLean, and M. H. Reilly, *Appl. Phys. Lett,* **88**, 123509 (2006).
3. P. K. Shin, Y. Aya, T. Ikegami, and K. Ebihara, *Thin Solid Films* **516**, 3767 (2008).
4. W. Lim, S. H. Kim, Y. L. Wang, J. W. Lee, D. P. Norton, S. J. Pearton, F. Ren, and I. I. Kravchenko, *J. Electrochem. Soc.* **155**, H383 (2008).
5. H. Q. Chiang, D. Hong, C. M. Hung, R. E. Presley and J. F. Wager, *J. Vac. Sci. Technol. B* **24**, 2702 (2006).
6. H. Q. Chiang, J. F. Wager, R. L. Hoffman, J. Jeong, and D. A. Keszler, *Appl. Phys. Lett.* **86**, 013503 (2005).
7. B. Bayraktaroglu, K. Leedy and R. Neidhard, *IEEE Electron Dev. Lett.* **29**, 1024 (2008).
8. Ü. Özgür, Y. I. Alivov, C. Liu, A. Teke, M. A. Reshchikov, S. Doğan, V. Avrutin, S.J. Cho, and H. Morkoç, *Appl. Phys. Rev.* **98**, 041301 (2005).
9. B. Bayraktaroglu, K. Leedy and R. Neidhard, *IEEE Electron Dev. Lett.* **30**, 946 (2009).
10. L. Bentes, R. Ayouchi, C. Santos, R. Schwarz, P. Sanguino, O. Conde, M. Peres, T. Monteiro, and O. Teodoro, *Superlattices and Microstructures* **42**, 152 (2007).
11. S. Amirhaghi, V. Craciun, D. Craciun, J. Elders and I. W. Boyd, *Microelectronics Engineering* **25**, 321 (1994).
12. L. Han, F. Mei, C. Liu, C. Pedro and E. Alves, *Physica E*, **40**, 699 (2008).
13. C.-F. Yu, C.-W. Sung, S.-H. Chen and S.-J. Sun, *Appl. Surface Sci.* **256**, 792 (2009).
14. Y. L. Wang, L. N. Covert, T. J. Anderson, W. Lim, J. Lin, S. J. Pearton, D. P. Norton, J. M. Zavada and F. Ren, *Electrochem. Sol. State Lett.* **11**, H60 (2008).

Poster Session II

Mater. Res. Soc. Symp. Proc. Vol. 1201 © 2010 Materials Research Society   1201-H10-01

Room Temperature Ferromagnetism in Powder Form of $Sn_{1-x}Cr_xO_2$

($x = 0.01, 0.02, 0.03, 0.04$ and 0.05) Solid Solution

Kun Xu[1], Mitsuru Izumi[1], Osami Yanagisawa[2], and Tetsuya Ida[3]

[1]Laboratory of Applied Physics, Tokyo University of Marine Science and Technology, Tokyo, Japan

[2]Maritime Technology Department, Yuge National College of Maritime Technology, Ochi-gun, Japan

[3]Department of Electronic Control Engineering, Hiroshima National College of Maritime Technology, Hiroshima, Japan

ABSTRACT

Structural and magnetic properties were investigated in the mixed powders of $Sn_{1-x}Cr_xO_2$ ($x = 0.01, 0.02, 0.03, 0.04$ and 0.05) in nominal composition. The lattice parameter observed in (110) x-ray diffraction indicates two step changes with increasing Cr content. The occupation of Cr ion at the interstitial position leads to elongation of the lattice parameter for x = 0.01 to x =0.03. Then, the Cr^{3+} ions are remarkably substituted into the Sn^{4+} ion site for x = 0.04 to x = 0.05, which results in shortening of the lattice. The lattice parameters for $x = 0.01$ and 0.02 are larger than $x = 0.03$ to 0.05. The room temperature ferromagnetism appeared in the sample with $x = 0.01$ and reaches maximum at the doping rate of $x = 0.02$; while the magnetization decreases for $x > 0.02$ was observed. Present study clearly shows the existence of correlation between appearance of ferromagnetism and the structural change.

INTRODUCTION

Diluted magnetic semiconductors (DMS), which are made by doping a few percentages of transition metal (TM) ions into semiconductors, attract much attention recently for their potential applications in the field of spintronics [1]. Tin dioxide (SnO_2) is a very attractive system for its wide band gap (~3.6 eV) and excellent transparency. Recently, many groups have reported the room temperature ferromagnetic (RT-FM) property of TM-doped SnO_2 theoretically [2] and experimentally [3,4]. Their results approved that the RT-FM originates from the TM doping and consequent oxygen vacancies instead of metallic clusters or any secondary phases. However, the magnetization results by different groups or using different preparing methods are different with each other. We try to clarify this problem from the viewpoint of correlation between structural and magnetic properties.

The ball milling is an effective sample preparing method to mix Cr^{3+} ions into SnO2 powders. However, few related works have been reported by using the ball milling method so far [5]. In this work, we studied the structural and magnetic properties of $Sn_{1-x}Cr_xO_2$ ($x = 0.01, 0.02, 0.03, 0.04$ and 0.05) synthesized from the mixture of SnO_2 and Cr_2O_3 powders prepared by ball milling. The effect of structural change on magnetic property is reported.

EXPERIMENTS

The $Sn_{1-x}Cr_xO_{2-\delta}$ samples were prepared by ball milling the Cr_2O_3 and SnO_2 powders. The starting materials of Cr_2O_3 (99.99%) and SnO_2 (99.9%) were mixed with x varying from 0.01 to

0.05. The mixture was ball milled for 15 minutes, and then thermally treated at 1000°C for 12

hours. Both the ball milling and annealing processes were done in the air. In order to prevent contamination, all of the processes were managed without using any kind of metallic tools.

The structural analysis of the samples was performed using a Bruker powder x-ray

diffractometer (XRD) with Cu K_α radiation, which is operated at 40 kV and 200 mA. All the

magnetic measurements were carried out by using a Quantum Design physical property measurement system- vibrating sample magnetometer (PPMS-VSM).

RESULTS AND DISCUSSION

Figure 1 (a) shows the x-ray diffraction patterns of $Sn_{1-x}Cr_xO_2$ ($x = 0.01, 0.02, 0.03, 0.04$ and 0.05) solid solution. The results show that the SnO_2 is in a single tetragonal phase. As seen in the fig. 1 (b), the peaks of Cr_2O_3 are identified for the sample of $x = 0.05$. There is no remarkable trace of the existence of Cr_2O_3 in other samples. This result indicates the successful substitution of Sn^{4+} ions by the Cr^{3+} ions at least $x < 0.05$.

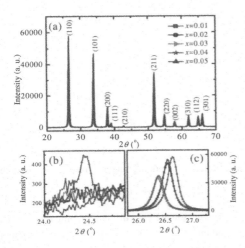

Figure 1. (a) Room temperature x-ray diffraction profiles in powder form of $Sn_{1-x}Cr_xO_2$ ($x = 0.01$, 0.02, 0.03, 0.04 and 0.05) in nominal composition. The enlarged view for (b) (012) peak of Cr_2O_3 in the sample with $x = 0.05$ and (c) SnO_2 (110) peaks.

The figure 1 (c) shows the enlarged view of SnO_2 (110) peaks for different Cr^{3+} doping rates. The SnO_2 (110) peak shifts to low diffraction angle associated with decreasing intensity from $x =$ 0.01 to 0.03, and shifts to high diffraction angle with increasing intensity for $x = 0.04$ and 0.05. This means that the amount of doped Cr^{3+} ions influence the structural property. The average grain size of $Sn_{1-x}Cr_xO_2$ powder, as shown in fig. 2, was calculated using the Scherrer Equation by the full width half maximum. For the sample $x = 0.02$, the average grain size is a little smaller than in the other doping rates (38.3 nm). In general, the grain size merely changes with varying the doping rate x. The uniformity approves the reliability of this experiment. Besides, the smallness of the doped SnO_2, compared to the undoped one (70 nm), suggests the doping would inhabit the particle growth [6].

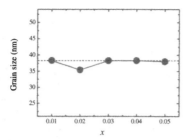

Figure 2. The variation of grain size as a function of doping rate x derived from the Scherrer Equation.

The variation of SnO_2 (110) peak position and intensity for different doping rates x was plotted in fig. 3. The parameters (position and intensity) of SnO_2 (110) peak show the similar trend. The position and intensity of SnO_2 (110) show a linear relationship as in the inset of fig. 3. The 2θ decreases from $x = 0.01$ to 0.03. And it becomes stable for $x = 0.03$ and 0.04. and increases finally for $x = 0.05$. The occupation of Cr ion at the interstitial position may lead to elongation of the lattice parameter for x = 0.01 to x =0.03. The further doping leads to both interstitial incorporation and substitution of Cr ion for Sn ion site. This resulted in decrease of lattice parameter. Thus, the 2θ slightly increases until $x = 0.03$ and 0.04. Finally, the appearance of Cr_2O_3 made the 2θ increase again. The reason for this explanation will be approved with magnetic results.

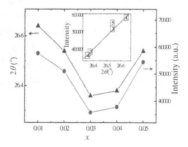

Figure 3. The variation of SnO_2 (110) peak position and intensity for different doping rates x. Inset shows the linear relationship between the SnO_2 (110) peak position and intensity, the squared numbers 1-5 mean the doping percentages as 0.01-0.05 at. %, respectively.

Magnetic hysteresis loop measurements were carried out between ±10 kOe at the room temperature. The resulted hysteresis loops of the $Sn_{1-x}Cr_xO_2$ ($x = 0.01, 0.02, 0.03, 0.04$ and 0.05) after subtraction of a paramagnetic component were plotted in the fig. 4 (a). All of the samples exhibit an obvious ferromagnetic behavior in spite of the diamagnetic and antiferromagnetic characters of SnO_2 and Cr_2O_3 respectively. From $x = 0.01$ to 0.02, variation of the ferromagnetic magnetization (M_{FM}), the paramagnetic magnetization (M_{PM}), and the d obtained from 2θ show that the amounts of both substituted and interstitial Cr^{3+} ions increase. The increase of M_{FM} and d should be related to the substituted and interstitial Cr^{3+} ions respectively. The interstitial Cr^{3+} ion would produce the oxygen vacancies which is the origin of ferromagnetism in the DMSs. The present role of the interstitial Cr ions is in contrast with the role of the substituted Cr ion in the interpretation with the nanoparticles made by the sol-gel methods [6,7]. The oxygen vacancies (V_O) prefer to form near the Cr^{3+} instead of Sn^{4+} site. The consequently Cr^{3+}-V_O-Cr^{3+} groups have a crucial effect to the ferromagnetism [8]. Furthermore, the M_{FM} of 0.4296 emu/g_{SnO_2} for $x = 0.02$ consists well with the result in ball milled Co_3O_4 and TiO_2 mixtures [5]. While for $x = 0.02$ ~ 0.04, the decrease of M_{FM} and increase of d might be induced by the decrease of substituted Cr^{3+} ion and increase of the interstitial one. The interstitial one will produce the oxygen vacancies as a result, the M_{FM} decrease. Finally, the appearance of Cr_2O_3 at $x = 0.05$ cause the increase of M_{PM} and decrease of d.

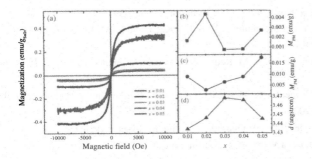

Figure 4. (a) The room temperature hysteresis loops for the ball milled $Sn_{1-x}Cr_xO_2$ ($x = 0.01$ ~ 0.05) powders. The variation of ferromagnetic (b), paramagnetic magnetization (c), and d (d) obtained from 2θ for different x.

CONCLUSIONS

In summary, we have investigated the structural and magnetic properties $Sn_{1-x}Cr_xO_2$ (x = 0.01, 0.02, 0.03, 0.04 and 0.05) powders in nominal composition. The structural property was studied by the x-ray diffraction pattern with emphasis on the SnO_2 (110) peak. Different mechanisms dominate for different doping rates. The occupation of Cr ion at the interstitial position may induce the oxygen vacancy associated with the elongation of the lattice parameter for x = 0.01 to x =0.03. The Cr^{3+}-V_O-Cr^{3+} groups may have a crucial effect to the appearance of the ferromagnetism for x = 0.02. The further doping enhances the substitution effect of Cr ion for Sn ion site. The magnetization decreases for $x > 0.02$ was observed. Present study shows the existence of correlation between appearance of ferromagnetism and the structural change.

ACKNOWLEDGMENTS

The work was financially supported by the Sasakawa Scientific Research Grant from The Japan Science Society. We acknowledge for the Shanghai University for provision of PPMS measurement.

REFERENCES

1. Y. Matsumoto, M. Murakami, T. Shono, T. Hasegawa, T. Fukumura, M. Kawasaki, P. Ahmet, T. Chikyow, S. Koshihara, and H. Koinuma, *Science* **291**, 854 (2001).
2. H. X. Wang, Y. Yan, Y. Sh. Mohammed, X. B. Du, K. Li, and H. M. Jin, *J. Magn. Magn. Mater.* **321**, 337 (2009).
3. S. K. Misra, S. I. Andronenko, K. M. Reddy, J. Hays, and A. Punnoose, *J. Appl. Phys.* **99**, 08M106 (2006).
4. S. K. Misra, S. I. Andronenko, K. M. Reddy, J. Hays, A. Thurber, and A. Punnoose, *J. Appl. Phys.* **101**, 09H120 (2007).
5. A. Serrano, E. Fernandes Pinel, A. Quesada, I. Lorite, M. Plaza, L. Perez, F. Jimenez-Villacorta, J. de la Venta, M. S. Martin-Gonzalez, J. L. Costa-Kramer, J. F. Fernandez, J. Llopis, and M. A. Garcia, *Phys. Rev. B* **79**, 144405 (2009).
6. C. Van Komen, A. Thurber, K. M. Reddy, J. Hays, and A. Punnoose, *J. Appl. Phys.* **103**, 07D141 (2008).
7. A. Thurber, K. M. Reddy, and A. Punnoose, *J. Appl. Phys.* **101**, 09N506 (2007).
8. J. M. D. Coey, A. P. Douvalis, C. B. Fitzgerald, and M. Venkatesan, *Appl. Phys. Lett.* **84**, 1332 (2004)

Mater. Res. Soc. Symp. Proc. Vol. 1201 © 2010 Materials Research Society   1201-H10-07

Room Temperature Ferromagnetism in Spin-Coated Anatase and Rutile Ti$_{1-x}$M$_x$ O$_2$ (M= Fe, Mn, Co) Films

Danilo Barrionuevo[1], Surinder P. Singh[2] and Maharaj S. Tomar[1]

[1]Department of Physics, University of Puerto Rico, Mayaguez, Puerto Rico, PR 00681
[2]Engineering Science and Materials, University of Puerto Rico, Mayaguez, PR 00681

ABSTRACT

Diluted magnetic semiconductors (DMS) have been explored extensively, because of their potential application in spintronic devices. We studied the structural optical and magnetic properties of Ti$_{1-x}$M$_x$O$_2$ (M= Fe, Mn, Co; x = 0.00, 0.05, 0.08, 0.10, 0.15, 0.20, 0.25, 0.30), thin films by sol-gel process and deposited using spin coating on Pt (Pt/Ti/SiO$_2$/Si) and quartz substrates. X-ray diffraction studies and Raman spectroscopy reveal anatase and rutile phases of the synthesized films when annealed at 500 and 1000^0 C, respectively. Optical transmission measurements show high degree of transparency that decreases with increase in transition metal ion concentration. The films show room temperature ferromagnetism, suggesting their potential in spin based heterojunction devices.

INTRODUCTION

Wide band gap II-VI and III-V compound semiconductors have attracted a great deal of interest as they could me made dilute magnetic semiconductor (DMS) by replacing host cation with magnetic impurity (Fe, Co, Ni, Mn, etc) up to few atomic percent [1-2]. These materials promises for simultaneous control of charge currents and spin polarized currents which may open new paradigms in information processing technologies and spintronics [3]. However, the fundamental condition imposed to these magnetic semiconductors to be used in spintronics is to have ferromagnetic properties at room temperature. Theoretical calculations predicted wide band metal oxide semiconductors (ZnO and TiO$_2$) as good candidates for the host materials, which after doping with transition metal ions could reveal room temperature ferromagnetism. Electronic structure calculations indicate that the alternative dopants such as Co, Fe and Mn may induce ferromagnetism [4] in TiO$_2$. It is known that TiO$_2$ has three different crystal structures i.e. rutile, anatase and brookite and excellent optical transmission in the visible and near-infrared regions with high n-type carrier mobility in material system for spintronics. Recently, dilute magnetic semiconductors such as Co doped TiO$_2$, Fe doped TiO$_2$, and Mn and Co doped ZnO have been reported to show the room temperature ferromagnetism [5]. However, the origin of ferromagnetism in magnetically doped TiO$_2$ at room temperature is still not clear. It has been argued that the origin of ferromagnetism is caused by the exchange interaction mediated by the vacancy-induced n-type carriers when magnetic ion replaces the Ti in host TiO$_2$ lattice. Also, the magnetic ions especially Co in Ti$_{1-x}$Co$_x$O$_2$ form clusters or Co incorporates into the interstitial position and/or forms Co–Ti–O complexes result in the high Curie temperature [6-8]. In this work, we prepared thin films of Ti$_{1-x}$M$_x$O$_2$ (M = Fe, Co, Mn; x = 0.00, 0.05, 0.08, 0.10, 0.15,

0.20, 25, and 0.30) in anatase and rutile phases by sol–gel spin coating. The films were
investigated for their structural, optical and magnetic properties.

EXPERIMENTAL

The precursor solution for the pure anatase and rutile TiO_2 samples was prepared by
homogeneously mixing titanium (IV) isopropoxide and 2-methoxyethanol. The iron nitrate,
cobalt chloride and manganese acetate were used as precursor to synthesize doped-TiO_2 films.
The corresponding solutions of transition metal ion doped $Ti_{1-x}M_xO_2$ (x = 0.05, 0.08, 0.10, 0.15,
0.20, 0.25 0.30) were prepared using ultrasonic bath. The thin films were deposited by spin
coating the solutions at 5000 rpm for 30 seconds onto platinum Pt ($Pt/Ti/SiO_2/Si$) and quartz
substrates, and annealed at 300^0 C for 5 min with every single coating. This process was repeated
several times to build up the desired thickness. The films were finally annealed at temperatures
500^0 C for development of anatase phase and 1000^0 C for rutile phase for 4 h. The crystal
structure has been investigated using powder x-ray diffraction (XRD) on Rigaku RU 2000
diffractometer and the vibrational spectra of the samples were analyzed by Raman spectroscopy
using a 109 mW and 532 nm laser spectra physics Renishaw Raman system. The magnetic and
optical properties were characterized using a vibrating sample magnetometer (VSM) and UV-
visible spectrophotometer (200 nm to 800nm).

DISCUSSION

Figure 1 shows X-ray diffraction (XRD) patterns of $Ti_{1-x}M_xO_2$ (M= Fe, Mn, Co) films annealed
at 500^0 C on quartz (Fig.1a) and platinum ($Pt/Ti/SiO_2/Si$) (Fig.1b) substrates. The observed
diffraction peaks reveal the formation of pure anatase phase. No segregated phase such as oxides
of transition metal ions or M_2TiO_5, $MTiO_3$ types are seen within the detection limit of x-ray
diffraction. Figure 2 shows XRD patterns of $Ti_{1-x}M_xO_2$ (M = Fe, Mn, Co) films annealed at
1000^0 C on platinum $Pt/Ti/SiO_2/Si$ substrate. All the peaks can be attributed to rutile phase of
TiO_2 without any impurity or segregated phase. It is to be noted from Fig.1b that we observe
impurity phase related to corresponding oxides of Mn at $2\theta = 29.15^\circ$ and Fe at $2\theta = 21.5^\circ$ for

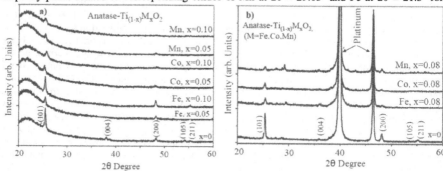

Figure 1. X-ray diffraction pattern for pure TiO_2 and $Ti_{1-x}M_x$ O_2 (M= Fe, Mn, Co) thin film
anatase phase a) quartz substrate and b) platinum $Pt/Ti/SiO_2/Si$ substrate.

248

x= 0.08 whereas no peak is observed in case of Co substitution suggesting the higher dissolution of cobalt in rutile phase of TiO$_2$ in comparison of Fe and Mn.

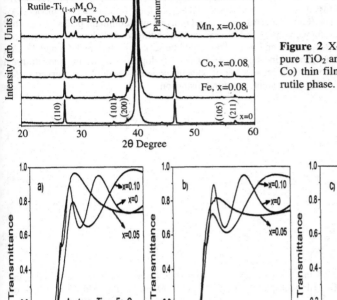

Figure 2 X-ray diffraction pattern for pure TiO$_2$ and Ti$_{1-x}$M$_x$ O$_2$ (M= Fe, Mn, Co) thin films on platinum substrate in rutile phase.

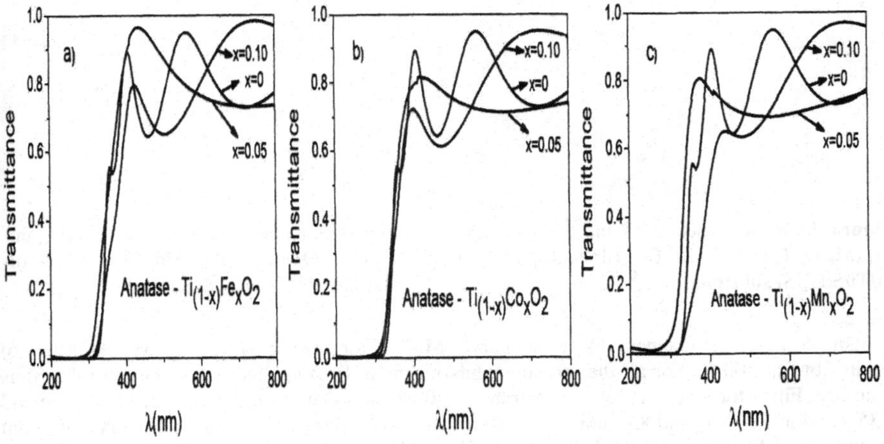

Figure 3. The UV transmission spectra for Ti$_{1-x}$M$_x$O$_2$ (M= Fe, Mn, Co) films deposited on quartz.

The optical transmission spectra of anatase Ti$_{1-x}$M$_x$O$_2$ (M= Fe, Mn, Co) films on quartz substrates are shown in **Fig. 3a, 3b, 3c,**. The peaks observed in wavelength region (λ) of 390–800 nm are due to thin film optical interference effects, whereas the dip at $\lambda < 390$ nm is due to electronic bandgap absorption. The optical band gaps for pure and doped films were obtained using the formula $\alpha_0 h\nu = C (h\nu - E_g)^2$ Where, h is Planck's constant and α_0 is the optical absorption coefficient. The band gap value obtained for anatase TiO$_2$ film is 3.3 eV and the values obtained for Ti$_{1-x}$M$_x$O$_2$ (M= Fe, Mn, Co) are 3.25 eV(5%), 3.23(10%) for Fe, 3.20 eV(5%), 3.23(17%) for Co, 3.28 eV(5%), 3.12 eV(10%) for Mn are in good agreement with other reports [9-11]. It could be clearly seen that the incorporation of magnetic ions results in decrease in band gap. Raman spectra of the anatase Ti$_{1-x}$M$_x$O$_2$ (M= Fe, Mn, Co) films are shown in **Fig. 4** and rutile $Ti_{1-x}M_x O_2$ (M= Fe, Mn, Co) in the **Fig. 5** on platinum $Pt/Ti/SiO_2/Si$ substrate. It is known that anatase TiO_2 has six Raman active modes: $A_{1g}+2B_{1g}+3E_g$ and the allowed bands are located at 144 cm^{-1} (E_g), 197 cm^{-1} (E_g), 399 cm^{-1} (B_{1g}), 513 cm^{-1} (A_{1g}), 519 cm^{-1} (B_{1g}), and 639 144 cm^{-1} (E_g) and rutile TiO_2 has four Raman active modes: A_{1g} +B_{1g}+B_{2g}+E_g [12–16]. The Raman spectra of rutile TiO_2 reveal two dominant peaks at 443 cm^{-1} (E_g) and 609 cm^{-1}(A_{1g}) along with a peak at 521 cm^{-1} corresponding to the substrate. The allowed Raman vibrational mode for anatase and rutile TiO_2 could be seen in both pure TiO_2 and $Ti_{1-x}M_x$ O_2 (M= Fe, Mn, Co) films, indicating that the (Fe, Co, Mn) dopants occupy the substitutional sites in the host lattice. However, a broadening of Raman mode at 400 cm^{-1} is observed in Mn doped TiO_2 anatase films suggesting some structural distortion.

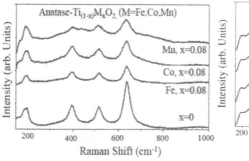

Figure 4. Raman spectra of anatase TiO_2 and $Ti_{1-x}M_x$ O_2 (M= Fe, Mn, Co) films on platinum $Pt/Ti/SiO_2/Si$ substrate.

Figure 5. Raman spectra of rutile TiO_2 and $Ti_{1-x}M_x$ O_2 (M= Fe, Mn, Co) films on platinum $Pt/Ti/SiO_2/Si$ substrate.

Figure 6 shows the magnetic hysteresis curve (M – H) for anatase $Ti_{1-x}M_xO_2$ (M= Fe, Mn, Co) thin films at 300 K. For all the measurements magnetic field applied was parallel to the films surface. Films for magnetic ion concentration x=0.08 shows small corecivities 22 Oe, 50 Oe and 35 Oe for Fe, Mn and Co doped films, respectively. Figure 7 shows M-H curve at room temperature for Mn doped rutile $Ti_{1-x}M_xO_2$ film with x= 0.08 showing small coercivity $H_c = 23$ Oe. These preliminary results indicate the possibility of ferromagnetism in magnetic ion doped TiO_2 films at room temperature.

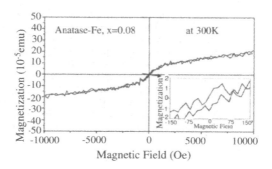

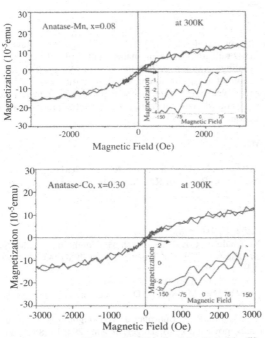

Figure 6. M–H curves for $Ti_{1-x}M_x O_2$ (M= Fe, Mn, Co) thin films on platinum Pt/Ti/SiO$_2$/Si substrate at room temperature.

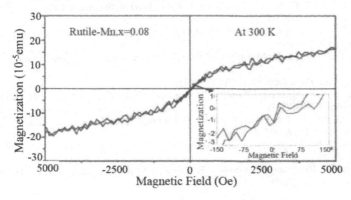

Figure 7. M–H curves for $Ti_{1-x}M_x O_2$ (M= Mn) thin film on platinum Pt/Ti/SiO$_2$/Si substrate at room temperature.

CONCLUSIONS

We have successfully prepared $Ti_{1-x}M_xO_2$ (M= Fe, Co, Mn) thin films on platinum ($Pt/Ti/SiO_2/Si$) and quartz substrates both in anatase and rutile phases by sol–gel process and optimized the post annealing conditions. XRD and Raman spectra of doped TiO_2 films suggest the possible incorporation of magnetic ion in the host lattice of TiO_2 on Ti ion sites. The magnetic measurements reveal weak ferromagnetism at room temperature and needs more systematic efforts with various doping concentrations and other process parameters. The work is in progress in this direction.

ACKNOWLEDGEMENT

The work is supported by DoE-EPSCoR Grant No. DE-SG02-08ER46526 and is gratefully appreciated. Thanks are due to Professor R.S Katiyar for providing measurement facility.

REFERENCES

1. A. P. Gary, *Science* **282**, 1660 (1998)
2. S. A. Wolf, D. D. Awschalom, R. A. Buhrman, J. M. Daughton, S. V. Molnar, M. L. Roukes, A. Y. Chtchelkanova and D. M. Treger, *Science* 294 1488–95 (1998)
3. H. Ohno, *Science* **281**, 951 (1998)
4. M.S. Park, B.I. Min, Phys. Rev. B **68**, 033202 (2003)
5. Pearton, S. J.; Heo, W. H.; Ivill, M.; Norton, D. P.; Steiner, T. Semicond. Sci. Technol, **19**, R59 (2004)
6. S. A. Wolf, D. D. Awschalom, R. A. Buhrman, J. M. Daughton, S. V. Molnar, M. L. Roukes, A. Y Chetchalkanova and D. M. Treger, *Science* **294**, 1488 (2001)
7. R. Suryanarayanan, J. Phys.: Condens. Matter **17**, 755–762 (2005)
8. Chunming Huang, Thin Solid Films **505**, 141 (2006)
9. W. Weng, M. Ma, P. Du, G. Zhao, G. Shen, J. Wang, G. Han, Surface & Coatings Technology **198**, 340 (2005)
10. Q. Deng, Y. Gao, X. Xia, R. Chen, L. Wan, G. Shao, Journal of Physics: Conference Series **152,** 012073 (2009)
11. C. Xu, R. Killmeyer, L. McMahan, U. Shahed, Electrochemistry Communications **8**, 1650 (2006)
12. C.R. Aita, Appl. Phys. Lett. **90**, 213112 (2007)
13. A. Orendorz, A. Brodyanski, J. Losch, L.H. Bai, Z.H. Chen, Y.K. Le, C. Ziegler, H. Gnaser, Surf. Sci. **601**, 4390 (2007)
14. T. Mazza, E. Barborini, P. Piseri, P. Milani, D. Cattaneo, A.L. Bassi, C.E. Bottani, C. Ducati, Phy. Rev. B **75**, 045416 (2007)
15. Y. Djaoued, S. Badilescu, P.V. Ashrit, D. Bersani, P.P. Lottici, J. Robichaud, Proc. SPIE **4469**, 70 (2001)
16. A. Orendorz, A. Brodyanski, J. Losch, L.H. Bai, Z.H. Chen, Y.K. Le, C. Ziegler, H. Gnaser, Surf. Sci. **601**, 4390 (2007)

Mater. Res. Soc. Symp. Proc. Vol. 1201 © 2010 Materials Research Society   1201-H10-21

Influence of O_2/Ar Ratio on the Properties of Transparent Conductive Titanium-Doped Indium Oxide Films by DC Radio-Frequency Sputtering

Lei Li, Li Li, Xin Yu, Hai Yu, Chen Chen, Zhenyu Song, Menglong Cong and Yiding Wang*

State Key Laboratory on Integrated Optoelectronics, College of Electronic Science and Engineering, Jilin University, Changchun, Jilin, China, 130012

ABSTRACT

This paper presents titanium doped indium oxide (TIO) thin films deposited on glass substrates by DC sputtering with different O_2/Ar gas ratios at 330 ℃. The effects of sputtering on the structural, morphologic, optical and electrical characteristics of TIO thin films were investigated by XRD, Hall measurements and optical transmission spectroscopy. The deposited films exhibited polycrystalline in the preferred (222) orientation, with higher mean grain size and lower resistivity 3.37 $\times 10^{-4}\Omega \cdot cm$ at O_2/Ar ratio of 1/10. The average optical transmittance of the films is over 90%, and the transmittance has no evident change with different O_2/Ar ratios.

INTRODUCTION

As a sort of transparent conductive oxide (TCO) thin films, the indium oxide thin film is attractive in many fields because of its high transmission, low resistivity, strong mechanical strength and stable chemical properties. The thin films are widely applied to transparent electrodes of flat panel displays, solar cells, and IR reflectors. As a kind of efficient TCO thin films, TIO (In_2O_3: Ti) thin films deposited onto glass substrates have been developed extensively and profoundly.

In_2O_3 -based TCO thin film can be deposited by chemical-vapor deposition (CVD), sol-gel, spray pyrolysis techniques, and so on. The film properties can be adjusted by controlling the doping materials' ratios, deposition parameters, post treatment, etc. However, the DC ratio-frequency (RF) sputtering technique with its own advantage makes it is a good candidate for the thin film preparation applications.

In this paper, highly transparent and conductive In_2O_3: Ti thin films are prepared in the case of depositing onto glass substrates at 330℃ with different O_2/Ar ratios by RF sputtering. The results indicate that In_2O_3: Ti can be used to fabricate high performance TCO thin films.

EXPERIMENT

DPS-III ultra-high vacuum against-target RF sputtering system is used to deposit the TIO thin film onto glass substrates with the substrate temperature at 330℃ and different O_2/Ar ratios. The dimensions of the glass substrate were prepared as $18 \times 16mm^2$ at beginning. A sintered

*Corresponding author: Tel.: +86 431 85168240; e-mail:wangyiding47@yahoo.com.cn.

ceramic 2wt % Ti-doped In_2O_3 target with 99.99% purity, commercially available was employed as the material. According to the requirement of the RF sputtering system, the diameter of the target was chosen as 60mm. At the beginning of sputtering, the vacuum chamber was evacuated with molecule pump to get a basic pressure of 1.0×10^{-4} Pa. In the following steps, the substrate temperature was controlled up to 330°C and then the separated mass flow-meters were opened high purity Ar(99.999%) and O_2(99.999%) injecting into the vacuum chamber at the same time. The O_2/Ar gas ratios were varied from 1/8 to 1/18.When the fingers of flow meter were steady, the sputtering current was launched and fixed at 150mA. Prior to the deposition, the target was pre-sputtered in O_2+Ar ambient condition for 3-5 minutes to remove the impurity on the surface of the target. In the end, forty minutes' sputtering time was employed for depositing the TIO thin film.

Bruker axs D8 advance X-ray system (XRD) was used to determine the component, the crystal structure, the lattice constant and the stable grainsize of the samples. It was conducted on a Rigaku D/Max-2500 X-ray diffractometer with Cu Kα radiation (λ = 1.5418Å) and an energy dispersion X-ray spectroscopy (EDX) equipped in scanning electron microscopy system (Hitachi S-4700). During the processing, the scanning angle was set from 20° to 80° with the speed was 0.02°.

The Hall-effects of the sample was measured by Bio-Rad Micro science HL5500 Hall System at room temperature. The resistivity, carrier concentrations, and Hall mobility of samples can be shown on the system. The UV-1700, Shimadu spectrophotometer was used to measure the optical transmittance. The scanning wavelength was chosen from 300nm to 1100nm. Morphologic photos were taken on a field emission scanning electron microscopy (FESEM) equipped with energy dispersive x-ray (EDX) spectroscopy (JEOL JEM-6700F). The scanning voltage and current were set at 15kV and 10μA, respectively [1-2].

DISCUSSION

The Ti content in the In_2O_3 thin film is about 2.5wt% which is detected by EDX. That is higher than target. The deposited thin films with different O_2/Ar ratios are measured by dispersive X-ray, and depicted in Figure 1. The (211), (222), (431) and (440) peaks can be mainly obtained at 2θ=21.32°,30.35°,45.44°and 50.76° respectively for all samples. The (222) orientation is as the preferred orientation for TIO film sputtered from In_2O_3: Ti target when the O_2/Ar ratio is 1/12 and 1/10 and a shift of the peaks to higher angles with the lower or higher ratio is observed. This result is agreed with other authors' reports. No more Ti diffraction peak is found from the XRD pattern, which indicates Ti^{4+} replaces In^{3+} in the hexagonal lattice or Ti-ion segregates to the non-crystalline region in the grain boundary. All the obtained films with thickness of about 500 nm that obey the polycrystalline with the hexagonal wurtzite structure and have a preferred orientation with the c-axis perpendicular to the substrates [3-5]. The maximum diffraction intensity of the (222) peak is obtained under the film deposited at the O_2/Ar ratio of 1/10. The tendency in the change of the FWHM is constant with the change of the (222) peak intensity. That is to say, stronger (222) direction can achieve better crystal quality. The crystalline size of 51.7 nm was calculated by the Scherrer equation from the full-width at half-maximum (FWHM) values of (222) peak.

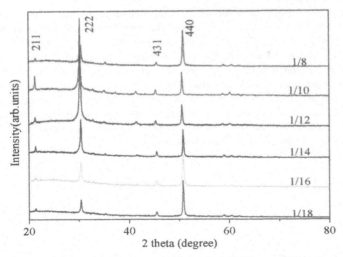

Figure 1 XRD patterns of TIO films deposited at different O_2/Ar ratios

Figure 2 FESEM images of TIO films deposited at O_2/Ar ratio (a) 1/8,
(b) 1/10, (c) 1/12, (d) 1/14, (e) 1/16 and (f) 1/18.

The Figure 2 (a)-(f) show the surface morphologies of the TIO films deposited at the O_2/Ar ratio from 1/18 to 1/8. The morphology of TIO grains is found to be continuing and dense. It can be obtained that the crystallite sizes increasing from 42.8nm to 51.7nm with reducing the ratio of the O_2/Ar from 1/8 to 1/10, then the sizes decrease to 38.7nm (1/12 to 1/6) .The films deposited at low oxygen flow (1/18-1/16) present roughness on the film surface (e), (f) which are responsible for the grains without enough oxidation to form non-stoichiometric ratio films. For an oxygen flow of 1/14 the sample is completely covered with spikes. When the oxygen flow is raised to 1/10, the sample's density is strongly reduced but shows a very flat surface morphology and the largest grains. For higher oxygen flow (1/8) the samples remain smoothness but change rapidly in the form of the grains on the films surface [6].

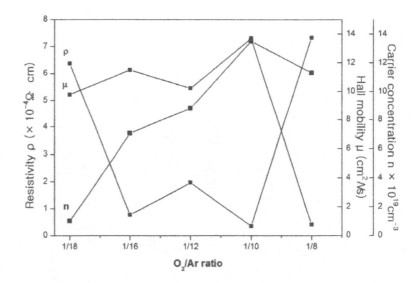

Figure 3 The resistivity, hall mobility and carrier concentrations of samples are functions of different O_2/Ar ratios.

The changes of electrical resistivity (ρ), Hall mobility (μ), and carrier concentration (n) are as a function of O_2/Ar gas ratio which is shown in Figure 3. The results indicate that all the films are degenerate doped N-type semiconductors. The lowest electrical resistivity of 3.37 × 10^{-4} $\Omega\cdot cm$ is achieved for the film deposited at the O_2/Ar ratio of 1/10. It is much higher than the film deposited without O_2 gas. It is because Ti^{4+} is very active at 330℃ and easier to react with O_2 to form TiO_2 molecular. The change of resistivity is mainly caused by the influence of carrier concentration and Hall mobility. It can be written as equation (1):

$$\rho = (nq\mu_n)^{-1} \tag{1}$$

where ρ is electrical resistivity, n is carrier concentration, μ_n is Hall mobility and q is electric charge constant. At lower O_2 conditions (<1/10), the reduced carrier concentration can be interpreted by the shallow donors or/and relative deep donors partially losing their dopant effect at this condition. When the O_2 content increased too much (>1/10), the number of O_2 vacancies decreases, Ti atoms and O_2 atoms probably combine with TiO_2, then substitution Ti^{4+} decreases, so the carrier concentration decreases[1, 4]. The Hall mobility of the film (1/10) is much larger than that of the films deposited at other O_2/Ar ratios, because the effect of grain boundary and scattering center on the film mobility [7-9].

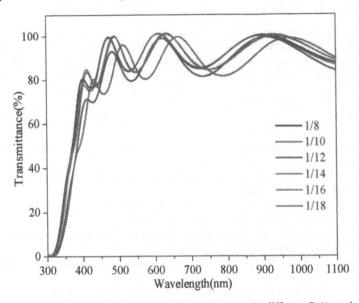

Figure 4 TIO thin films' transmittance dependence on the different O_2/Ar ratios

The Figure 4 depicts that the average transmittance of TIO films with different O_2/Ar ratios. The average optical transmittance in the visible region from 400 nm to 800 nm reaches 90.3% for all films. For all of the O_2/Ar ratios, the transmittance of the films increase sharply with the wavelength ranging from 300 to 400 nm and get the highest transmittance of – 94.3% between 450 nm and 500 nm, then the transmittance decrease slightly from 500 nm to 550 nm and then increase stably again (550–620nm), at last vary gently (620-800nm) [10]. The transmittance at 1/10 O_2/Ar ratio is an exception with maximum transmittance appears around 520 nm. The wavelength shift of the transmittance was caused by a different thickness of the TIO films [11].

CONCLUSIONS

TIO thin films are deposited on glass substrates by DC Radio-Frequency Sputtering at 330 ℃ and the influences of O_2/Ar ratio on the film properties are studied. The as-deposited films are polycrystalline with the cubic phase of bixbyite-type In_2O_3 structure and have a preferred orientation with the (222) perpendicular to the substrates. The lowest resistivity of 3.37 $\times 10^{-4} \Omega \cdot cm$ is obtained at the O_2/Ar ratio of 1/10. The average optical transmittance of the films is over 90 %.

ACKNOWLEDGEMENTS

This research are supported by the National High Technology Research and Development Programs of China ("863" Programs), No.2007AA03Z112, No.2007AA06Z112, No. 2009AA03Z442, the Program of Jilin Provincial Science and Technology Department of China, No. 20070709, No. 20090422 and the Program of Ministry of Education of China, No.20060183030.

REFERENCE

1. Feng Cao, Yiding Wang, Lei Li, Baojia Guo and Yupeng An, Scripta Materialia 61,231–233 (2009).
2. Cao Feng, Wang Yiding, Liu DaLi, Yin Jingzhi, Guo Baojia, Li Lei, An Yupeng, Chin. Phys. Lett. , Vol. 26, No. 3,034210(2009).
3. K.H. Kim, K.C. Park, D.Y. Ma, J. Appl. Phys. 81, 7764(1997).
4. C. Guillén, J. Herrero, Materials Chemistry and Physics 112,641–644(2008).
5. J.B. Chu, S.M. Huang ,H.B. Zhu , X.B. Xu , Z. Sun , Y.W. Chen , F.Q. Huang, Journal of Non-Crystalline Solids 354,5480–5484 (2008).
6. V. Sittinger, F. Ruske, W. Werner, C. Jacobs, B. Szyszka, D.J. Christie, Thin Solid Films 516, 5847–5859(2008).
7. Qun Zhang, Xifeng Li, Guifeng Li, Thin Solid Films 517, 613–616(2008).
8. Y.C. Lin, W.Q. Shi, Z.Z. Chen, Thin Solid Films 517, 1701–1705(2009).
9. H.T. Cao, Z.L. Pei, J. Gong, C. Sun, R.F. Huang, L.S.Wen, Surf. Coat. Techno. 184, 84 (2004).
10. Chang S. Moon, Jeon G. Han, Thin Solid Films 516, 6560–6564 (2008).
11. Ren Bingyan , Liu Xiuoping, Wang Minhua , and Xu Ying, RARE METALS, Vol. 25, Spec. Issue, 137-140(2006).

Mater. Res. Soc. Symp. Proc. Vol. 1201 © 2010 Materials Research Society   1201-H10-25

Wet Chemical Etching of Zn-Containing Oxides and HfO$_2$ for the Fabrication of Transparent TFTs

Jae-Kwan Kim, Jun Young Kim, Seung-Cheol Han, Joon Seop Kwak, and Ji-Myon Lee*

Department of Materials Science and Metallurgical Engineering, Sunchon National University, 315 Maegok, Sunchon, Chonnam 540-742, Korea

ABSTRACT

The etch rate and surface morphology of Zn-containing oxide and HfO$_2$ films after wet chemical etching were investigated. ZnO could be easily etched using each acid tested in this study, specifically sulfuric, formic, oxalic, and HF acids. The etch rate of IGZO was strongly dependent on the etchant used, and the highest measured etch rate (500 nm/min) was achieved using buffered oxide etchant at room temperature. The etch rate of IGZO was drastically increased when sulfuric acid at concentration greater than 1.5 molar was used. Furthermore, etching of HfO$_2$ films by BF acid proceeded through lateral widening and merging of the initial irregular pits.

INTRODUCTION

Oxide-based thin film transistors attract much attention due to their advantageous properties such as high mobility, high electrical conductivity, and high visible transmittance [1]. Zn-containing oxide thin film transistors (TFTs) are a promising large-area backplane electronic device for flat panel displays such as liquid crystal displays and organic light-emitting displays [2,3]. Zn-containing oxide and hafnium oxide (HfO$_x$) have been investigated for use in the fabrication of transparent thin film transistors (TTFTs). A number of studies have investigated Zn-containing oxide TFTs that used ZnO [4], In-Zn-O (IZO) [5], or In-Ga-Zn-O (IGZO) [6] as the active channel material. In addition, considerable effort has been made to reduce power consumption, such that flexible and mobile applications of oxide-based TFTs with a high-k gate dielectric can be realized. High-dielectric-constant (high-k) materials have received significant attention as gate dielectrics because they enable greater physical gate oxide thicknesses (EOT) that are equivalent to those of the conventional gate dielectrics SiO$_2$ and SiO$_x$N$_y$, which results in significant reductions in gate leakage [7]. Among the potential candidates to replace SiO$_2$ and SiO$_x$N$_y$ as gate dielectrics, hafnium oxide is one of the most promising materials. Hafnium oxide has a high dielectric permittivity with low leakage current due to a reasonably high barrier height [8]. Etching is key step. It is very important to develop well-defined etching processes for various transparent oxides that are used in such devices [1], because the thickness of the channel layer is less than a few tens of nanometers. Moreover, because Zn-containing oxides and HfO$_2$ might be used as the channel and gate dielectric material, respectively, it is very important to develop well-defined wet-etching processes for various transparent oxides [9] that result in low etch damage and reliable pattern transfer. In the present study, various acidic solutions were used to investigate the etch-rate and etched surface morphology of both Zn-containing oxides (i.e., ZnO, IZO, and IGZO) and HfO$_2$.

EXPERIMENTAL

In this work, 600-nm thick films were deposited on Si (001) substrates using radio-frequency magnetron sputtering [10]. HfO_2 films were deposited using an Hf target in Ar/O_2 ambient at 3 mTorr. Each film was chemically cleaned using acetone, isopropyl alcohol and de-ionized water. Wet-etching experiments were performed using various acidic solutions such as sulfuric acid, buffered oxide etchants (BOE), 85% formic acid ($H_2C_2O_4$), oxalic acid (CH_2O_2) and a mixed solution (BF) of BOE and formic acid, respectively. A photo-resist etch mask with 20 μm circular patterns was formed on the films using standard lithography. A 1.2 μm thick photo-resist (AZ1512) was deposited by 3-step spin coating (400 rpm/ 15 s, 800 rpm/ 30 s, and 1200 rpm/45 s). After baking at 95 °C for 5 min, the samples were loaded into the mask aligner. Patterned photo-resisted films were developed by immersion in the developer (AZ300MF) for 60 s.

Etch rates were determined using a surface profilometer and averaging 10 points for each sample. The surface morphology, chemical composition, and crystalline quality of the samples were characterized by field emission-scanning electron microscope (FE-SEM), energy dispersive spectroscopy (EDS), and X-ray diffraction (XRD), respectively.

RESULTS and DISCUSSION

Table I. Etch depths of ZnO-containing oxides and HfO_2 resulting from use of various acid solutions. (The etching duration for the Zn-containing oxides and HfO_2 were 1 and 10 min, respectively.)

	ZnO	a-IZO	c-IZO	IGZO	HfO_2
		Etch time : 1 min			Etch time : 10 min
Sulfuric acid (2.0 mole)	590 nm	480 nm	<10 nm	346 nm	<10 nm
Sulfuric acid (1.0 mole)	590 nm	325 nm	<10 nm	85 nm	<10 nm
Formic acid	590 nm	<10 nm	<10 nm	<10 nm	<10 nm
Oxalic acid (0.1 mole)	589 nm	<10 nm	<10 nm	105 nm	<10 nm
Oxalic acid (0.05 mole)	578 nm	<10 nm	<10 nm	<10 nm	<10 nm
BOE	600 nm	<10 nm	<10 nm	506 nm	326 nm
BF acid	598 nm	<10 nm	<10 nm	64 nm	28 nm

Table.1 shows the etch depth of the Zn-containing oxide and HfO_2 films that resulted from etching in the acidic solutions for 1 and 10 min, respectively. We found that ZnO was easily etched by each of the acidic aqueous solutions. As for IZO, the crystalline quality affected the etch resistance of the film. Figure 1 shows the X-ray diffraction spectra of a-IZO and c-IZO thin film. As can be seen in the figure, In_2O_3-related peaks such as (222) and (440) were observed in the c-IZO XRD spectrum. As shown in Table 1, amorphous IZO (a-IZO) was easily etched by sulfuric acid. However, the crystalline IZO (c-IZO) films were mostly resistant to etching by the acidic aqueous solutions. The result indicates that the In_2O_3 crystallite may suppress the chemical reaction required for wet chemical etching by an acidic solution [11].

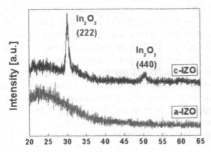

Figure 1. X-ray diffraction spectra of c-IZO and a-IZO thin film.

The IGZO was easily etched using sulfuric acid or buffered oxide etchant (BOE). This result is discussed in detail below. It is also of interest that IZO and IGZO were highly resistant to etching by an organic acid solution. This observation indicates that not all Zn-containing oxides can be etched by acidic solutions, despite the fact that ZnO can be easily etched using conventional acidic solutions such as HCl or H_2SO_4 [12]. Thus, the etch rate of Zn-containing oxide materials is strongly influenced by the chemical composition of the material, such as the inclusion of In or Ga.

By contrary, the HfO_2 film was resistant to etching by acidic aqueous solutions with the exception of the BOE and BF solutions.

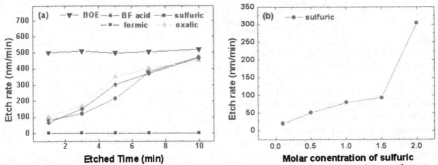

Figure 2. Etch rates of IGZO films using (a) various acidic solutions and (b) various molar concentrations of sulfuric acid.

Figure 2(a) shows the etch rate of IGZO produced by the various acidic solutions. As shown in the figure, the etch rate of IGZO that resulted from the use of BOE was constant, regardless of the etching time. This result is indicative of a surface-reaction-rate-limited regime [13]. By contrast, the etch rates that resulted from the use of either oxalic (0.1 M), sulfuric (1.0

M), or BF acid increased with longer etching durations. After ten minutes, the etch rates for oxalic, sulfuric, and BF etchant approached that of BOE. This result indicates that this etching process is a diffusion-rate-limited regime [13]. However, IGZO could not be etched by the formic acid solution, even when the etching time was increased to longer than 10 min. Although the exact reason for the lag in etching of IGZO by oxalic, sulfuric, and BF etchants remains unknown, the authors believe that surface contamination or a surface oxide might play an important role. Fig. 2 (b) shows the relationship between the etch rate of IGZO and the molar concentration of sulfuric acid. When lower concentrations of sulfuric acid were used (i.e., less than 1.5 M), the etch rate was reduced. However, use of 2 M sulfuric acid drastically increased the etch rate of IGZO, which reached approximately 320 nm/min. These results indicate that the etch rate is mainly dependent on the molar concentration of the active etchant. In addition, the pH of the solution also appears to be responsible for the increase in the etch rate of IGZO. In other words, as the molar concentration of sulfuric acid increased, the pH decreased. Thus, the surface reaction might have been enhanced by the highly acidic solution.

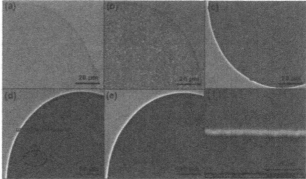

Figure 3. FE-SEM images of the IGZO film etched using different concentrations of sulfuric acid at room temperature. (a) 0.1 M for 5 min, (b) 0.5 M for 5 min, (c) 1.0 M for 5 min, (d) 2.0 M for 5 min, (e) 2.0 M for 10 min, and (f) tiled-view of the SEM images of the etched side wall.

Figure 3 shows the FE-SEM images of IGZO films etched by different concentrations of sulfuric acid at room temperature. The surface of the film that was etched using 0.1 M sulfuric acid was smooth, as shown in Fig. 3(a). As the molar concentration was increased, the roughness of the surface increased, leaving a high density of submicron-sized island residues on the surface, as shown in Fig. 3(b). As the molar concentration of sulfuric acid was increased further, the surface roughness was reduced due to removal of the submicron-sized island residues, as shown in Fig. 3(c) and (d). The surface morphology was smooth after complete removal of the submicron-sized island residues, as shown in Fig. 3(e). The morphology of the etched sidewall was also smooth, as shown in Fig. 3(f). These results indicated that residues could be eliminated by controlling the molar concentration of the active etchant. Tasi [12] reported that the residue remaining on ITO after etching by oxalic acid could be removed via the agitation or sonication during etching.

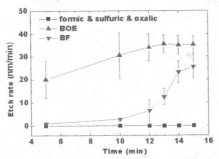

Figure 4. Etch rate of HfO_2 films resulting from use of various acidic solutions.

Figure 4 shows the etch rate of HfO_2 produced by the various acidic solutions. The HfO_2 was found to be etched by using BOE and BF acid solutions with maximal etch rates of 35 nm/min and 25 nm/min, respectively. Additionally, it was also observed that the etch rate of IGZO using formic, sulfuric, and oxalic solutions showed extremely low etch rates of less than 10 nm. These results indicate that the F atoms/ions in the etchant solution play a critical role in the etching of HfO_2. Although the etch rate of IGZO by using BOE was relatively constant regardless of etching durations, the etch rate using BF solution was increased with increasing the etching time, maybe due to the consumption of formic acid which hampered the surface reaction of active etchant F in the BF solution. These also indicate that the etching by using BOE was governed by reaction-rate-limited and that the BF by diffusion-rate-limited.

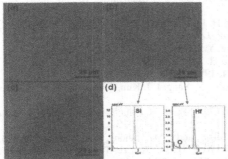

Figure 5. FE-SEM image of the HfO_2 film etched using BF acid at room temperature, for (a) 10 min, (b) 12 min, and (c) 15 min. (d) EDX spectra of the etched and unetched regions.

Figure 5 shows both the morphology of etched surface by SEM and the surface atomic concentration of the etched surface by EDX spectra of HfO_2 after etching in BF etchant. After etching the HfO_2 for 10 min in BF acid, irregularly shaped pits (dark areas) were observed on the surface at high density, as can be seen in Fig. 5(a). As the etching time was increased, the lateral dimension of the irregular pits increased, and the density of the pits decreased, as shown in Fig. 5(b). The increased in pit size might be due to merging or agglomeration of pits on the surface.

The EDX spectra, which are shown in Fig. 5(d), confirmed that the dark areas that resulted from pit merging were due to complete removal of HfO_2. This result indicates that the pits have the highest etch rate and/or act as channels to transfer the active etchant. Finally, etching duration longer than 20 min resulted in smooth etched surfaces with bare Si substrate exposed due to complete removal of the HfO_2, as shown in Fig. 5(c). Base on these results, we believe that etching of HfO_2 by BF acid solutions proceeds via lateral widening and merging of the initial irregular pits. The pits also play a role in active transfer of the etchant.

CONCLUSIONS

The etch rate and surface morphology of Zn-containing oxide and HfO_2 films after wet chemical etching were investigated. We found that the ZnO could be easily etched using each of the acidic aqueous solutions that were tested. Amorphous IZO (a-IZO) was easily etched by sulfuric acid. However, the crystalline IZO (c-IZO) films were highly resistant to etching by the acidic aqueous solutions that were tested.

The etch rate of IGZO was highly dependent on the etchant, and the highest etch rate 500 nm/min was achieved using BOE acid at room temperature in a reaction-rate-limited regime. The etch rates achieved using sulfuric, BF, and oxalic acid were similar. The initial surface morphology of IGZO was rough due to the formation of submicron island residues, but disappeared after continued dipping in the solution. HfO_2 could be etched by BOE or BF acidic solutions with maximal etch rates of 35 nm/min and 25 nm/min, respectively. Furthermore, the etching of HfO_2 film by BF acid proceeded through lateral widening and merging of the initial irregular pits.

REFERENCES

1. K. Nomura, H. Ohta, A. Takagi, T. Kamiya, M. Hirano and H. Hosono, *Nature*. **432**, 288 (2004).
2. M. Kim, J. H. Jeong, H. J. Lee, T. K. Ahn, H. S. Shin, J. S. Park, J. K. Jeong, Y. G. Mo, and H. D. Kim, *Appl. Phys. Lett.* **90**, 212114 (2007).
3. J. S. Park, J. K. Jeong, Y. G. Mo, and H. D. Kim, *Appl. Phys. Lett.* **90**, 262106 (2007).
4. H. Jeon, K. Noh, D.-H. Kim, M. Jeon, V. P. Verma, W. Choi, D. Kim and J. Moon, *J. Korean Phys.* Soc. **51**, 1999 (2007).
5. J. M. Park, J. S. Hong, J. Y. Yang, J. J. Kim, S. H. Park, H. M. Kim and J.-S. Ahn, *J. Korean Phys.* Soc. **48**, 1530 (2006).
6. K. Nomura, A. Takagi, T. Kamiya, H. Ohta, M. Hirano and H. Hosono, *Jpn. Appl. Phys.* **45**, 4303 (2006).
7. G. D. Wilk, R. M. Wallace, and J. M. Anthony, *J. Appl. Phys.*, **89**, 5243 (2001).
8. S. J. Lee, H. F. Luan, W. P. Bai, C. H. Lee, T. S. Jeon, Y. Senzaki, D. Roberts, and D. L. Kwong, *Tech. Dig. Int. Electron Devices Meet.* 31 (2000).
9. J. M. Lee, K. K. Kim, C. K. Hyun, H. Tampo, and S. Niki, *J. Nanosci. and Nanotechnol.* **6**, 1 (2006).
10. Y. J. Jo, J. K. Kim, S. C. Han, J. S. Kwak, and J. M. Lee, *J. Kor. Inst. Met. & Mater.* **47**, 44 (2009).
11. G. Gonçalves , P. Barquinha, L. Raniero, R. Martins, and E. Fortunato, *Thin Solid Films*. **516** 1374(2008).
12. T.H. Tsai and Y. F. Wu, *Microelectron. Eng.* **83**, 536 (2006).
13. S. J. Pearton, J. J. Chen, W. T. Lim, F. Ren, and D. P. Norton, *ECS trans.* **6**, 501 (2007).

Mater. Res. Soc. Symp. Proc. Vol. 1201 © 2010 Materials Research Society   1201-H10-27

ZnO Thin Film Transistors Fabricated by Atomic Layer Deposition Method

Yumi Kawamura[1], Nozomu Hattori[2], Naomasa Miyatake[2], Kazutoshi Murata[2],
and Yukiharu Uraoka [1,3]

[1] Graduate School of Materials Science, Nara Institute of Science and Technology,
8916-5 Takayama, Ikoma, Nara, 630-0192, Japan
[2] Mitsui Engineering & Shipbuilding Co., Ltd.,
3-16-1 Tamahara, Tamano, Okayama 706-0014, Japan
[3] CREST, Japan Science and Technology Agency,
4-1 Honcho, Kawaguchi, Saitama 332-0012, Japan

ABSTRACT

In this study, we deposited zinc oxide (ZnO) thin films by atomic layer deposition (ALD) as an active channel layer in thin film transistor (TFT) using two different oxidizers, water (H_2O-ALD) and oxygen radical (PA-ALD). The fabricated TFTs were annealed at various temperatures, in oxygen ambient gas. The electrical properties of TFTs with PA-ALD ZnO film annealed at the temperature up to 400 °C improved without any degradation of the subthreshold swing or any large shift of the threshold voltage. Through this study, we found that the high performance ZnO TFTs is possible to obtain by using PA-ALD at low temperature, and the electrical properties are dependent on the annealing temperature.

INTRODUCTION

The most commonly used materials as an active channel layer in thin film transistors (TFTs) have been amorphous (a-Si:H) and polycrystalline silicon (poly-Si).[1] However, there are a number of drawbacks for these materials, such as issue of uniformity, their high temperature fabrication process, and it is difficult to fabricate on large substrates.[1-3] In recent years, the application of zinc oxide (ZnO) thin films as an active channel layer in TFTs has become of great interest owing to their specific characteristics. ZnO is transparent to visible wavelengths because of its wide band gap (~3.37eV), and the ability to fabricate good quality films over large areas at low temperature suggests the compatibility of these films with plastic or flexible substrates [4-6]. Higher field-effect mobility of ZnO-TFTs than a-Si:H TFTs has recently been demonstrated.[4-10] However, there is still critical issue that has to be solved. Reliability for electrical stress is serious problem in their mass production.

An atomic layer deposition (ALD) method is a thin film preparation technology that attracts much attention in the LSI industry. ALD thin films are deposited with alternating exposures of a source gas and an oxidant. ALD films have additional features of accurate thickness control, high conformity, and uniformity over large areas, because of their layer-by-layer growth. [11] Further, it is reported that TFTs with ALD thin film as the channel layer demonstrate high mobility. [12-14]

In this study, we fabricated TFTs using ZnO thin film as the channel layer deposited by ALD (ALD-ZnO), and the electrical properties were measured. The dependences of TFT characteristics on annealing conditions are evaluated, and the dependence of their electrical properties on the composition of the ZnO films was characterized.

EXPERIMENT

A schematic structure of the bottom-gate-type ZnO TFTs fabricated in this study is shown in Fig. 1. 50-nm-thick SiO₂ gate insulator was prepared by thermal oxidation. 30-nm-thick ZnO thin films were deposited on p-type Si(100) substrates at 100 °C. We used alkyl Zn as a metal precursor. Furthermore, we used two different oxidizers, water (H₂O-ALD), and oxygen radical by plasma assisted ALD (PA-ALD). Ti metal was deposited and patterned by a lift-off technique to serve as the source/drain (S/D) electrodes. The Si substrate was used as the gate electrode.

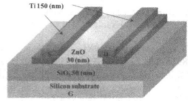

Fig.1.   Bottom-gate ZnO TFT

The fabricated TFTs were annealed at 100 -500 °C for 1 h in O_2 (O_2 = 20 %, N_2 = 80 %) ambient. [15] The electrical properties were measured using a semiconductor parameter analyser (Agilent 4156C).

The channel length (L) and width (W) used in this study were 10 and 20 µm, respectively. Threshold voltage (V_{th}) was defined by the gate voltage (V_g), which induces a drain current of 1 nA at a drain voltage (V_d) of 5 V. On/off current ratio was measured from the ratio of maximum to minimum drain current (I_d) on the gate voltage axis, and on-current (I_{d_on}) was defined as the drain current measured with V_g = 10V. The field effect mobility was defined by the maximum value calculated using the conventional measurement of I_d.

RESULTS and DISCUSSION

Electrical characteristics

The transfer characteristics of the fabricated TFTs were measured in a single-sweep mode of gate voltage at V_d = 5 V. The variations in the transfer characteristics of the TFTs non-annealed and annealed at 300 °C in O_2 ambient are shown in Figs 2(a) and 2(b), respectively. The TFTs with H₂O-ALD ZnO films did not exhibit the switching characteristics before the annealing as shown in Fig. 2(a). Further, no switching characteristics were also obtained at the H₂O-ALD ZnO TFTs annealed at 200 °C. After the annealing over 300 °C, the switching characteristics were obtained as shown in Fig. 2(b). However, poor TFT device characteristics such as low I_{d_on} and mobility were observed. On the other hand, the TFTs with PA-ALD ZnO films exhibited the clear TFT behaviours in the case of without annealing as shown in Fig. 2(a). For the PA-ALD ZnO TFTs annealed at 300 °C in O_2 ambient, the off-current (I_{d_off}) of 4×10^{-14} A, and the on/off ratio of 1×10^9 was obtained. The field effect mobility was approximately 1.5 cm²/Vs, the V_{th}, and the subthreshold swing (S.S.) was 1.2 V and 0.3 V/decade, respectively. Figure 3 shows the I_{d_on} of the TFTs as a function of the O_2 annealing temperature. The PA-ALD ZnO TFTs annealed over 450 °C showed lower I_{d_on} than that annealed at 400 °C. We suppose

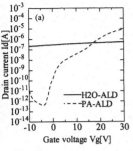

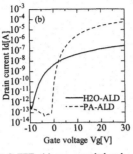

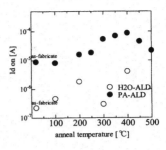

Fig. 2 Transfer characteristics of ZnO TFTs (a) non-annealed and (b) annealed at 300 °C. (V_d=5 V)

Fig. 3 Temperature dependence of I_{d_on} (V_d=5 V)

that the decrease of the I_{d_on} of PA-ALD ZnO TFTs annealed over 450 °C are due to the oxidation of Ti used for S/D electrodes.

These results suggest that the high performance ZnO TFTs is possible to obtain by using ZnO films deposited by PA-ALD at low process temperature.

XPS measurement

X-ray photoelectron spectroscopy (XPS) analysis were performed to investigate analyze the composition of the ALD-ZnO films. Figure 4 shows the atomic concentration of oxygen and carbon in the ZnO films as a function of annealing temperature. The H$_2$O-ALD ZnO films exhibited the lower atomic concentration of oxygen than that of the PA-ALD ZnO films. In addition, the carbon concentration in the H$_2$O-ALD ZnO films was higher than that of the PA-ALD ZnO films. In the H$_2$O-ALD samples annealed over 300 °C, the oxygen concentration was increased, and the carbon concentration was decreased. It is assumed that the number of oxygen deficiencies caused by remaining carbon in the films was reduced by the introduction of oxygen during the annealing.

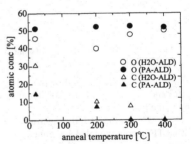

Fig. 4. Temperature dependence of O and C atomic concentration in ALD-ZnO films.

To evaluate the bonding state of the oxygen in the films, the results of the O 1s peaks of non-annealed and annealed in O$_2$ ambient at 300 °C ZnO films prepared by PA-ALD and H$_2$O-ALD are shown in Figs. 5(a) - 5(d), respectively. The O 1s peaks have the shoulders at higher binding energy, and it decreases obviously after the annealing. The lowest energy peak detected at lower than 530 eV corresponds to O-Zn bonding, the middle energy peak detected at approximately 531 eV is attributed to O-C bonding, and the highest energy peak is attributed to O-H bonding. [16] The O-C and O-H bonds seem to be caused by the alkyl Zn as a metal precursor when the ZnO films were deposited. Figures 5(a) and 5(b) show the results of the non-annealed and annealed PA-ALD ZnO films, respectively. The O-C bonding in the PA-ALD ZnO

267

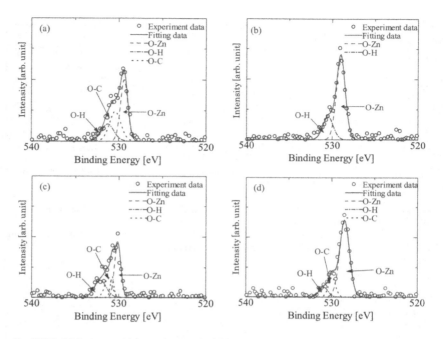

Fig. 5 XPS of O 1s spectra with two or three resolved O bonding components for (a) non-annealed PA-ALD, (b) 300 °C -annealed PA-ALD, (c) non-annealed H₂O-ALD, and (d) 300 °C -annealed H₂O-ALD ZnO films.

films vanished after annealing as shown in Fig. 5(b). In the H₂O-ALD ZnO films shown in Figs. 5(c) and 5(d), the O-H and O-C bonding decreased after annealing, however, the O-C bonding remained.

Compared with the electrical properties of the ALD-ZnO TFTs, those of the H₂O-ALD ZnO TFTs non-annealed and annealed at lower than 200 °C did not show switching characteristics. It is considered to be caused by the high career concentration because the films were zinc rich. Furthermore, the insufficient TFT device characteristics such as low I_{d_on} are caused by remaining carbon in ZnO films.

CONCLUSIONS

We investigated TFTs with ZnO channel layers deposited by ALD using two different oxidizers, at low temperatures. Fabricated ZnO TFTs were annealed at various temperatures in O₂ ambient, and the changes of their electrical properties were measured. The TFTs with ZnO films prepared by PA-ALD stably exhibited clear transistor operations with high on/off current ratio. On the contrary, the ZnO TFTs prepared by H₂O-ALD did not exhibit any switching properties in the case without or with lower than 200 °C annealing. Further, the TFT device performances of the H_2O-ALD ZnO TFTs annealed over 300 °C exhibited insufficient properties, such as low on-current and mobility.

XPS analysis revealed the low oxygen concentration and the large remaining carbon in ZnO films deposited by H_2O-ALD compared with that of PA-ALD ZnO films. The carbon in the PA-ALD ZnO films vanished after annealing over 300 °C in O_2 ambient, however, that of the H_2O-ALD ZnO films was remained. Through this study, we found that the electrical properties of ZnO TFTs are dependent on the annealing temperature, and the low on-current is attributed to the remaining carbon in the ZnO films. Furthermore, these results suggest that high performance and low temperature processed ZnO TFTs is possible to obtain by using PA-ALD.

REFERENCES

1. Y. Kuo, Thin Film Transistors: Materials and Processes (Kluwer Academic, New York, 2004), Vol. 1, p. 6.
2. T. Sameshima, J. Non-Cryst. Solids 227–230, 1196 (1998).
3. H. Kakinura, Phys. Rev. B 39, 10473 (1989).
4. R. B. M. Cross and M. M. D. Souza, Appl. Phys. Lett. **89**, (2006) 263513.
5. Ü. Özgür, Y. I. Alivov, C. Liu, A. Teke, M.A. Reshchikov, S. Doğan, V. Avrutin, S.J. Cho and H. Morkoç, J. Appl. Phys. **98** (2005) 041301.
6. P. F. Carcia, R. S. McLearn, M. H. Reilly, and G. Nunes, Appl. Phys. Lett. **82** (2003) 1117.
7. R. L. Hoffman, B. J. Norris, and J. F. Wager, Appl. Phys. Lett. **82**, (2003) 733.
8. S. Matsuda, K. Kitamura, Y. Okumura, and S. Miyatake: J. Appl. Phys., **93** (2003) 1624.
9. P. F. Carcia, R. S. McLearn, M. H. Reilly, and G. Nunes: Appl. Phys. Lett. **82** (2003) 1117.
10. E. Fortunato, P. Barquinha, A. Pimentel, A. Goncalves, A. Marques, L. Pereira, and R. Martins: Thin Solid Films **487** (2005) 205.
11. K. Murata, K. Washio, N. Miyatake, Y. Mori, H. Tachibana, Y. Uraoka, and T. Fuyuki, ECS Transactions, 11(7) (2007) 31.
12. S. Kwon, S. Bang, S. Lee, S. Jeon, W. Jeong, H. Kim, S. C. Gong, H. J.Chang, H. Park and H. Jeon, Semicond. Sci. Technol. **24** (2009) 035015.
13. D. H. Levy, D. Freeman, S. F. Nelson, P. J. Cowdery-Corvan, and L. M. Irving, Appl. Phys. Lett. **92** (2008) 192101.
14. S. J. Lim, S. J. Kwon, H. G. Kim and J. S. Park, Appl. Phys. Lett. **91** (2007) 183517.
15. Y. Kawamura, Y. Uraoka, H. Yano, T. Hatayama and T. Fuyuki, 2009 IMFEDK abstract pp. 84, (2009).
16. J. F. Moulder, W. F. Stickle, P. E. Sobol, and K. D. Bomben, Handbook of X-ray Photoelectron Spectroscopy, edited by J. Chastain, and R. C. King Jr. (Physical Electronics, Inc. publishers, Minnesota, 1995) p.40.

Mater. Res. Soc. Symp. Proc. Vol. 1201 © 2010 Materials Research Society

Photocurrent measurements of $Mg_xZn_{1-x}O$ epitaxial layers of different x

R. K. Thöt[1], T. Sander[1], P. J. Klar[1], and B. K. Meyer[1]
[1] I. Physikalisches Institut, Justus-Liebig Universität, Heinrich-Buff-Ring 16, 35392 Giessen, Germany

ABSTRACT

The $Mg_xZn_{1-x}O$ alloy in wurtzite structure can be grown with Mg contents x up to 0.4. The band gap of the alloy increases with x. Furthermore, $ZnO/Mg_xZn_{1-x}O$ quantum well structures are of type I and thus are of interest for the active region of opto-electronic devices.
We report on in-plane photocurrent measurements of $Mg_xZn_{1-x}O$ epitaxial layers with x up to about 0.4 in the temperature range from 80 K to 300 K. Epitaxial films are either grown by plasma-assisted molecular beam epitaxy on c-plane sapphire substrates with a thin MgZnO buffer layer and by chemical vapor deposition on a-plane ZnO substrates. We map the evolution of the band gap transitions as a function of the Mg composition at different temperatures for the c-plane samples and as a function of polarization of the incoming light for an a-plane sample. The contributions of A, B and C interband transitions to the band gap signals are analysed and discussed.

INTRODUCTION

One of the key issues in modern solid-state electronics is energy band engineering for design and fabrication of heterostructures and quantum wells. ZnO can be alloyed with MgO to form high quality single-crystal $Mg_xZn_{1-x}O$ films with the Mg content up to almost 40%, while retaining the wurzite crystal structure [1, 2]. Alloying ZnO with MgO increases the direct bandgap of ZnO from 3.3 eV to about 4.0 eV (for 33% Mg incorporation). $ZnO/ Mg_xZn_{1-x}O$ forms a type-I heterostructure with reported conduction band offsets $\Delta E_c \sim 0.7 - 0.9\ \Delta E_g$ of the bandgap difference ΔE_g [1, 3, 4], ideal for electrical and optical confinement. The small lattice mismatch and the high bandgap ensure the effective use of $Mg_xZn_{1-x}O$ as a barrier layer for ZnO based heterostructures and quantum wells.

It is important to note that the wurtzite crystal structure of ZnO does not have inversion symmetry perpendicular to [0001], the crystallographic c-axis causing ZnO and $Mg_xZn_{1-x}O$ alloys to be polarized along the [0001] c-axis. Due to presence of equal numbers both of Zn as well as of O atoms on the surface, a-plane ZnO and $Mg_xZn_{1-x}O$ exhibit no net surface charges and a net zero dipole moment normal to the surface in this non-polar plane, thus preventing divergence of surface energy [5, 6]. While most ZnO based devices have been reported on c-axis oriented ZnO films and heterostructures, the active layers of the devices grown with this orientation suffer from undesirable spontaneous and piezoelectric polarizations. These polarizations produce electrostatic fields, which gives rise to the quantum-confined Stark effect (QCSE) that causes a spatial separation of the electrons and holes, thus reduces their overlap, lowering of the quantum efficiency in optoelectronic devices. This problem can be overcome for non-polar layered structures where spontaneous and piezoelectric polarization fields are avoided. On the other hand, layered structures grown on a-plane substrates exhibit an in-plane anisotropy which is reflected in their optical properties via the selection rules for optical transitions between the A, B, and C valence bands with the conduction bands. The anisotropy of the optical properties may lead to novel device concepts [7-9].

EXPERIMENT

$Mg_xZn_{1-x}O$ layers have been grown on c-plane sapphire by plasma-assisted molecular beam epitaxy (PAMBE) using conventional double-zone effusion cells for the evaporation of Mg ($4N$) and Zn ($6N$) and a radio-frequency plasma source to provide oxygen radicals [10]. A nominally undoped ZnO layer was grown on an a-plane ZnO substrate by chemical vapor deposition [11]. For the photocurrent measurements, the samples two In-contacts were prepared on the sample surface at a distance of about 2 mm. The photocurrent measurements were performed at normal incidence using light of a Xe-lamp dispersed by a single-grating monochromator with a spectral resolution of about 2 Å. The light spot illuminated the entire area between the contacts and its intensity was modulated using a mechanical chopper working at 220 Hz. A voltage of about 5 V was applied between the contacts. The AC photocurrent at the chopper frequency was detected using lock-in technique. The spectra acquired between 200 and 400 nm were corrected for the spectral response of the optical system. The samples were mounted in a cold-finger cryostat cooled with liquid nitrogen for the temperature-dependent measurements. Using a linear polarizer, the polarization direction of the incoming light was varied with respect to the crystallographic axes in case of the samples grown on a-plane substrates.

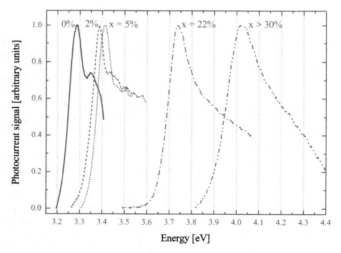

Figure 1. Photocurrent spectra of the series of $Mg_xZn_{1-x}O$ samples obtained at 295 K. The Mg content x is indicated in the figure.

272

DISCUSSION

Figure 1 depicts normalized photocurrent spectra for the series of $Mg_xZn_{1-x}O$ samples of different x obtained at 295 K. All spectra show a clear feature related to the interband transition which shifts to higher energy with increasing x. At this temperature no finestructure of the interband transition related to the A, B, and C valence bands can be observed. The peak-like feature in all the spectra is shifted to lower energies with respect to the actual band gap. This is confirmed by room-temperature absorption measurements (not shown here) which reveal rather step-like bandgap absorptions without sharp features. The peak-like feature in each photocurrent spectrum and its shift towards lower energies with respect to the band gap is due to the complex interplay of absorption, penetration depth of the probe light, and higher recombination rates close to the surface. Absorption of light is essential to photo-excite carriers. Thus, at first, the number of photo-excited carriers increases with increasing absorption until the penetration depth of the light is somewhat between layer thickness and the width of the surface depletion region. In the surface region less of the photo-excited carriers actually contribute to the measured photocurrent as the recombination rates are higher than in the bulk of the layer. Therefore, the photocurrent spectrum may show a maximum although the absorption is still increasing. Nevertheless, the peak in the photocurrent spectrum is related to band gap absorption. On cooling the samples to 80 K, the photocurrent spectra shift to higher energies as the band gap increases and, in particular for the samples of low x, distinct features can be observed, which we tentatively relate to the finestructure of the valence band at the Γ-point. The photocurrent spectra of the samples with $x =$ 0%, 2% and 5%, respectively, obtained at 80 K are given in Figure 2.

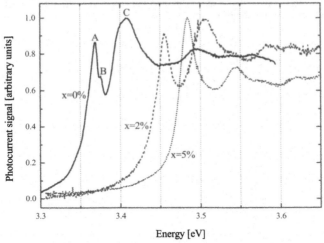

Figure 2. Photocurrent spectra of the $Mg_xZn_{1-x}O$ samples of low x measured at 80 K.

For example, the spectrum of the ZnO sample shows a sharp peak at 3.369 eV with a small satellite at 3.376 eV and an additional broad peak at 3.407 eV. The separation of these

273

features is very close to those of the A, B and C valence band edges of bulk ZnO of 7 meV between A and B and 50 meV between A and C. Furthermore, it is well known that the splitting between the A and B is rather independent of biaxial strain whereas that between A and C depends strongly on biaxial strain [12]. Therefore, we consider these peaks as band gap related, of course, affected by the dependence of the penetration depth on the magnitude of the absorption and by the dependence of the recombination rates on the distance from the surface. Surprisingly, the linewidths of the A and B-related peaks are much smaller than that of the C-related feature. The narrowing of the A and B peaks may be related to the distinct increase of the excitonic absorption feature at low temperatures. The reason of the much broader C-related feature is currently unknown. Several effects may contribute such as the much lower penetration depth due to the onset of the C-edge absorption which restricts the penetration of the light to a region where the recombination rates change rapidly with the distance from the surface or the high sensitivity of the C-edge position on biaxial strain and thus on relaxation effects as a function of penetration depth. Further studies are required to clarify this point. An A as well as a C-related feature are also observed for the samples with $x = 2\%$ and 5%. A B-related feature cannot be resolved in the corresponding spectra due to the alloy broadening of the spectral features. The splitting between the A and C related feature is found to increase with x from 38 meV for 0%, to 49 meV for 2% and 60 meV for 5%. The magnitude of the shifts again confirms that the feature at higher energies is rather of electronic origin than related to phonon-assisted absorption, i.e. C-related and not A-related in conjunction with creation of phonons, as the changes of phonon frequencies with x are much smaller than the observed increase of the splitting between the two spectral features. Figure 3 summarizes the energy positions of the A-related feature versus Mg content x and as an inset the variation of the splitting between A and C-related features as a function of x.

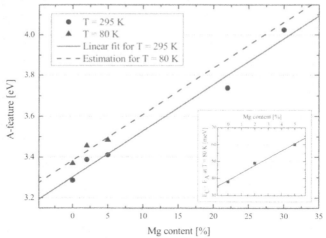

Figure 3. Position of the A-related feature in the photocurrent spectra as a function of Mg content x at 295 K and 80 K, respectively. Inset: Splitting between A and C-related features versus x at 80 K.

Figure 4 depicts a series of photocurrent spectra in the band gap region of ZnO obtained using linearly polarized light at normal incidence. The spectra are taken on the sample grown on the a-plane ZnO substrate. Thus, normal incidence corresponds to an a-direction and the linear polarization of the light can be continuously varied between $E\|c$ and $E\|a$. The spectra obtained at room temperature at different angles with respect to the crystallographic axes are shown in figure 4. In contrast to the spectra obtained on the c-plane samples, A and C related features can be observed at room temperature. This is related to the strong dependence of the interband transitions on the incidence direction of the probe light with respect to the crystallographic axes in the wurtzite structure [13].

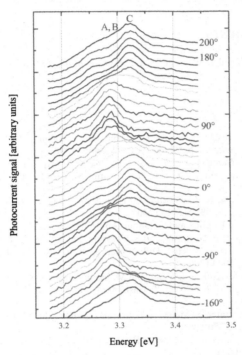

Figure 4. Series of photocurrent spectra in the band gap region of ZnO obtained using linearly polarized light at normal incidence. The spectra are taken at 295 K on the sample grown on an a-plane ZnO substrate.

In particular, it should be noted that the A-related interband transition is forbidden for $E\|c$. This explains the observed variation of the spectral features as a function of the polarization angle and even allows one to determine the direction of the c-axis of the ZnO layer (e.g. for this sample polarizer angle of 125°). Furthermore, it confirms that the spectral feature at higher energy is not related to phonon-assisted transition involving the A-valence band edge because, if this was the case, both features would show the same angular dependence of the intensity.

CONCLUSIONS

We have demonstrated that the photocurrent spectra of $Mg_xZn_{1-x}O$ layers grown on c-oriented and a-oriented substrates may yield useful information about the valence band structure of the layers despite the complexity of the rather complicated interplay between magnitude of the absorption, penetration depth and surface recombination effects.

ACKNOWLEDGEMENTS

We thank M. Eickhoff and T. Wassner for the provision of the PAMBE samples used in the experiments.

REFERENCES

1. A. Ohtomo, M. Kawasaki, T. Koida, K. Masubuchi, Y. Sakurai, Y. Yoshida, T. Yasuda, Y. Segawa, and H. Koinuma, Appl. Phys. Lett. **72**, 2466 (1998)
2. S. Muthukumar, J. Zhong, Y. Chen, T. Siegrist, and Y. Lu, Appl. Phys. Lett. **82**, 742 (2003)
3. A. Ohtomo, M. Kawasaki, I. Ohkubo, H. Koinuma, T. Yasuda, and Y. Segawa, Appl. Phys. Lett. **75**, 980 (1999)
4. H. Tampo, H. Shibata, K. Matsubara, A. Yamada, P. Fons, S. Niki, M. Yamagata, and H. Kanie, Appl. Phys. Lett. **89**, 132113 (2006)
5. U. Diebold, L. Vogel Koplitz, and O. Dulub, Appl. Surf. Sci. **237**, 336 (2004)
6. M. Du, S. B. Zhang, J. E. Northrup, and S. C. Erwin, Phys. Rev. B, **78**, 155424 (2008)
7. C.J. Kao, Y.W. Kwon, Y.W. Heo, D.P. Norton, F. Ren, G.C. Chi, and S.J. Pearton, J. Vac Sci. Technol. B **23**, 1024 (2005)
8. S. Sasa, M. Ozaki, K. Koike, M. Yano, and M. Inoue, Appl. Phys. Lett. **89**, 053502 (2006)
9. K. Koike, D. Takagi, M. Kawasaki, T. Hashimoto, and T. Inoue, Jpn J. Appl. Phys. **46**, L865 (2007)
10. T.A. Wassner, B. Laumer, S. Maier, A. Laufer, B.K. Meyer, M. Stutzmann, and M. Eickhoff, J. Appl. Phys. **105**, 023505 (2009)
11. S. Graubner, C. Neumann, N. Volbers, B. K. Meyer, J. Bläsing, and A. Krost, Appl. Phys. Lett. **90**, 042103 (2007)
12. A. Schleife, C. Rödl, F. Fuchs, J. Furthmüller, and F. Bechstedt, Appl. Phys. Lett. **91**, 241915 (2007)
13. S.I. Gubarev, phys. stat. sol. (b) **134**, 211 (1986)

Mater. Res. Soc. Symp. Proc. Vol. 1201 © 2010 Materials Research Society   1201-H10-34

Structural and Magnetic Properties of Zn$_{1-x}$Co$_x$O Nanoparticles Prepared by a Simple Sol-Gel Method at Low Temperature

Segundo R. Jáuregui-Rosas[1], O. Perales-Perez[2], S. Urcia-Romero[3], M. Asmat-Uceda[3] and E. Quezada-Castillo[1]

[1] Lab. Física de Materiales - Departamento de Física, Universidad Nacional de Trujillo, Av. Juan Pablo II S/N, Trujillo - Peru.
[2] Department of Engineering Science and Materials, University of Puerto Rico, Mayaguez, Puerto Rico 00681-9044, USA
[3] Department of Physics, University of Puerto Rico, Mayaguez, Puerto Rico, 00681-9016, USA

ABSTRACT

Pure and Zn$_{1-x}$Co$_x$O nanoparticles have been synthesized by a simple sol-gel method at low temperature where neither a chelating agent nor subsequent annealing was required. The effect of Cobalt atomic fraction, 'x' $\leq$ 0.0625, on the structural and magnetic properties of the doped ZnO powders was evaluated. X-ray diffraction and Fourier-transform infrared spectroscopy analyses evidenced the exclusive formation of the ZnO-wurtzite structure; no isolated Co-phases were detected. The linear dependence of cell parameters a and c with 'x', suggested the actual replacement of Zn by Co ions in the oxide lattice. Micro Raman spectroscopy measurements showed a band centered at 534cm^{-1}, which can be assigned to a local vibrational mode related to Co species, in addition to the normal modes associated with wurtzite. The intensity and broadening of this band at 534 cm^{-1} were enhanced by increasing 'x'. In turn, the other bands corresponding to A$_1$ (E$_2$, E$_1$) and E$_2^{High}$ modes were red shifted at higher Co contents. Room-temperature magnetization measurements revealed the paramagnetic behavior of the Co-doped ZnO nanoparticles.

INTRODUCTION

ZnO, exhibiting a direct wide bandgap of 3.37eV and an exciton binding energy of ~60meV at room temperature [1], has emerged as a very promising material for different applications. The expected room temperature ferromagnetism in wide band gap semiconducting-based diluted magnetic semiconductors (DMS) [2,3] has opened new possibilities for these materials in spintronic applications. However, the synthesis of a DMS system with practical magnetic ordering temperatures is still a challenging task. Regarding candidate DMS materials, wurtzite-phase semiconductor ZnO doped with rare earth [4] or transition metal species [5-10] has been the focus of the research efforts. ZnO has been synthesized as nanorods [9], thin films [10] or nanoparticles [4,5,8] through diverse routes including autocombustion [6], hydrothermal processing [8] or sol-gel [4,9]. Despite of the progress made on the production of this material, experimental results of inherent ferromagnetism in doped ZnO are still controversial. In particular, it has been found that magnetism in ZnO containing Co^{2+} ions is strongly sensitive to synthesis conditions [11,12]. On this basis, the present work was focused on the systematic investigation of the incorporation of Co species into host ZnO lattice and the corresponding structural and magnetic properties at room temperature. The selected synthesis approach for

ZnO-based nanocrystalline powders was based on a simple low-temperature sol-gel approach that does not require the thermal treatment of the samples to develop the oxide structure.

EXPERIMENTAL

Materials synthesis

Suitable amounts of Zn(II) acetate (Sigma Aldrich, 99.99%) and Co(II) acetate (Sigma Aldrich, 99.995%) were dissolved separately in 2-ethylhexanoic from Alfa-Aesar, 98% (boiling point 227°C, viscosity at 20°C, 7.5mPa.s) at 200°C to obtain a 0.02mol/L solution. No chelating agent was used. The weights of each precursor salt were calculated to give the desired Co-atomic fractions, 'x', according to $Zn_{1-x}Co_xO$ stoichiometry ($0.00 \leq$ 'x' ≤ 0.0625). Obtained solutions were stirred for 10 min and heated to 200°C for 72 hours to eliminate the solvent. As-synthesized solids were pulverized in an Agate mortar and submitted for characterization.

Characterization

The solid products were structurally characterized by X-Ray Diffraction (XRD) using the Cu-Kα radiation in the 2Ω range between 15° and 80°; the average crystallite size was estimated using the Scherrer's equation. These structural analyses were complemented by Fourier-transform infrared (FTIR) spectroscopy measurements using a MIRacleTM ATR FTS 1000 spectrometer in the transmittance mode. The Raman spectra were taken at room temperature by using a Renishaw micro-Raman system with a 514.5nm (2.41eV) excitation line from an Ar$^+$-ion laser. Magnetic measurements were curried out at room temperature by using a Vibrating Sample Magnetometer (VSM).

RESULTS AND DISCUSSION

Structural characterization

As-synthesized Co-doped powders exhibited a green tone that became darker with increasing cobalt content. The average crystallite size in the powders was found to be 12±1.3nm, for all doping levels. Figure 1.a shows the XRD patterns of $Zn_{1-x}Co_xO$ powders synthesized at various atomic fractions of Co, 'x'. The diffraction lines for bare ZnO (P6$_3$mc) are shown at the bottom. All peaks correspond to ZnO-wurtzite structure; no peaks related to isolated Co phases were detected. Two intense fcc-CoO peaks would be expected at 36.5° and 42.4°; although the peak at 36.5° might be hidden by the (101) ZnO peak, the peak at 42.4° should be clearly defined if appreciable fcc-CoO was formed. As suggested by the detailed analysis of the XRD data in the 36°-44° 2Ω range, shown in logarithmic scale (figure 1.b), there is no evidence of the presence of this compound in co-existence with ZnO.

The variation of the lattice parameters ('a' and 'c') and cell volume 'V' of hexagonal wurtzite with 'x' is shown in figures 1.c-e. The linear variation of the lattice parameters with 'x' suggests the actual incorporation of Co into the ZnO lattice. This behavior is in good agreement with previously published results [13,14], and was explained in terms of the small size of Co^{2+} (0.60Å) in tetrahedral coordination in the oxide lattice [15]. In an octahedral coordination, Co^{2+}

ions would be characterized by a significant increase in both lattice parameters because of this larger ionic radious (0.65 Å, low spin, or 0.74 Å, high spin).

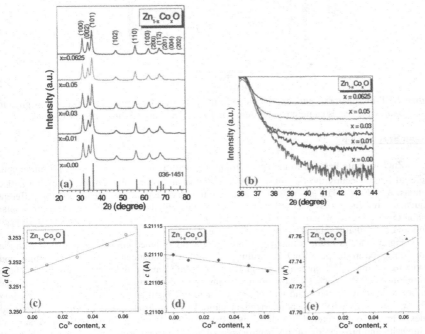

Figure 1 (a) XRD patterns for $Zn_{1-x}Co_xO$ powders. (b) Detail of (a) for 2θ between $36°$ and $44°$ showing the absence of the strong peak of CoO at 2Ω $42.4°$. (c-e) Variation of lattice parameters a and c, and cell volume of ZnO as a function of Co^{2+} atomic fraction, 'x', in ZnO.

FT-IR analyses

As-synthesized powders were also analyzed by FT-IR spectroscopy to complement the structural information provided by XRD. Figure 2 shows the spectra for the solids synthesized at different 'x' values. The sharp band at $524cm^{-1}$ is assigned to Zn-O bound in the wurtzite lattice. Broad bands centered at $1419cm^{-1}$ and $1573cm^{-1}$ were clearly observed for all synthesized samples. These bands are assigned to asymmetric and symmetric C=O stretching vibration modes, respectively, of acetate species that would remain as adsorbed species and/or as intermediate that could not be converted into the oxide phase. No bands associated to Co oxides, [16], were detected.

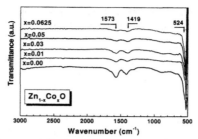

Figure 2 FT-IR spectra of pure and Co-doped ZnO powders. The band at 524cm^{-1} is assigned to Zn-O bond, while bands centered at 1573cm^{-1} and 1419cm^{-1} correspond to acetate species.

Raman Spectroscopy

ZnO-wurtzite belongs to the space group C_{6v}^4, with two formula units per primitive cell, where all atoms occupy the C_{3v} sites. Therefore, the group theory predicts the following lattice phonons: $A_1+2E_2+E_1$. The phonons of A_1 and E_1 symmetry are polar and, hence, exhibit different frequencies for the transverse-optical (TO) and longitudinal-optical (LO) modes. Non-polar phonon modes with symmetry E_2 have two frequencies: E_2^{High}, which is associated with oxygen atoms and E_2^{Low} that is related to the Zn sub-lattice.

Figure 3.a shows the Raman spectra of pure and Co-doped powders. Compared with the modes observed in the pure ZnO spectrum ('x' = 0.0), a pronounced red-shift, broadening, and weakening of the E_2^{High} phonon band in Co-doped ZnO nanoparticles were observed. These features in the Raman spectra have been attributed to the presence of structural defects in the host lattice [17]. A similar trend was observed for the A_1 band (figure 3.b). The corresponding processes and symmetry are summarized in table 1. In addition to the previously discussed modes, a broad band centered at 534cm^{-1} was detected for 'x' values above 0.01. Interestingly, the intensity and broadening of this band increased as Co content increases but no any shift was observed, which agrees well with previous reports [10,14,17]. The presence of this band would be to due to a local vibrational mode of Co associated to donor defects and would play a crucial role in mediating the high-temperature ferromagnetism [17].

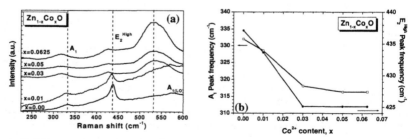

Figure 3 a) Room temperature micro-Raman spectra of $Zn_{1-x}Co_xO$ powders at various Co atomic fractions, 'x'. b) Variation of A_1 and E_2^{High} band positions with 'x'.

Table 1 Room temperature symmetries and processes in pure and Co-doped ZnO powders. The modes were assigned according to reference [18].

Symmetry	Process	Brillouin zone points/lines	Frequency (cm^{-1})				
			ZnO	x=0.01	x=0.03	x=0.05	x=0.0625
A₁ (E₂,E₁)	E₂^{High}-E₂^{Low}	Γ	332	330	320	316	315
A₁	A₁(TO)	Γ	382	382			
E₂	E₂^{High}	Γ	437	433	425	425	425
A₁	2B₁^{Low};2LA	Γ; L, M, H		534	534	534	534
A₁	A₁ (LO)	Γ	576	571			
A₁	TA + LO	M	665		665		

Magnetic properties

Figure 4 shows the room temperature M-H loops for as-synthesized $Zn_{1-x}Co_xO$ powders. Similar to reported for Fe (0.01)-doped ZnO nanoparticles prepared by co-precipitation method [19], the powders were diamagnetic in nature. However, unlike previous reports [14,17], where the presence of a Raman band at 534cm^{-1} was observed, no ferromagnetism was evidenced at higher Co contents; instead, a paramagnetic behavior is clearly observed. This behavior agrees with previous studies [7,9] and would be consistent with Spaldin's *et al.* results [20], who claimed that carriers are indispensable to induce ferromagnetism in ZnO-based DMSs. Our ongoing work addresses the co-doping of ZnO with Cu and Co and the corresponding results will be a matter of a forthcoming publication.

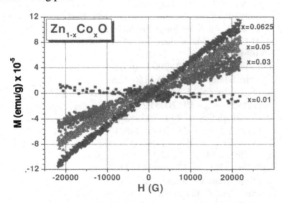

Figure 4 Room temperature M-H measurements for as-synthesized $Zn_{1-x}Co_xO$ ('x' = 0.01, 0.03, 0.05, 0.0625) powders.

CONCLUSIONS

Nanocrystalline $Zn_{1-x}Co_xO$ powders were successfully synthesized by a simple sol-gel method at low temperature with no need for thermal treatment. XRD analyses confirmed the exclusive development of the host ZnO-wurtzite structure for all levels of Co in the atomic fraction, 'x', range between 0.00 and 0.0625. No isolated Co-based compounds were detected, which was also confirmed by FT-IR measurements. The linear dependence of the lattice parameters with 'x' suggested the actual replacement of Zn by Co species in the host oxide lattice. Micro Raman spectroscopy analyses revealed a red shift of the bands corresponding to A_1 (E_2, E_1) and E_2^{high} that was dependent on the 'x' value. Also, the Raman band at 534cm^{-1}, observed for 'x' values above 0.01, was assigned to a local vibrational mode associated to Co. The intensity and broadening of this band increased as Co content did. Room temperature magnetic measurements evidenced the paramagnetic nature of the Co-doped ZnO powders.

ACKNOWLEDGMENTS

This material is based upon work supported by the Presidencia Consejo de Ministros-Peru, through the Programa de Ciencia y Tecnologia-FINCyT (Contract N° 163-2009-FINCyT-BDN). OPP acknowledges the support from the DoE-Grant N° FG02-08ER46526.

REFERENCES

[1]   A. Janotti and Ch. G. Van de Walle, Rep. Prog. Phys. **72** (2009) 126501
[2]   T. Dietl, et al., Science **287** (2000) 1019-1022
[3]   K. Sato and H. Katayama-Yoshida, Jpn. J. Appl. Phys. Part 2, **39** (2000) L555-L558
[4]   S. Jáuregui-Rosas, et al., in *Functional Metal-Oxide Nanostructures*, edited by J. Wu, W. Han, A. Janotti, H-C. Kim (Mater. Res. Soc. Symp. Proc. **Volume 1174**, Warrendale, PA, 2009), p. **1174-V09-07**
[5]   B. Martinez, et al., Phys. Rev. B **72** (2005) 165202
[6]   S. Deka, et al., Chem. Mater. **16** (2004) 1168-1169
[7]   C. N. R. Rao and F. L. Deepak, J. Mater. Chem. **15** (2005), 573-578
[8]   X. Qiu, et al., Nanotechnology **19** (2008) 215703
[9]   J.H. Yang, et al., J. Alloys Comp. **473** (2009) 543-545
[10] K. Samanta, et al., Phys. Rev. B **73** (2006) 245213
[11] C.J. Cong, et al., Mater. Chem. Phys. **113** (2009) 435-440
[12] R. Boubekri, et al., Chem. Mater. **21** (2009) 843–855
[13] Y.B. Zhang, et al., Phys. Rev. B **73** (2006) 172404
[14] L.B. Duan, et al., Solid State Commun. **145** (2008) 525-528
[15] R.D. Shannon, Acta Cryst. A **32** (1976) 751-767
[16] Ch.-W. Tang, et al., Thermochim. Acta **473** (2008) 68-73
[17] X. Wang, et al., Adv. Mater. **18** (2006) 2476-2480
[18] Ramon Cuscó, et al., Phys. Rev. B **75** (2007) 165202
[19] P.K. Sharma et al., J. Magn. Magn. Mater. **321** (2009) 2587–2591
[20] N.A. Spaldin, Phys. Rev. B **69** (2004) 125201

Mater. Res. Soc. Symp. Proc. Vol. 1201 © 2010 Materials Research Society   1201-H10-39

Reaction Anisotropy and Size Resolved Oxidation Kinetics of Zinc Nanocrystals

Xiaofei Ma[1] and Michael R. Zachariah[1,2]

[1]Department of Mechanical Engineering, University of Maryland-College Park,
College Park, MD 20742, U.S.A.

[2]Department of Chemistry and Biochemistry, University of Maryland-College Park,
College Park, MD 20742, U.S.A.

Abstract

In this work, size-classified substrate-free Zn nanocrystals (NCs) are prepared and investigated for their oxidation kinetics using an in-flight tandem ion-mobility method. The first mobility characterization size selects the NCs, while the second mobility characterization measures changes in mass resulting from a controlled oxidation of the NCs. This method allows for a direct measurement of mass change of individual particles and thus enables us to explore the intrinsic reactivity of NCs while minimizing the sampling error introduced by mass and heat transfer. Two reaction regimes were observed for Zn NC oxidation. A shrinking core model is used to extract the size-dependent oxidation activation energies. We also observed a strong anisotropy effect in the oxidation process as imaged by electron microscopy. An oxidation mechanism is proposed that qualitatively explains the oxidation rate anisotropy and its relationship to the surface energy of the Zn NCs.

Introduction

Zn is an attractive fuel and has been employed in powder form as a rocket propellant when mixed with sulfur[1]. Zinc is also used as the anode or fuel in the zinc-air fuel cell which forms the basis of the theorized zinc economy[2]. More recently Zn/ZnO redox reactions has been considered for thermo-chemical two-step water-splitting cycles for hydrogen generation[3-5]. The cycle involves a Zn hydrolysis reaction to produce hydrogen, followed by the solar reduction of zinc oxide. In a recent study[6], Zn nanoparticles have been employed so as to take advantage of their high surface to volume ratio and the ability to operate a continuous flow process.

As a wide-band-gap semiconductor, bulk ZnO has band gap energy of 3.37 eV at room temperature, and also exhibits piezoelectric properties. ZnO nanostructures such as nanowires/belts have attracted considerable attention due to their novel application in optical and electrical devices, sensors and medical devices[7-9]. While there exist many different techniques to prepare ZnO, the thermal-oxidation from Zn is the most basic and is widely employed to create a variety of ZnO nanostructures[10-14]. Despite the widening application of Zn based nanostructures, very little attention has been paid to the quantitative kinetics of Zn nanocrystal (NC) oxidation.

In this paper we use electrical mobility classification and characterization to explore the size resolved reactivity of Zn NCs in free-flight, and determine quantitative Arrhenius reaction parameters. We will also observe that oxidation is anisotropic as imaged from electron-microscopic analysis. What distinguishes this study from previous bulk sample studies, are first, free NCs are suspended in an inert gas environment, which means free of any potential substrate effects, and thus all crystal planes are co-existent and exposed to the same experiment conditions simultaneously. Secondly, NCs are nanometer in size, which means there are a larger portion of atoms on the edges. Those edges are sharp interfaces between the crystal planes. Thus, edge effect may play an important role in the NC oxidation process.

Experimental Approach

The basic strategy for studying the reactivity of Zn NC, is to size select NC's and then quantify their size resolved reactivity. The experiment system consists of three components. Preparation of monodisperse Zn NCs, exposure of size selected Zn NCs into a controlled temperature region, and finally, measurement of the mass change resulting from oxidation. A complete schematic of the experimental system with temperature and flow rate control is shown in Figure 1. Our experiment consists of two different ion-mobility schemes in series[15]. The first mobility characterization is to size select NCs with a differential mobility analyzer[16-18]. The second mobility characterization employs an aerosol particle mass analyzer[19] and measures changes in mass resulting from a controlled oxidation of the Zn NCs.

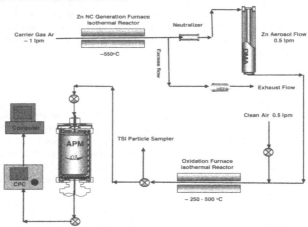

Figure 1 Experimental system for Zn, generation, size selection by DMA, oxidation and subsequent mass analysis with the APM

The Zn NCs are generated using an evaporation/condensation method in argon. Since our objective is to study size resolved reactivity we use a DMA as a band pass filter to generate monodisperse NCs. The DMA selects particles based on their electric mobility, which is related to the drag force and charge on a particle. Our previous experiments have used the DMA-APM technique to measure the inherent density of nanoparticles, as well as to study the mechanism of aluminum and nickel oxidation[15,20]. Furthermore, our DMA/APM mass measurements of NIST SRM 60 and 100 PSL spheres were within about 1.4% and 5.6%, respectively.

In our experiment, Zn NCs of initial ion-mobility sizes of 50nm, 70nm and 100nm were selected in sequential experiments to study the size-dependent oxidation. The temperature in the oxidation furnace was set between 250-500°C in increments of 25°C. The particle mass distribution was then measured for each reactor temperature after the system reached steady state. The room temperature particle mass distribution was also taken, and set as the base of the mass measurement. Samples for electron-microscopic analysis were collected exiting the evaporation furnace by electrostatically precipitating the aerosol onto a TEM grid.

Results and Discussion

Figure 2 shows the normalized particle mass distributions for Zn NCs of initial size 50nm oxidized in air at different temperatures, obtained from APM-CPC measurements. The peak mass of Zn NCs at each oxidation temperature is obtained by fitting the experimental data using a Gaussian distribution. The results in the figure show that the peak mass remains unchanged at low temperatures, and then increases in mass at elevated temperatures reflecting the increase rate of oxidation.

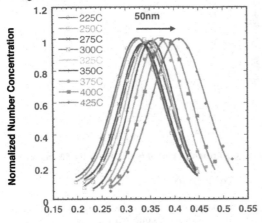

Figure 2 Normalized particle mass distributions for 50nm Zn NC at different oxidation temperatures

It is believed that metal oxidation is a diffusion controlled process. This suggests that the classic diffusion controlled shrinking core model could be applied in our study as a relatively straightforward way to extract reaction kinetics data[21]. Following Carter's steady state analysis for spherical nanoparticles, we derived equation:

$$\ln(\frac{\Delta m}{\tau}) = -E_a / RT + \ln(4\pi M_{O_2} C_{O_2} A \frac{r_c r_p}{r_p - r_c}) \tag{1}$$

to relate the mass change rate of NCs with reaction temperature. Figure 3. shows an Arrhenius plot of the reaction rate for 50nm NCs. Two different regimes of oxidation represented by the two linear fits can be observed from the plot. A slower reaction region occurs at lower temperatures followed by a faster oxidation regime occurring at higher temperatures. The intersection of the two straight lines represents the transition temperatures between the two regimes. Similar trends are observed for the other sizes of Zn NCs. For the case of the slower reaction, the activation energy decreases from 46.1kJ / mol for 100nm size NCs, to 34.0kJ / mol for 50nm NCs. While for the faster reaction, the activation energy decreases from 63.1kJ / mol for 100nm NCs to 49.3kJ / mol for 50nm NCs. In general, the smaller particles have lower activation energy.

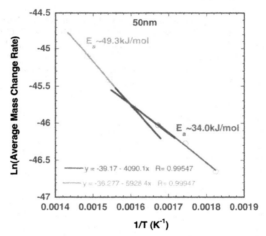

Figure 3 Arrhenius plot of reaction rate for 50nm Zn NCs

Oxidation Rate Anisotropy

To further explore the nature of the two oxidation regimes observed in the Arrhenius plots, samples of 100nm Zn NCs were collected for electron microscopic analysis after oxidation at 350°C, 400°C and 450°C, which are below, close to and above the transition temperature, respectively. The high resolution SEM images are shown in Figure 4.

As we can see from the figure, there is a preferable surface of oxidation during the initial stages. From the images collected at 350°C (4(a) and 4(b)), we see that there are band-shaped oxide layers formed around the six side surfaces of the NCs, while the top and bottom surfaces of NC are flat and remain unchanged. The Zn NCs show a strong oxidation anisotropy with the rate of oxidation on Zn $\{1\bar{1}00\}$ planes much faster than on the $\{0001\}$ planes. As the oxidation temperature increases to 400°C, the Zn NC undergoes restructuring with the edges between the top surfaces and side surfaces becoming blurred. The NC deforms and the original hexagonal-prism shape can not be distinguished. As oxidation temperatures well above the transition temperature (as shown in Arrhenius plots), the NC exhibits a flower-shaped morphology. Given the observed oxidation anisotropy we propose the following reaction mechanism: at the lower oxidation temperatures, only the side surfaces of the NCs are activated and oxide layer are formed first around those surfaces, while at higher oxidation temperatures, both the side and top surfaces are activated, which enhances the oxidation kinetics but requires a higher activation energy. This at least is consistent with the observed kinetic regimes.

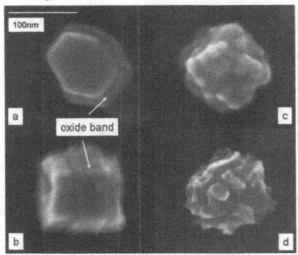

Figure 4 HR-SEM pictures of partially oxidized Zn NCs at different oxidation temperature (a) top view of the NC at 350°C (b) side view of the NC at 350°C (c) 400°C (d) 450°C

As discussed in the introduction, our NCs are free of any potential substrate effects, and their are a larger portion of atoms on the edges of our NCs. Those edges are sharp interfaces between the crystal planes. As such evaporation and oxidation are taking place simultaneously from all faces, and thus at the interface of these planes either a morphology change with time, or alternatively a change in reaction rate must occur. No doubt these edge atoms are at a higher potential energy, due to their lower coordination number and high stain, thus potentially more reactive. This perhaps, is a more important distinction between larger crystals and NCs.

287

Conclusion

Size-classified substrate-free Zn NCs are explored for their oxidation kinetics using an in-flight tandem ion-mobility DMA-APM method. This method allows for a direct measure of mass change of individual particles and thus enables us to explore the intrinsic reactivity of NCs while minimizing the sampling error introduced by mass and heat transfer occurring in traditional methods. Two reaction regimes were observed for Zn NC oxidation, a slower reaction regime at lower temperatures followed by a faster oxidation regime occurring at higher temperatures. A shrinking core model is used to extract the size-dependent oxidation activation energies. Using electron microscopic analysis, we observed a strong effect of oxidation anisotropy. An oxidation mechanism based on surface energy anisotropy and edge-enhanced oxidation effects was proposed to explain the observed oxidation behavior of Zn NCs.

Acknowledgments

Support for this work comes from the National Science Foundation, and the Army Research Office. In addition, the authors wish to acknowledge the microscopy support through the University of Maryland Nanocenter, and the University of Maryland NSF-MRSEC.

References

(1) Sutton, G. P. *Rocket Propulsion Elements*, 7th ed.; Wiley-Interscience, 2000.

(2) Laboratory, L. L. N. *Science and Technology Review* **1995**.

(3) Steinfeld, A. *International Journal of Hydrogen Energy* **2002**, *27*, 611.

(4) Steinfeld, A. *Solar Energy* **2005**, *78*, 603.

(5) Abu Hamed, T.; Davidson, J. H.; Stolzenburg, M. *Journal of Solar Energy Engineering-Transactions of the Asme* **2008**, *130*.

(6) Ernst, F. O.; Tricoli, A.; Pratsinis, S. E.; Steinfeld, A. *Aiche Journal* **2006**, *52*, 3297.

(7) Wong, E. M.; Searson, P. C. *Applied Physics Letters* **1999**, *74*, 2939.

(8) Tang, Z. K.; Wong, G. K. L.; Yu, P.; Kawasaki, M.; Ohtomo, A.; Koinuma, H.; Segawa, Y. *Applied Physics Letters* **1998**, *72*, 3270.

(9) Pan, Z. W.; Dai, Z. R.; Wang, Z. L. *Science* **2001**, *291*, 1947.

(10) Wang, Z.; Harris, R. *Materials Characterization* **1993**, *30*, 155.

(11) Kim, S.; Jeong, M. C.; Oh, B. Y.; Lee, W.; Myoung, J. M. *Journal of Crystal Growth* **2006**, *290*, 485.

(12) Wang, Y. G.; Lau, S. P.; Lee, H. W.; Yu, S. F.; Tay, B. K.; Zhang, X. H.; Hng, H. H. *Journal of Applied Physics* **2003**, *94*, 354.

(13) Lu, H. B.; Li, H.; Liao, L.; Tian, Y.; Shuai, M.; Li, J. C.; Fhu, M.; Fu, Q.; Zhu, B. P. *Nanotechnology* **2008**, *19*.

(14) Nakamura, R.; Lee, J. G.; Tokozakura, D.; Mori, H.; Nakajima, H. *Materials Letters* **2007**, *61*, 1060.

(15) Zhou, L.; Rai, A.; Piekiel, N.; Ma, X. F.; Zachariah, M. R. *Journal of Physical Chemistry C* **2008**, *112*, 16209.

(16) Liu, B. Y. H.; Pui, D. Y. H. *Journal of Colloid and Interface Science* **1974**, *49*, 305.

(17) Liu, B. Y. H. P., D. Y. H. *J. Colloid Interface Sci.* **1974**, *47*, 155.

(18) Knutson, E. O.; Whitby, K. T. *Journal of Colloid and Interface Science* **1975**, *53*, 493.

(19) Ehara, K.; Hagwood, C.; Coakley, K. J. *Journal of Aerosol Science* **1996**, *27*, 217.

(20) Rai, A.; Park, K.; Zhou, L.; Zachariah, M. R. *Combustion Theory and Modelling* **2006**, *10*, 843.

(21) Levenspiel, O. *Chemical Reaction Engineering*, 3rd ed. ed.; John Wiley & Sons, 1999.

Mater. Res. Soc. Symp. Proc. Vol. 1201 © 2010 Materials Research Society   1201-H10-42

I. A. Al-Omari[1, *], S. H. Al-Harthi[1], M. J. Al-Saadi[1], K. Melghit[2]

[1]Department of Physics, Box 36, Sultan Qaboos University, PC 123, Muscat, Sultanate of Oman
[2]Department of Chemistry, Box 36, Sultan Qaboos University, PC 123, Muscat, Sultanate of Oman
*Corresponding author: e-mail: ialomari@squ.edu.om or ialomari@yahoo.com

ABSTRACT

Nanoparticles Fe (x wt. %)-doped Zn-TiO_2 rutile powders, with x between 0 an 10 wt. %, were prepared using a solution chemistry route based on the wet-gel stirring method. Using the TEM images we found that the powder samples exhibit nanorods and nanosheets with nanorods oriented in different directions and accompanied by an amorphous Zn on the surface. The average length of these nanorods is about 60 nm and they have an average diameter of 7 nm. The x-ray diffraction patterns revealed the formation of the nanocrystalline particles with the rutile phase, which is characterized by the (101) diffraction peak. The magnetic properties of the samples were studied using a vibrating sample magnetometer (VSM) in magnetic filed up to 13.5 kOe and in the temperature range of 100 K to 300 K. We found that the magnetization of the samples does not saturate in the maximum available field. The magnetization (M) at an applied magnetic field of 13.5 kOe is found to increase with increasing the Fe percentage at room temperature and at 100 K. TEM measurements and atomic-force microscopy (AFM) were used to image the samples.

INTRODUCTION

Oxide duiluted magnetic semiconductors, in which nonmagnetic oxide semiconductors are doped with low percent magnetic element such as Fe, Co, and Ni, have an important role in the development and the applications of semiconductor spintronics photovoltaic materials, gas sensors, solar energy conversion …etc [1-9]. Fe^{3+} doping of TiO_2 was shown to enhance the photo activity efficiency of this material. Since the discovery of the ferromagnetic state in the Co-doped TiO_2 semiconductors at room temperature and with a curie temperature of 400 K [7] the attention of many researchers have been drawn to understand the effect of doping a transition metal on the magnetotransport properties of these systems. Among the well-known three crystalline structures (Rutile, Anatase, and Brookite) of TiO_2, Rutile is the most thermodynamically stable phase. Zhao et al. [9] prepared and studied the optical properties of TiO_2 nanoparticles using a facile gas flame combustion method. They were able to produce an ultrafine TiO_2 particles with a diameter of about 9 nm. They also found that the as-prepared nanoparticles crystalline with the anatase phase. Resently, Wie et al. [10] studied the structural and the magnetic properties of TiO and TiO_2 films prepared by cluster deposition method. They were able to produce nanoclusters with an average size between 15 nm and 40 nm. This study showed that all the samples of TiO and TiO_2 anatase phase or TiO_2 rutile phase are ferromagnetic with room temperature saturation magnetization of 12.32 emu/cc for the rutile phase, 0.12 emu/cc for the anatase phase, and 0.14 emu/cc for the TiO pahse. They also investigated the moment of TiO_2 rutile phase as a function of cluster size and showed that the moments are located near the surface of the clusters, which is due to the creation of oxygen defects and magnetic moments by oxygen diffusion at the cluster surfaces.

Duhalde et al. [3] studied the role of dopants transition metals (TM) in the origin and significance of room temperature ferromagnetism in the transition-metal-doped TiO_2 thin films, where TM = Mn, Fe, Co, Ni, Cu. They found that the films are ferromagnetic at room temperature in the cases of Fe, Co, Ni, and Cu but not in the case of Mn. Shinde et al. [11] claimed that the ferromagnetism is due to the formation of Co clusters within TiO_2 structure, while others like Griffin et al. [12] provided experimental evidence of intrinsic ferromagnetism in insulating Co doped anatase TiO_2 films. The knowledge of the local environment of the dopant plays an important role in understanding the mechanism of magnetic order in the different materials.

In this work, we prepared and characterized the properties of nanoparticles Fe (x wt. %)-doped Zn-TiO_2 rutile powders, with x between 0 an 10 wt. %, to study the effects of Fe and Zn dopant on the magnetic properties of these nanoparticles.

EXPERIMENTAL DETAILS

Nanoparticles Fe (x wt. %)-doped Zn-TiO_2 rutile powders, with x between 0 an 10 wt. %, were prepared using a solution chemistry route based on the wet-gel stirring method. First, we synthesized zinc-containing TiO_2 nanoparticles using Titanium (III) chloride ($TiCl_3$) which was already doped with Zn (II) precursor. About 10 ml of $TiCl_3$ (15% weight per volume (W/V)) was diluted with 40 ml of distilled water and stirred for about 65 hours at room temperature until the color of the solution turned from dark purple to white colloidal solution. 500 ml of distilled water was then added to the solution followed by 20 ml of Ammonium hydroxide (NH_4OH) with (10% wt, molar weight of 35.05 g/mol and density equal to 0.89 g/cm³) to precipitate the Zn-TiO_2 at the bottom of the beaker. The white formed gel was then easily separated from the solution by decantation and washed with distilled water several times until the silver nitrate $AgNO_3$ (molecular weight of 169.87 g/mol and density equal to 1.180 g/cm³) test indicates the absence of chloride ions in the solution. The wet gel obtained was dried at room temperature for one day then weighed and stored in a clean bottle. To dope the Zn-TiO_2 nanoparticles with Fe, Iron nitrate $Fe(NO_3)_39H_2O_{(s)}$ with the molar weight of 404.0 g/mol was used. The required amount of $Fe(NO_3)_39H_2O_{(s)}$ was weighed and dissolved in 100 ml of distilled water and heated up to 100 °C. Simultaneously the Zn-TiO_2 gel which was prepared as mentioned above was heated in a different beaker at 100 °C. The two solutions were mixed and stirred for about 15 minutes. The final product was separated by decantation and washed several times by distilled water then left to dry at room temperature for one day and stored in clean bottles. Using the same doping procedure several Fe-doped Zn-TiO2 nanoparticle samples were prepared with Fe nominal weight concentrations ranging from 0.1 to 10 %.

RESULTS and DISCUSSION

X-ray diffraction patterns of the samples were measured using a Phillips Analytical X-ray 1710 diffractometer (with Cu-K$_\alpha$ radiation) fitted with graphite diffracted-beam monochromator. Figure 1 shows the x-ray diffraction patterns of the different samples. This figure show that all the samples crystalline in the rutile phase, which is characterized by the (101) diffraction peak. The broading of the diffraction peaks is due to the nano-size effects. The shift in the peaks toward lower angles can be attributed to the iron doping. No characteristic peaks were detected for Zn, Fe or any impurities, either because of the Fe small size crystal or less concentration or for the Zn being on the surface of the specimens.

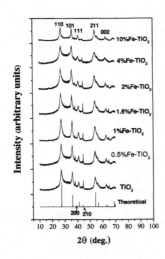

Figure1. X-ray diffraction pattern for the different samples of nanoparticles Fe (x wt. %)-doped Zn-TiO$_2$.

The size and morphology of the undoped and the Fe doped Zn-TiO$_2$ samples were characterized by Transmission Electron Microscopy (TEM, JEOL 1230). The TEM instrument has a lattice resolution of about 0.2 nm. The transmission electron microscopy (TEM) images for two representative samples undoped (x = 0) and Fe-doped with (x = 4%) are shown in Fig. 2. In the undoped sample we can see that there are many nanorods with an average length of 60 nm and an average diameter of 7 nm while in the doped sample we can see that the density of the nanorods decrease due to the breakdown to sheets, which is due to the increase of the acidity of Fe(NO$_3$)$_3$9H$_2$O solution while sample preparation [13].

Figure 2. Transmission electron microscopy (TEM) images for two representative samples undoped (x = 0) and Fe-doped with (x = 4 wt. %).

Atomic-Force Microscopy (AFM) was used to investigate the topography and the particle distribution of the undoped and doped Zn-TiO_2 nanoparticles. To minimize the sample damage the AFM imaging were performed in tapping mode. Figure 3 shows the tapping mode AFM images for the Fe-doped with (x = 4%) sample. These images confirm the coexistence of the nanorods and nanosheets and the increase of the number of nanosheets compared with the number of nanorods for the doped sample.

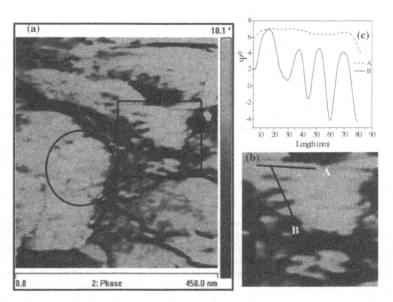

Figure 3. (a) AFM phase image of the Fe-doped with (x = 4%) doped Zn-TiO_2 nanoparticles indicating nanorods and nanosheets in the area enclosed by square and circle respectively. (b) Enlarge area enclosed by square in (a) showing the lengths and thickness of the nanorods. (c) Two cross section profiles taken from (b) along the growth direction of a nanorod (line A), across the four nanorods (line B).

The magnetic properties of the samples were studied using a vibrating sample magnetometer (VSM) in magnetic filed up to 13.5 kOe and in the temperature range of 100 K to 300 K. We found that the magnetization of the samples does not saturate in the maximum available field, 13.5 kOe, see for example a representative magnetization loop for x = 4 wt. %, Fig.4. The magnetization (M) at an applied magnetic field of 13.5 kOe is found to increase with increasing the Fe percentage at room temperature and at 100 K. Figure 5 shows the room temperature magnetization values as a function of Fe weight percentage. We also found that the magnetization of each sample increases by about 10-times when we reduce the temperature from 300 K to 100 K.

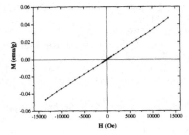

Figure 4. Magnetization loop for Fe (x wt. %)-doped Zn-TiO$_2$ sample, at a temperature of 300 K.

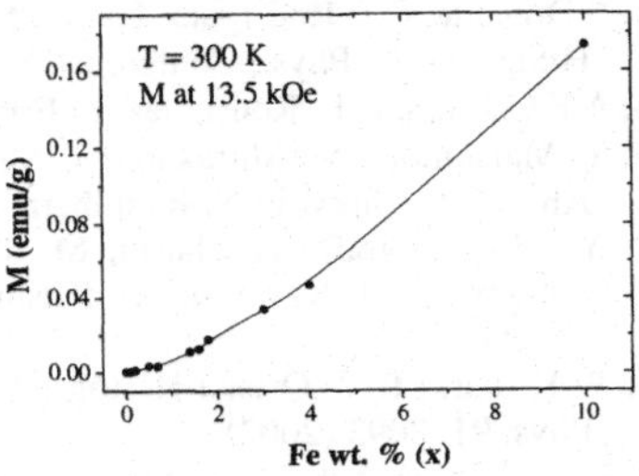

Figure 5. Dependence of the magnetization values, at an applied magnetic field of 13.5 kOe, for the different samples of nanoparticles Fe (x wt. %)-doped Zn-TiO$_2$ on the Fe weight percentage, x. The solid line represents a guide to the eye.

CONCLUSIONS

A solution chemistry route based on the wet-gel stirring method has been employed to synthesis the nanoparticles Fe (x wt. %)-doped Zn-TiO$_2$ rutile powders. We found that all the samples crystalline in the rutile phase. This study showed that we have many nanorods with an average length of 60 nm and an average diameter of 7 nm in the undoped sample, while in the doped samples the density of the nanorods decrease due to the breakdown to sheets. The magnetization values at the maximum applied magnetic field were found to increase with increasing the Fe weight percentage at room temperature and at 100 K.

ACKNOWLEDGMENTS

The authors would like to thank Sultan Qaboos University for the support under Grant number IG-SCI-PHYS-07-05.

REFERENCES

1. W.T. Geng, Kwang S. Kim, Phys. Rev. B **68**, 125203 (2003).
2. S.V. Chong, K. Kadowaki, J. Xia, and H. Idriss, Appl. Phys. Lett. **92**, 232502 (2008).
3. S Duhalde, C E Rodriguez Torres, M F Vignolo, F Golmar, C Chillote, A F Cabrera, F H Sanchez, J. Phys.: Conference Series **59**, 479 (2007).
4. A F Cabrera, C.E. Rodriguez, L. Errico, F.H. Sanchez, Physica B 384, 345 (2006).
5. Y. Matsumoto, M. Murakami, T. Shono, T. Hasegawa, T. Fukumura, M. Kawasaki, P. Ahmet, T. Chikyow, S. Koshihara, H. Koinuma, Science **291**, 854 (2001).
6. Y. Matsumoto,R. Takahashi, M. Murakami, T. Koida, X. J. Fan, T. Hasegawa, T. Fukumura, M. Kawasaki, S. Koshihara, H. Koinuma, Jpn J. Appl. Phys. **40**, L1204 (2001).
7. W.K. Park, R. J. Ortega-Hertogs, J. S. Moodera, A. Punnoose, M.S. Seehra, J. Appl. Phys. **91**, 8093 (2002).
8. T.L. Thompson, J.T. Yates, J. Chem. Rev. 106, 4428 (2006).
9. Y. Zhao, C. Li, x. Liu, F. Gu, H. Jiang, w. Shao, L. Zhang, Y. He, Matter. Lett. **61**, 79 (2006).
10. X. Wei, R. Skomski, B. Balamurugan, Z. G. Sun, S. Ducharme, D. J. Sellmyer, J. Appl. Phys. 105, 07C517 (2009).
11. S Shinde, S ogale, S Sarma, J Simpson, H Drew, S Lofland, C Lanci, J Buban, N Browing, V Kulkarni, J Higgins, R Sharma, R Green, T Venkatesan, Phys. Rev B **67**, 115211 (2003).
12. K Griffin, A Pakhomov, C Wang, S Heald, M Krishnan, Phys. Rev. Lett. **94**, 157204 (2005).
13. Salim Al-Harthi, Mubarak Al-Saadi, Imad A. Al-Omari, Husein Sitepu, Khalid Melghit, Issa Al-Amri, Ashraf T. Al-Hinai, and Senoy Thomas, Appl. Phys. A (in press DOI 10.1007/s00339-009-5508-4), (2009)

Mater. Res. Soc. Symp. Proc. Vol. 1201 © 2010 Materials Research Society

Optimising the Low Temperature Growth of Uniform ZnO Nanowires

Nare Gabrielyan, Shashi Paul[1] and Richard B.M Cross

Emerging Technologies Research Centre, De Montfort University, Leicester, LE1 9BH, UK.

ABSTRACT

Zinc oxide (ZnO) nanowires have been widely investigated and various different methods of their synthesis have been suggested. This work is devoted to the optimisation of the growth conditions for uniform and evenly distributed ZnO nanowire arrays. The nanowire growth process includes two steps: 1. Radio-frequency (RF) magnetron sputtering of a ZnO nucleation layer onto a substrate; 2. A hydrothermal growth step of ZnO nanowires using the aforementioned sputtered layer as a template. The optimisation process was divided into two sets of experiments: (i) the deposition of different thicknesses of the ZnO nucleation layer and the subsequent nanowire growth step (using the same conditions) for each thickness. The results revealed a strong dependence of the nanowire size upon the seed layer thickness and structural properties; (ii) the second set of experiments were based on growth solution temperature variation for the nucleation layers of the same thicknesses. This also showed nanowire size and distribution change with solution temperature variation.

INTRODUCTION

ZnO has numerous applications in optics, optoelectronics, sensors, solar cells, etc. due to its transparency, electrical conductivity, ferromagnetic, piezoelectric [1] and short-wavelength light-emitting properties [2]. And it is an important member of the II-VI group semiconductors with a direct band gap energy of ~3.37eV and strong exciton binding energy of 60 meV at room temperature.

As the structural morphology is an important factor in determining the properties of the material, significant research has been focused on synthesizing different types of ZnO nanostructures such as nanowires, nanobelts, nanotubes and nanorods. Recently, devices based on nanostructured ZnO have been studied extensively such as room temperature lasers [3], gas sensors [4], transistors [5] and field emitters [6]. Furthermore, certain properties of ZnO, such as its nontoxicity, chemical stability, electrochemical activity, high electron transfer features and high isoelectric point (~9.5), render nanostructures of this material highly promising for applications such as biosensors.

A number of methods for synthesising aligned one dimensional (1D) ZnO nanostructures have been reported previously. These include: the vapour–liquid–solid process (VLS) [7], metal–organic chemical vapour deposition (MOCVD) [8], thermal decomposition [9] and thermal evaporation [10]. Though there are a variety of the methods that have been developed, these often use metal catalysts to facilitate growth, which may cause unintentional doping of the nanostructures thereby affecting their physical properties. To negate this possibility, recent

---

[1] Corresponding author. Email: spaul@dmu.ac.uk

research has attempted to produce 1D ZnO structures using a two-step growth process [11]: i.e. the deposition of a thin nucleation layer of nanostructural ZnO by room temperature radio frequency (RF) magnetron sputtering, followed by a hydrothermal growth step using equimolar aqueous solutions of hexamine and zinc nitrate. The suggested method is promising as no metal catalysts are used in the growth process. Indeed, the nanowires grown by this technique show zinc and oxygen to be the only elements present with no contamination by any impurity. However, an extensive optimisation of the nanowire arrays produced by this method has yet to be reported.

In this regard, this work describes the results of optimisation experiments for the large-area, low temperature uniform growth of ZnO nanowires using a two-step approach.

EXPERIMENTAL PROCEDURE

Nanowires studied in this work were grown by the method described in Cross et al. [11]. The optimisation was carried out by varying two experimental parameters: the variation of the ZnO seed layer thickness and the growth solution temperature.

The first set of experiments involved the deposition of between 5 – 200 nm of ZnO as nucleation layers using RF magnetron sputtering where the system base pressure was <10–7 mbar. The sputtering was carried out using an 8 inch ZnO (99.99% purity) target in an argon (Ar) (99.999% purity) gas atmosphere at room temperature; The RF power was 150W at a frequency of 13.56 MHz. The Ar flow rate was kept such that the sputtering pressure was 1.3 × 10–2 mbar. The distance between the target and the substrate was approximately 7 cm. The substrates used were p-type silicon wafers (100) that had been pre-cleaned using acetone and buffered HF (20:1) before being given an ultrasonic rinse in 18.2 MΩ deionised (DI) water. In addition, soda-lime glass substrates were used for the UV-Vis and some XRD investigations of the ZnO sputtered layers. Following this, equimolar aqueous solutions (0.025 M) of zinc nitrate hexahydrate and hexamine were combined in a temperature controlled bath at 75 °C. After the period of temperature stabilisation, the samples were suspended upside down in the growth solution for 2h. The samples were then removed from the growth vessel, rinsed with DI water and dried in air. The second set of experiments involved using the same ZnO seeding layer thickness (75nm), but varying the growth temperature between 40 and 90 °C using the same concentration of equimolar solutions as before; the growth time was again set at 2h. The structural properties of the as-grown ZnO nanowires were studied using scanning electron microscopy (SEM) and X-ray diffraction (XRD), and the nucleation layers properties (such as surface roughness, grain size and optical bandgap) were studied using UV-Vis spectroscopy, atomic force microscope (AFM) and XRD.

RESULTS AND DISCUSSION

Varying the nucleation layer thickness results in a change in diameter of the nanowires Figure 1((a)-(d)) shows SEM planar views of as-grown ZnO nanowire arrays grown from 5 nm, 50 nm, 100 nm and 200 nm nucleation layers respectively.

Figure 1. SEM images of ZnO nanowires grown from 5nm, 50nm, 100nm and 200nm ZnO sputtered layers in 75 °C growth solution (the measurement bar is the same for each scan).

The diameter of the nanowires was found to vary between approximately 10 nm for the nanowires grown from the 5 nm seed layer to approximately 40 nm for the 200nm seed layer.

Thin film layers obtained by sputtering have rough surfaces. E. Makarona et al. [12], which suggests that surface roughness and large grain formations improve the initiation and hydrothermal growth process of the nanowires. In this regard, contact–mode AFM surface measurements have been carried out on ZnO nucleation layers with thickness of 25 nm, 50 nm, 75 nm, 100 nm , 150 nm and 200 nm. The results obtained showed the root mean square (RMS) roughness variation for the abovementioned samples Table I.

Table I. RMS Roughness values of the nucleation layers

Thickness of the seed layer (nm)	25	50	75	100	150	200
RMS Roughness (nm)	0.59	0.96	1.75	2.34	4.23	5.27

In order to investigate further the nucleation layer morphology and its influence on the nanowire growth process, 5 nm, 25 nm, 100 nm and 200 nm thick seed layer samples and corresponding nanowire arrays have been subjected to XRD analysis. The results show that the nanowire growth direction is perpendicular to the substrate (ZnO (002) was observed in the results). In addition, they show no crystallinity in the thin layer of 5 nm, but with increasing thickness the crystallinity increases. We suggest that this property of the seeding layer is directly involved in the nanowire growth process the following way; the thicker the seed layer (and the higher the crystallinity), the larger the diameter of the nanowires.

The crystal sizes of the seed layers are calculated (Table II) from the XRD data using the Scherrer equation [13];

$$L = \frac{K\lambda}{B(2\theta)\cos\theta} \tag{1}$$

where L is the average crystal size, K is the Scherrer constant that depends on the crystal shape, λ is the X-ray wavelength, $B(2\theta)$ is the FWHM (Full Width at Half Maximum) of the diffraction peak calculated in radians where θ – angle corresponding to the peak. Subsequent values are all calculated using the following values: $K = 0.9$, $\lambda = 0.154$nm.

Table II. Effect of thickness of the ZnO sputtered layer on crystal size

Film Thickness	25 nm	100 nm	200 nm
FWHM (002)(radians)	0.0136	0.0104	0.0091
Crystal size (nm)	12.5	16.3	18.7

It has been shown previously by C.C. Yang et al that the optical band gap of semiconductor quantum dots and wires increases with decreasing nanocrystal size for groups IV, III – V and II – VI semiconductors [14]. As nanocrystal size decreases, the continuous energy bands break into discrete levels which causes the optical band gap to widen [15]. In order to investigate the optical bandgap of the nucleation layers, UV-Vis spectroscopy has been used. This is determined using the absorption coefficient (α) as a function of the incident photon energy (hν) for allowed indirect transitions. Using the Tauc relation[16], the optical band gap (E$_g$) is calculated using,

$$\alpha = \frac{A(h\nu - E_g)^{\frac{1}{2}}}{h\nu} \tag{2}$$

Where A is a function of the refractive index of the material, reduced mass and speed of light[16]. It can be seen from Figure 2 that the optical band gap of the nanostructured nucleation layer increases slightly with the thickness. This property of the seed layer is currently the topic of further investigation.

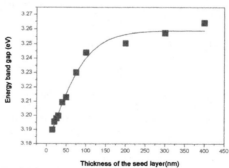

Figure 2. The Optical band-gap dependence of the ZnO layer on its thickness

The SEM images of ZnO nanowires grown when varying the temperature of the growth solution from 40 °C to 70 °C are shown in Figure 3. As can be seen, for a solution temperature of 40 °C, there is no evidence of nanowire growth. Nanowires grown at 70 °C and 75 °C were the most uniform in terms of structure and most evenly distributed on the substrate, with an average nanowire diameter of approximately 30 nm.

Figure 3- SEM images of ZnO nanowire arrays grown from 75nm ZnO sputtered layer in 40 °C (left) and in 70 °C (right) growth solution temperatures respectively (the measurement bar is the same for both scans).

The maximum temperature of the nanowire growth solution was set to 90 °C; the SEM images of those samples can be seen in Figure 4. The nanowire arrays were found to be very non-uniform, with circular formations seen on the sample surfaces at low magnifications (Figure 4 (left)). At higher magnification, (Figure 4 (right)) it is noticeable that the circles are formed by larger nanowires with diameters of up to 100 nm, whereas the nanowires growing inside the circles were of approximately 30 nm in diameter. This effect could be attributed to bubbles forming in the higher temperature solutions. However the exact cause is not yet known.

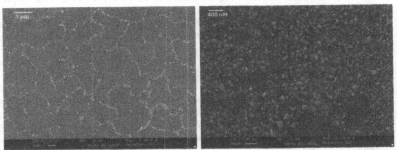

Figure 4- SEM images of ZnO nanowire arrays grown from 75nm ZnO sputtered layer in 90 °C temperature bath (4.000 and 20.000 magnifications respectively).

CONCLUSIONS

In summary, it has been demonstrated that the nucleation layer plays a vital role in the initiation and hydrothermal growth of ZnO nanowires. XRD analysis demonstrates that for the initiation of the hydrothermal growth of the nanowires to take place, a crystalline seed layer is required. This characteristic of the sputtered layer increases with thickness. The optical band gap of the nucleation layers also increases with nucleation layer thickness. The nanowire growth solution temperature also has a significant influence on the size of the nanowires and their distribution on the substrate. It has been shown that nanowires grown at approximately 70 °C form the most uniform and evenly-distributed arrays.

REFERENCES

1. Kong, X.Y. and Z.L. Wang, *Spontaneous Polarization-Induced Nanohelixes, Nanosprings, and Nanorings of Piezoelectric Nanobelts.* Nano Letters, 2003. **3**(12): p. 1625-1631.
2. Service, R.F., *Will UV lasers beat the blues?* Science, 1997. **276**(5314): p. 895.
3. Sun, X.W., et al., *Room-Temperature Ultraviolet Lasing from Zinc Oxide Microtubes.* Japanese Journal of Applied Physics, Part 2: Letters, 2003. **42**(10 B): p. L1229-L1231.
4. Fan, Z. and J.G. Lu, *Zinc oxide nanostructures: Synthesis and properties.* Journal of Nanoscience and Nanotechnology, 2005. **5**(10): p. 1561-1573.
5. Fan, Z. and J.G. Lu, *Gate-refreshable nanowire chemical sensors.* Applied Physics Letters, 2005. **86**(12): p. 1-3.
6. Xu, C.X. and X.W. Sun, *Field emission from zinc oxide nanopins.* Applied Physics Letters, 2003. **83**(18): p. 3806-3808.
7. Yang, P., et al., *Controlled growth of ZnO nanowires and their optical properties.* Advanced Functional Materials, 2002. **12**(5): p. 323-331.
8. Park, W.I., et al., *Metalorganic vapor-phase epitaxial growth of vertically well-aligned ZnO nanorods.* Applied Physics Letters, 2002. **80**(22): p. 4232.
9. Xu, C.X., Guoding; Liu, Yingkai; Wang, Guanghou, *A simple and novel route for the preparation of ZnO nanorods* Solid State Communications 2002. **122**(3-4): p. 175-179.
10. Yao, B.D., Y.F. Chan, and N. Wang, *Formation of ZnO nanostructures by a simple way of thermal evaporation.* Applied Physics Letters, 2002. **81**(4): p. 757.
11. Cross, R.B.M., M.M. De Souza, and E.M. Sankara Narayanan, *A low temperature combination method for the production of ZnO nanowires.* Nanotechnology, 2005. **16**(10): p. 2188-2192.
12. Makarona, E., et al. *ZnO nanorod growth based on a low-temperature silicon-compatible combinatorial method.* in *Physica Status Solidi (C) Current Topics in Solid State Physics.* 2008.
13. Elizabeth, E., et al. *Influence of iron dopant on structure, surface morphology and optical properties of ZnO nanoparticles.* in *Advanced Materials Research.* 2009.
14. Yang, C.C. and S. Li, *Size, dimensionality, and constituent stoichiometry dependence of bandgap energies in semiconductor quantum dots and wires.* Journal of Physical Chemistry C, 2008. **112**(8): p. 2851-2856.
15. Wang, J., et al., *The Al-doping contents dependence of the crystal growth and energy band structure in Al:ZnO thin films.* Journal of Crystal Growth, 2009. **311**(8): p. 2305-2308.
16. Pamu, D., P. Dharama Raju, and A.K. Bhatnagar, *Structural and optical properties of Ba (Zn1 / 3 Ta2 / 3) O3 thin films deposited by pulsed laser deposition.* Solid State Communications, 2009. **149**(43-44): p. 1932-1935.

AUTHOR INDEX

SUBJECT INDEX

Printed in the United States
by Baker & Taylor Publisher Services